ASTE Series in Science Education

More information about this series at http://www.springer.com/series/11803

Alandeom W. Oliveira • Molly H. Weinburgh
Editors

Science Teacher Preparation in Content-Based Second Language Acquisition

Springer

Editors
Alandeom W. Oliveira
State University of New York
Albany, NY, USA

Molly H. Weinburgh
Texas Christian University
Fort Worth, TX, USA

ISSN 2214-5435 ISSN 2214-5443 (electronic)
ASTE Series in Science Education
ISBN 978-3-319-43514-5 ISBN 978-3-319-43516-9 (eBook)
DOI 10.1007/978-3-319-43516-9

Library of Congress Control Number: 2016953008

Printed on acid-free paper

This Springer imprint is published by Springer Nature
The registered company is Springer International Publishing AG Switzerland

The Project

The Inheritances Books are a collaborative project between Ichabod Crane High School's Illustration and ELL students. It is made possible by a grant from the Berkshire Taconic Community Foundation.

The Student-Artist

Colleen Curry, Grade 12, 18 years old.

Acknowledgments

On behalf of all the authors in the book, we wish to thank the leadership team of the Association for Science Teacher Education (ASTE) for their generous support throughout the planning and preparation process. The publication of this collection of chapters would not have been possible without the hard work of the ASTE Publications Committee and endorsement of the board. In addition, we (Alan and Molly) wish to thank the reviewers who were recruited due to their expertise in the field of second-language acquisition in science. Each reviewer provided insightful and provocative feedback to the authors in ways that clarified and strengthened the chapters. Special thanks needs to be given to Rory Glass, Autumn Joy Florencio-Wain, Jerusalem Rivera-Wilson, and Monika Kirner-Ludwig. Their thoughtful input was greatly appreciated.

We are also indebted to the authors who contributed to this publication. The geographic distribution of the authors provides a wide diversity of views on science teacher preparation in contexts in which language acquisition is an issue. The authors' generosity and willingness to provide us with a glimpse of their efforts to make science accessible to all learners (regardless of linguistic or cultural background) illuminate how science teacher educators can promote social justice and equity at a global scale.

Lastly, we would like to express our gratitude to Gayle Abrams (arts teacher in Upstate NY) for providing us with the student artwork. Produced as part of *The Inheritances Books*, this artwork is the result of a collaborative project between Ichabod Crane High School's illustration and second-language learners made possible by a generous grant from the Berkshire Taconic Community Foundation. Thanks for their generosity in showcasing the artistic talents of our second-language learners.

Contents

Chapter 1
Introduction: Science Teacher Preparation in Language and Content

Alandeom W. Oliveira and Molly H. Weinburgh

Background Issues and Rationale

In the US and around the world, educational reform documents are calling for increased communication skills in science for all students (NRC 2012) as classrooms experience growing numbers of second language learners (Camarota 2007; U.S. Census 2010). The population of English language learners in US public schools has nearly doubled in just one decade between 1994 and 2004 (Peregoy and Boyle 2013). Further, the well documented achievement gap that exists between English language learners and non-English language learners across all content areas (August et al. 2011; Echevarria et al. 2011) makes it clear that more needs to be done to meet the educational needs of English learning students in content-area classes. Therefore, it is necessary for science teacher educators who work with pre-service and in-service teachers to know more about how to meet the needs of second language learners.

Many K-12 teachers face the task of helping students develop content knowledge while acquiring language skills (Bunch 2013). To address the issue, pedagogical models of content-based language instruction have become the focus of much educational research. Centered on metaphorical notions such as "sheltering" (Fritzen 2011) and "accommodation" (Glass and Oliveira 2014), these pedagogical models emphasize the need for science teachers to be able to make their classrooms safer and more inclusive by providing non-native speakers with comprehensible input (i.e., understandable information) while ensuring accessibility to rigorous academic

A.W. Oliveira (✉)
State University of New York, Albany, NY 12222, USA
e-mail: aoliveira@albany.edu

M.H. Weinburgh
Texas Christian University, Fort Worth, TX 76129, USA
e-mail: m.weinburgh@tcu.edu

A.W. Oliveira, M.H. Weinburgh (eds.), *Science Teacher Preparation in Content-Based Second Language Acquisition*, ASTE Series in Science Education,
DOI 10.1007/978-3-319-43516-9_1

content (as opposed to a "watered-down" curriculum that could disadvantage speakers of other languages). Despite these recent advances in pedagogical methodology, more research is needed to determine how teachers of subject matters like science can be effectively prepared to accomplish content-language integration (Bunch 2013; Lee et al. 2013). Teacher educators need professional training models that can help science instructors develop an improved awareness and more sophisticated understanding of why language matters to science instruction.

To date, there appears to be a lack of professional development models that can help practitioners and researchers build professional communities in which practice is co-created to improve science instruction for second language learners. One interesting example is the *teacher collaboration and co-teaching model* (Honigsfeld and Dove 2010) wherein language specialists and content-area teachers receive training on how to effectively combine their expertise and maximize their ability to collectively meet the needs of second language learners. However, this model is generalist as it is not tailored specifically to science teachers. Teacher educators and researchers of science education still need a venue to share professional development strategies and teacher training models that are working.

As members of the Association for Science Teacher Education (ASTE), we recognized the need for members and non-members to share models of professional development that included strong science and language components. Therefore, the chapters in this book serve to fulfill several goals: to provide literature beyond methods books; to extend the research on multilingualism; and provide evidence of language-focused science standards being implemented.

Terminology

One particularly challenging aspect of writing this book was with regard to the terminology used to identify the target student population with whom science teachers were being prepared to work. Our options were somewhat constrained by the fact that this book is being published in English by the Association for Science Teacher Education (ASTE), which has an English-speaking membership that is predominately based in the US. As such, our main focus was on how to prepare science teachers to effectively work at the intersection of science content and English language acquisition. Although English is privileged in this book, we acknowledge that the issue of effectively preparing science teachers to overcome language obstacles to content learning is of worldwide interest, and reaches far beyond the English-speaking world. Therefore, we felt the need to emphasize our recognition that English is not the only language that students must learn in order to also learn science, that careful consideration must also be given to a multitude of other languages (often simultaneously to English).

We purposely did not use English language learner (ELL), English learner (EL), limited English proficient (LEP), Non-Native Speaker (NNS) or L2 Speaker in the title. Part of the reason is that there is little agreement as to what name best describes these students. Each designation has different connotations and problems, with different terms being favored by researchers within distinct research traditions depending upon one's philosophical commitment, sociopolitical orientation, and unique focus (e.g., English as a Foreign Language rather than English as a Second Language). We recognize that by selecting 'second language acquisition' for the title we utilized a label that may cause reservation and discomfort to some readers. We acknowledge that the students who will be taught science bring a rich heritage of culture and language. They simply do not speak the language of the school in which they find themselves. At the very least, the students are learning a second language and are emerging bilinguals (Garcia et al. 2008). Many of the students are, in fact, already bilingual or multilingual.

Literature Gap

The present book attends to an important gap in the science teacher education literature. The current literature available to science educators is limited to teaching methods books that provide classroom practitioners with a variety of pedagogical strategies that integrate science and language and introduce teachers to longstanding pedagogical models of content-based instruction such as SIOP (Sheltered Instruction Observation Protocol) (Echevarria et al. 2003) and SDAIE (Specially Designed Academic Instruction in English) (Cline and Nicochea 2003). For instance, NSTA Press recently released *Science for English Language Learners: K-12 Classroom Strategies* (Fathman and Crowther 2006); and *Teaching Science to English Language Learners: Building on Students' Strengths* (Rosebery and Warren 2008) – both written for an audience of science educators. Similar books have also been published by Heimen such as *English Learners and the Secret Language of School: Unlocking the Mysteries of Content-Area Texts* (Pilgreen 2010), *Academic Language for English Language Learners and Struggling Readers: How to Help Students Succeed across Content Areas* (Freeman and Freeman 2009) and *The New Science Literacy: Using Language Skills to Help Students Learn Science* (Their 2002) targeted mainly at language/literacy educators. However, to the best of our knowledge, there is currently no publication written specifically for an audience of science teacher educators. This is particularly problematic given the recent advent of more sophisticated and content-specific pedagogical models such as the 5R Instructional Model (Silva et al. 2013; Weinburgh et al. 2012). Designed specifically for teaching science, the 5R provides teachers with a more flexible and reflective approach to language-science integration without reducing science instruction to a linear and fixed sequence of steps or phases that are mechanically followed. A book with professional development models such as the 5R that could help teacher educators effectively prepare school teachers to teach science to language learners and support second language learners through science was yet to be published.

Research on Multilingualism

This book is also aimed at addressing the dearth of research on multilingualism that currently exists in the science education literature. Despite growing interest on language-related issues among science educators (e.g., argumentation, questioning, writing, reading, cooperative discussions), surprisingly little research has been conducted on how to effectively prepare teachers to meet the linguistic demands of science teaching. This shortcoming is particularly problematic given the primacy of the conceptual dimensions of science education. Science teaching is a profession typically linked to the overcoming of science misconceptions and progression toward science experts' conceptions. Because language issues are often overlooked in science preparation, many science teachers encounter difficulty when dealing with classroom situations involving language such as supporting second language learners and incorporating language-focused pedagogy into their teaching practices (Luft 2002). In fact, only four states currently require that content area teachers complete coursework in teaching second language learners (NCATE 2007). As a result, many teachers fall short of recognizing the importance of language as a critical mediating tool to science teaching and learning and even develop negative attitudes toward the integration of language with science – as documented in the article: "I am a Science Teacher, not a Reading Teacher: An Unfortunate Categorization" (Rush 2002). These issues underscore the need to problematize language in the specific context of science teacher professional development. As emphasized by scholars of classroom discourse, teachers need to develop *metalinguistic awareness* (Andrews 2010) – explicit knowledge of the underlying systems of language that enables teachers to teach effectively; *rhetorical consciousness* (Swales 1990) – an awareness of the rhetorical functions of language; and, *pragmatic awareness* (Oliveira et al. 2007) – an understanding of the multifunctionality of science classroom discourse.

Language-Focused Standards

The present book is also aligned with the advent of more rigorous, language-focused set of educational standards. This is particularly evident in the 2012 *English Language Proficiency Development (ELPD) Framework*, a document that identifies the language demands of the Common Core State Standards and the Next Generation Science Standards as well as a pedagogical shift in which a shared responsibility for English language development must exist between ESL and content-area teachers. The ELPD Framework (Council of Chief State Schools 2012) states:

> At present, second language development is often seen as the primary responsibility of the ESOL teacher, while content development (particularly in grades 6–12) as that of the subject area teacher. Given the diverse range of program design and explicitness [in the new standards] regarding how language must be used to enact disciplinary knowledge and skills, such a division of labor is no longer viable (p. 3).

This framework also identifies a new direction in science education; one in which language and content will need to be taught to second language learners in a deeply integrated manner, and one in which all teachers bear the responsibility of developing all students' academic language related to their discipline. Science teachers need to be able to recognize and target the key language and literacy practices inherent in their discipline – such as explaining and arguing with evidence – to enhance the engagement of second language learners with science content. Helping science teacher educators accomplish such a complex and challenging endeavor is our main hope for the present book.

The primary purpose of this book is to provide science teacher educators with exemplars of professional development programs designed to prepare school teachers to effectively help language learners in science classrooms simultaneously gain language proficiency and conceptual understanding. It is envisioned that this book will serve as a valuable resource for science teacher educators seeking to identify language-focused professional development activities that can be used to introduce science teachers to content-based approaches to second language instruction such as "sheltered science" as well as pedagogy that can be used to ensure accessibility and comprehensibility of science content to students with limited proficiency in the language of instruction such as translanguaging (Wei and Garcia 2013) and academic conversations (Wright 2014). Lastly, it is expected that the book can help support collaboration between science and language educators by increasing their familiarity of specialized terminology across the two scholarly fields. Together, the chapters will assist readers develop a stronger grasp of jargon commonly utilized by experts in language education (e.g., realia, model, wordwall, and Frayer Model) as well as in science education (e.g., inquiry, argumentation, scientific modeling, and nature of science). This interdisciplinary focus can help educators who read the book to be in a better position to collaborate more effectively across the two content areas. As an initial step, attention is now given to what we perceive as a major obstacle to effective science teacher preparation in content-based second language acquisition, namely the content-language divide that permeates many teacher education programs.

The Content-Language Divide

Traditionally, student conceptual development and language acquisition have been addressed separately in teacher preparation programs. On the one side, science teacher educators have given primacy to conceptualization by focusing their professional development efforts mainly on issues related to student cognitive development such as informing teachers about misconceptions or alternative conceptions that students commonly bring with them to the science classroom, providing instructors with pedagogical approaches and strategies to effectively improve students' science conceptions, and increasing teacher awareness of the developmental pathways followed by students in the course of acquiring more sophisticated science conceptual understandings in science (i.e., their science learning progression). On the other side, language educators have sought to help teachers better understand how students learn language, how to improve students' communicative competence (i.e., their knowledge of what constitutes contextually successful and culturally appropriate language use), and how to effectively engender learning outcomes such as mastery of vocabulary and grammar, ability to felicitously perform speech acts (e.g., asking questions, giving directives, disagreeing, etc.), and academic literacy.

This divide between language and content is particularly apparent in current educational research and policy. A good example is the parallel advent of science learning progressions and language progressions in distinct educational circles. In recent years, there has been an increased interest in learning progressions in the field of science education. Learning progressions map out student conceptual development across time (grade-levels). More specifically, students are viewed as progressing along a developmental continuum or pathway toward scientific understanding as they achieve higher (more sophisticated) levels of understanding of a big idea in science. The pathway of conceptual development is typically composed of a series of levels (often six) ranging from naïve knowledge to incomplete or disconnected mastery of isolated ideas, and ultimately sophisticated scientific understanding of complex conceptual connections. Learning progressions have recently been developed for a number of big ideas in science, including Earth's seasons (Plummer and Maynard 2014), genetics (Shea and Duncan 2013), water in environmental systems (Gunckel et al. 2012), and the particular nature of matter (Smith et al. 2006). Across these progressions, epistemic sophistication overshadows linguistic skill, with little attention given to how students' conceptions relate to their ability to verbally encode ideas in a particular language.

At the same time, language educators have witnessed the introduction of language progressions by educational policies both nationally and internationally (Council of Europe 2011; Interagency Language Roundtable 2011; WIDA 2012). Language progressions map out nonnative speakers' trajectories of linguistic development. Additional language acquisition is conceived as a progression across a series of levels of increasing communicative ability or proficiency. Higher proficiency levels are characterized in terms of student acquisition of a specific set receptive skills (listening and reading abilities) and productive skills (speaking and writing abilities). For instance, New York State recently adopted the *New Language Arts Progressions*, a learning trajectory that organizes student acquisition of a foreign language in terms of a sequence of five levels of proficiency, namely Entering, Emerging, Transitioning, Expanding, and Commanding (NYSED 2012a, b). Based on recent research on second language learners and the Framework for English Language Proficiency (Council of Chief State Schools 2012), each level identifies a distinct developmental stage in students' acquisition of a new language (other than the one spoken at home). Across these progressions, linguistic skill overshadows epistemology, with little attention given to how student fluency may relate to their knowledge of the topics under discussion.

The content-language divide in the above literature is readily apparent. Having separate progressions for language and content not only makes systematically tracking student development more difficult but it is also inconsistent with the holistic nature of student development. Using unidimensional roadmaps to track student progress along bi-dimensional trajectories seems ineffective and counterintuitive, particularly in classrooms that favor content-based pedagogical approaches to second language acquisition. In such settings, progression toward scientific understanding is expected to parallel progression toward fluency in an additional language. So, rather than being detached and unrelated, conceptual and linguistic development are closely interrelated and mutually influential. Yet, theoretical consideration of student progression along both conceptual and linguistic developmental pathways remains to be given, an issue that we tackle next.

Theorizing the Issue

Overcoming the content-language divide requires theoretical sophistication and clarity with regard to the role of language in cognition. Teacher preparation efforts need to be informed by theoretical models with clear and explicit articulation of how conceptions are mapped onto linguistic forms (i.e., the verbal encoding of science concepts), the cognitive foundation that underlies acquisition of a new language, and the pedagogical scaffolding of student development (conceptual and linguistic). Recent theoretical and empirical work in various fields of scholarship can provide helpful and novel insights in this regard.

Cognitive aspects of language have been the object of a considerable body of work in the field of cognitive linguistics. A common theme in this literature is the potential of language to bias and even transform thinking (Bowerman and Levinson 2001). Foreshadowed by Whorf's (1956) seminal work on language and thought, evidence is provided that linguistic forms (words and grammatical structures) provide us with a framework of understanding that predispose us to see things in particular ways. More specifically, recent studies have shown that a persons's way of speaking influences his or her ways of conceptualizing and perceiving natural phenomena such as time (Boroditsky 2001; Gentner et al. 2002; McGlone and Harding 1998).

For instance, English speakers tend to think of time horizontally (unfolding from left to right), whereas Mandarin Chinese conceive of it in terms of vertical spatial relations (up/down), and members of the Hopi tribe conceive of time duration in terms of ciclycity (successive recurrences of events). Such differences can be the source of added cognitive load for speakers who set out to learn a new language. Similar points have been made by scholars of education such as Sutton (1992) who argues that "[science] words in a sense are theories" (p. 11) and Pimm (1987) who emphasizes the important epistemic role of metaphoric terms in enabling the extrapolation of ideas and expansion of disciplinary knowledge in fields such as mathematics. Language plays a central role in the shaping of human perception and creation of collective vision (i.e., socially organized ways of seeing the natural world). As such, language constitutes an important epistemic tool whose influence on disciplinary knowledge-building is undeniable.

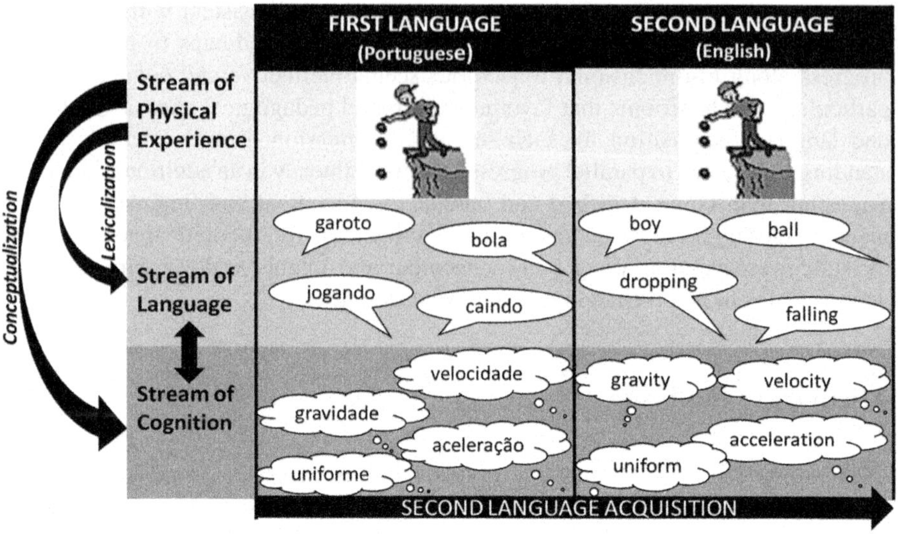

Fig. 1.1 Lexicalization and conceptualization of physical experience

A potentially useful way of theorizing the content-language relationship is provided by Gentner and Boroditsky (2001) who conceives of language acquisition as learning to perform nominal reference (i.e., refer to objects, processes, or entities by pointing and naming them). A central premise of this perspective is that second language learners need to have a clear concept of what the referent is (i.e., clearly understand the nature of the entity or phenomenon being referred to) in order to be able to accurately pinpoint and name it in the local context wherein their physical experiences take place. Language learning entails emergent mastery of both *lexicalization* (particular ways of verbalizing physical experiences) and *conceptualization* (particular ways of understanding physical experience). As shown on Fig. 1.1, this process can be visualized in terms of three parallel streams: experiential, linguistic, and cognitive. Lexicalization and conceptualization are parallel and mutually influential processes. Language learners learn to lexicalize as they learn to conceptualize the world in new ways.

Linguistic aspects of cognition have been examined in the field of sociology of education where evidence abounds that students have difficulty grasping grammatically encoded scientific knowledge when reading written science texts (Maton 2013a). Student comprehension science texts is often made difficult due to the predominance of a complex discursive style characterized by high semantic density or degree of meaning condensation (high number of content words per sentence) and low semantic gravity or degree of meaning dependence on context (predominance of context-independent generalizations). Decoding and interpreting such complicated texts usually requires downward semantic shifts (re-articulation into less abstract and more contextualized meanings) through pedagogical scaffolds such as literature circles, collaborative concept mapping and other classroom activities involving *transmediation* (Short 2004), that is, transfer of meanings across representational systems. Further, educators can more effectively promote cumulative knowledge-building, Maton (2013b) argues, by systematically tracking the extent to which student language performance (oral or written) fluctuates between abstract ideas (e.g., general scientific principles behind experiments) and concrete notions (e.g., specific procedural details of experiments). To build knowledge, students need to learn to encode their thoughts into abstract and generalized linguistic forms.

There is also increasing recognition among scholars that classroom language is a complex system that is both dynamic and nonlinear. As emphasized by Larsen-Freeman (1997), "language grows and organizes itself from the bottom up in an organic way, as do other complex nonlinear systems" (p.148). So, rather than assuming a direct causality (univariate cause-effect link), it is more productive to conceive of language and content as a contextually-dependent outcome inevitably characterized by a certain degree of indeterminacy, uncertainty and unpredictability.

Emphasized in the above literature is the dialogic and nonlinear manner whereby language and thought are interrelated. Although our ways of speaking can influence and bias our ways of conceptualizing perceptual experience and vice-versa, one does not determine the other. Therefore, it seems unwarranted and simplistic to presume the existence of a deterministic relationship between language and thought (e.g., treating language as a prerequisite for content learning) as language is not simply a determinant of or precursor to conceptual thinking ability. Instead, it is more productive to theorize language and content as sharing a dialogic, mutually influential relationship. Recognition of the complex and highly fluid nature of this relationship is essential to the success of professional development efforts aimed at preparing science teachers to effectively promote content-based second language acquisition.

Book Format and Organization

This book examines seventeen science teacher preparation programs that span a wide variety of grade levels (elementary, middle, and secondary), countries (Italy, Luxemburg, Spain, UK, and US), and linguistic contexts (English as a Second Language, English as a Foreign Language, trilingual classrooms, and teaching deaf children science through sign language). The book is divided into three main parts. Each part consists of chapters that illustrate a common, cross-cutting theme in science teacher preparation in content-based second language acquisition, namely pre-service teacher preparation, in-service teacher preparation, and international perspectives. Each part provides many insights on the similarities and differences in the professional development approaches used to prepare science teachers with varied amounts of instructional experience to help students in different parts of the world overcome linguistic barriers while simultaneously learning concepts central to science.

Chapters in Part I of the book focus on a set of US-based programs designed to prepare pre-service teachers to effectively help immigrant students whose first language is not English learn science in an additional language. Throughout these programs, pre-service teachers are provided with foundational knowledge and skills needed to seamlessly integrate science and language instruction (for a comparative overview of the chapters, see Table 1.1). Novices to the teaching profession gain expertise in making mainstream science instruction accessible to second language learners by learning to design and implement sheltered lessons that are supportive, yet academically rigorous. Three of the chapters deal specifically with pre-service elementary teachers. **Bravo** describes how two courses – in second language acquisition and science teaching methods – were systematically redesigned with a common focus on content-language integration to ensure coherence and convergence across coursework aimed at preparing preservice teachers to effectively work with ELs in elementary science. **de Oliveira** introduces pre-service teachers to LACI (Language-based Approach to Content Instruction) – a pedagogical framework or model designed to guide preservice teachers on how to effectively use texts to support second language learners' mastery of science content. And, **Hernández** describes a school-based program in which pre-service teachers are offered workshops at a bilingual elementary classroom where they observe first-hand the practices of an experienced teacher and reflect about content-language integration in an authentic setting.

Two other chapters examine pre-service teacher preparation at the high-school level. In an attempt to overcome disciplinary compartmentalization and fragmentation, **Roberts, Bianchini, and Lee** resort to a capstone course designed specifically for the purpose of coherently bridging language and content, and fostering in science teachers an adaptive disposition to support second language learners (as opposed to simply providing them with a set of pedagogical "tricks and tools"). **Stoddart, Solis, Tolbert, and Lyon** share a similar concern for ensuring coherence across courses and field experiences aimed at preparing preservice teachers to work with second language learners in secondary science. Their efforts take the form of a practice-based model wherein pre-service teachers systematically experience, deconstruct, and then attempt to approximate content-language integration practices through sustained engagement with videos, cases, and feedback from coaches.

The remaining two chapters focus on programs targeting wider pre-service teacher audiences that span the entire K-12 grade-level range. **Viesca, Mahon, Carson, et al.** take an online approach to science teacher professional development. More specifically, K-12 teachers are supported by means of a series online modules designed to foster expertise in linguistically responsive science teaching. And, **Smetana & Heineke** take a field-based approach to pre-service science teacher preparation for culturally and linguistically diverse classrooms. Through a 4-year B.S.Ed. program implemented mainly in authentic settings such as schools and museums, teachers are introduced to culturally responsive, language rich teaching practices and are encouraged to develop an asset-based mindset for ELs (focused on their strengths rather than deficits).

In Part II, attention then shifts to in-service teacher education. Also comprised of a total of seven chapters, this second part examines US-based professional development programs aimed at improving practicing science teachers' abilities to work with second language learners and promote content-based second language acquisi-

Table 1.1 Overview of chapters in Part I pre-service teacher preparation

Chapter	Authors	Preparation model & activities	Aims & principles
2	Bravo	Two teacher education courses (language acquisition and science methods); Meta-pedagogical discussions about areas of convergence and synergy between science & language (coherence & integration).	Integrated approach to preparing teachers (avoidance of disciplinary silos and encapsulated approach); Cultivating knowledge of how synergistically teach **elementary** science & language; Boosting teacher self-efficacy to simultaneously scaffold conceptual development and vocabulary development; Conceiving of words as concepts, and science processes as productive/receptive language.
3	de Oliveira	Provision of examples of language-based **elementary** science instruction; Provision of templates for lesson planning; Provided of framework for classroom implementation (questions and prompts to identify linguistic topics, strategies, etc.)	Teaching content through language, not language through content; Helping teachers "see" key linguistic features of science texts; Helping teachers make informed text selections; EL access to content is through mastery of discipline-specific discourse (not simplification of content); "Six C's of support" principle: Connection, Code-breaking, Culture, Challenge, Classroom interactions, & Community and Collaboration.
4	Hernández	School workshop in a bilingual classroom; Modeling of content-language integration practices; Classroom observations, daily reflections, and guided pedagogical discussions.	School-based approach to preservice teacher preparation; Emphasizes promotion of bilingualism through integration of language and **elementary** science; Uses Guided Language Acquisition Design (GLAD); Combines collaborative science inquiry (small student groups) with comprehensible input; Visual support of learners (posters, charts, cards, etc.).

(continued)

Table 1.1 (continued)

Chapter	Authors	Preparation model & activities	Aims & principles
5	Roberts, Bianchini, & Lee	Capstone methods course bridging **secondary** science and language acquisition;	Developing adaptive disposition (beyond pedagogical "tricks & tools");
			Appreciating ELs' diversity and need for differentiation;
		Emphasis on trying out and scaffolded reflection;	Seeing scientific language as sense-making (not just vocabulary);
		Investigating teaching and learning (videos and observations).	Building on EL's funds of knowledge through cognitively demanding, language-rich tasks;
			Engaging ELs in the disciplinary language of science.
6	Stoddart, Solis, Tolbert, & Lyon	**Secondary** methods course and practicum;	Preparing teachers to implement NGSS and CCSS;
		Modeling of content-language integration practices;	Training teachers to foster contextualized and authentic use of language in science through systematic content-language integration;
		Analysis of videos and cases (noticing ability);	Encourage science teachers to become reflective practitioners and to see themselves as "language planners";
		Approximation assignments (implementation of content-language integration practices).	Practice-based model of teacher education wherein content-language integration is experienced, deconstructed, and then approximated;
		Intensive feedback, coaching, and support.	Four practices: scientific sense-making, scientific discourse, language development, and contextualized instruction.
7	Viesca, Mahon, Carson, et al.	Online modules that support science-language integration;	Preparing **K-12** teachers for NGSS and WIDA;
			Helping teachers create opportunity for language acquisition in science;
		Teachers are guided by "essential questions";	Fostering linguistically responsive science teaching;
		Teachers Explore, Make it Work, and Share;	Asset-based approach (building on the strengths of teachers and students);
		Collaborative Communities.	Classrooms as sites of intentional investigation.

(continued)

Table 1.1 (continued)

Chapter	Authors	Preparation model & activities	Aims & principles
8	Smetana & Heineke	Four-year program (B.S.Ed.).	Field-based apprenticeship model (not in university classrooms);
		Phases: Exploration, concentration, and specialization;	Contextualized/embedded approach centered on schools and community spaces;
		Introduction to culturally responsive pedagogy (WIDA) and inquiry-based teaching (interdisciplinary, language-rich);	Aimed at preparing **K-12** teachers to positively influence students;
			Strong emphasis on community partnerships;
		"Science talks" with students, museum visits, and collaboration	Fosters asset-based mindsets for ELs, awareness of policy enactment by educators, and expertise on language-intensive practices;
			Sociocultural perspective on teacher training as authentic participation in cultural activities.

tion. Throughout these chapters, emphasis is placed on encouraging school teachers to recognize the essential role that language plays in science instruction and to improve their knowledge of language concepts, language pedagogies, and multilingualism (see Table 1.2 for a comparative overview). The first three chapters deal specifically with the professional development of teachers in the lower-grade levels (elementary and middle school). **Gomez-Zwiep & Straits** take an "inquiry/language-development" blended approach wherein elementary teachers receive sustained training in science-language integration by means of summer institutes and school-year lesson studies offered as part of a 3-year professional development program. **McDonald, Miller, Lord, & Cook** describe a 2-day professional development effort wherein elementary and middle school teachers are provided with curricular support materials (posters, scaffolds, etc.) and training on how to effectively facilitate classroom discussions that are inclusive of ELs as sense-makers. Emphasis is also placed on encouraging teachers to conceive of ELs in more productive ways (in terms of assets they bring to the classroom). **O'Hara, Pritchard, Pitta, Newton, Do, & Sullivan** describe a capacity-building program wherein middle-school teachers and administrators are jointly offered workshops on science-language integration and form school-based design teams that plan, implement, and assess contextually informed interventions in support of emergent bilinguals' science learning.

The next two chapters examine various aspects of the *Language-rich Inquiry Science with English language learners through Biotechnology* (LISELL-B), a teacher preparation program for secondary and middle school teachers. In addition to discussing the iterative design-based approach behind LISELL, **Buxton, Allexsaht-Snider, Rodriguez, Aghasaleh, Gaibisso, & Kirmaci** describe how participating

Table 1.2 Overview of chapters in Part II in-service teacher preparation

Chapter	Authors	Preparation model & activities	Aims & principles
9	Gomez-Zwiep & Straits	Three-year program (summer institute and school-year lesson studies);	Inquiry-based approach;
			Inquiry as context for authentic language use (peer collaboration);
		Collaborative development and implementation of inquiry/ELD blended lessons;	Sessions on science content, science pedagogy, and second language acquisition (increasing integration over the years);
		Provide **elementary** teachers with lesson plan templates (5E lessons).	Preparing teachers to merge science and language instruction;
			Model lessons to teachers.
			Develop teachers' abilities to differentiate instruction to different levels of student language proficiency (i.e., make language accommodations).
10	McDonald, Miller, Lord, & Cook	Provision of curricular support materials: posters and examples of productive discourse patterns (Teacher and Student Moves);	Changing interaction from IRE to sense-making;
			Teacher appropriation of facilitation tools – moves to scaffold deeper reasoning and create meaning-making;
		Guidance on NGSS;	Focus on ELs' assets (strengths and capacities);
		Creation and adaptation of classroom activities and units.	Seeing ELs as sense-makers, questioners, and thinkers;
			Replacing **elementary** and **middle** school teachers' atomistic views of language (accumulation of vocabulary) with an experiential language perspective.
11	O'Hara, Pritchard, Pitta, Newton, Do & Sullivan	Provides video examples of instructional practices;	Capacity-building approach centered on the formation of leadership teams and mentorship;
		Provides guiding framework;	Aimed at strengthening the STEM learning "ecosystem" of **middle** schools;
		Creates a network website;	Professional learning communities that inquire about instructional practices integrating science and language;
		Facilitates workshops;	
		School-based design teams.	

(continued)

Table 1.2 (continued)

12	Buxton, Allexsaht-Snider, Rodriguez, Aghasaleh, Gaibisso, & Kirmaci	Summer Institute (**middle** and **secondary** teachers) and Summer Academy (students);	Focused on designing language-rich science inquiry lessons;
			Making teachers aware of the "language of NGSS";
		Teachers develop language-rich investigations for student academy;	Students need to acquire mastery over the language of scientific investigations (e.g., verbalizing hypothesis, observations, claims, explanations, etc.) in order to participate in inquiry activities;
		"Grand rounds" (classroom observation and online logging);	
		Bilingual family workshops;	Language development can be scaffolded through various curricular support materials (concept cards, lab role cards, general academic vocabulary cards, and language frames);
		Teacher workshops for exploring students' writing.	
			Promotes translanguaging (English + home language).
13	Dominguez & Allexsaht-Snider	Summer Institute (**middle** and **secondary** teachers) and Summer Academy (students);	Teachers as architects of learning environments conducive to emergent bilingualism;
		Teachers develop language-rich investigations for student academy, and receive Spanish lessons;	Language as important expressive component of learning environment;
			Expression (meaningful language use and communication) leads to domestication of the learning environment and territoriality (ownership).
		Student presentations to family members, and visits to labs.	Teachers promote meaningful student-space interactions through incorporation of familiar cultural practices (native language, soccer).
14	Hernandez, Baker, Reyes, & Uribe-Flórez	One-year preparation program;	Standards-based professional development approach (CCSS, NGSS, CREDE);
		Online course work, face-to-face professional development sessions, site-based meetings.	Preparing STEM career changers to integrate content and language, and to implement culturally responsive pedagogy (language-rich pedagogy);
		Mentoring, coaching, co-teaching, video-based reflection, meetings with EL students, and peer collaborations.	Science as authentic context for language use
			Teachers design and implement 5E lessons with language strategies (vocabulary supports, word roots, text sets, visual, cues, etc.)
			Training **secondary** teachers to support and meet the needs of diverse students.

(continued)

Table 1.2 (continued)

15	Zhang	Two-year certification program;	Aims at prepare K-12 teachers to integrate language and science literacy;
		Online courses, face-to-face activities, and practicum;	Focused on language-based content instruction;
		Collaborative course projects on SFL text analysis (online vocabprofiler, nominalization, lexical density, theme-rheme).	Combines SIOP with System Functional Linguistics (SFL);
			Making texts accessible to English learners through games (long word detective) and graphic organizers (Frayer model, long noun group scramble).

science teachers are made more aware the language of classroom investigations and supported when designing and implementing language-rich science inquiry lessons aligned with the Next Generation Science Standards. **Dominguez & Allexsaht-Snider** examine the professional development activities of LISELL (teacher summer institutes, bilingual family workshops, student summer academies, etc.) from an architectural perspective wherein science teachers as viewed as "architects" being prepared to build "spaces" that are conducive to EL science learning.

This second part then ends with two chapters that share a particularly strong orientation to current educational standards. **Hernandez, Baker, Reyes, & Uribe-Flórez** describe their standard-based approach to preparing STEM career changers to integrate science and language and respond to the needs of diverse learners in secondary science. Through a 1-year program that combines online coursework with in-person sessions, teachers are encouraged to align their classroom practices with multiple sets of standards (CCSS, NGSS, and CREDE), simultaneously. **Zhang** resorts to a SIOP/SFL model of professional development in an effort to prepare K-12 school teachers for science-language integration in alignment with the CCSS, NGSS, and Science for All Americans: A Project 2061. Aimed at promoting functional science literacy among second language learners, teachers enroll in a 2-year TESOL endorsement program with both online and face-to-face components.

In Part III of the book, the focus shifts to international perspectives on science teacher preparation in content-based second language acquisition. The four chapters in this last part of the book examine the professional development efforts of science teacher educators in different parts of Europe. The extremely diverse range of linguistic contexts across these chapters provide a glimpse of the many complexities and nuances that varied patterns of language distribution across different countries add to science teacher preparation endeavors. As emphasized by Mackey (1970), "a learner who speaks only one language at home and the same language in school, even though it may not be the language of the community, is in quite a different position from that of a learner who uses two languages at home and the same two at school" (p. 415). Teaching in such multilingual learning environments requires careful consideration to be given to the unique patterns of home-school-area-nation relations that make up the local educational landscape. Toward this end, **Espinet, Valdés, Carrillo, Farro, Martínez, López, & Castillón** describe their triadic partnership approach to teacher professional development in the unique, multilingual context of Catalonia (Spain). As part of this program, elementary teachers are introduced to the Content and Language Integrated Learning (CLIL) approach, provided with lesson plan templates, and encouraged to go beyond vocabulary learning. Similarly, **Rasulo, De Meo, & De Santo** report a collaborative, CLIL-for-Science practicum model of teacher preparation on science and language issues in an Italian context. **Wilmes** resorts to inquiry-based workshops in an effort to prepare primary teachers to integrate science content and literacy in the unique and complex trilingual landscape of Luxembourg. In the last chapter in Part III, **Cameron, O'Neill, & Quinn** (UK) seek to prepare science teachers to promote English/sign language bilingualism in science classrooms with deaf children by offering workshops where teachers are provided BSL science glossaries that show how to translate technical terms in science into sign language (Table 1.3).

Finally, in the concluding chapter, we provide an epilogue. More specifically, we articulate the main themes that emerged out of the preceding chapters, summarize how they can provide guidance on preparing science teachers in content-based second language acquisition, and draw out implications (both theoretical and practical) for science teacher education. We then end by identifying areas in need of further research and practical consideration.

Table 1.3 Overview of chapters in Part III international perspectives

Chapter	Authors	Preparation model & activities	Aims & principles
16	Espinet, Valdés, Carrillo, Farro, Martínez, López, & Castillón (Spain)	Scaffolds to integrate scientific practices and discourse practices;	Content and language integrated learning approach (CLIL) to **elementary** science;
		Template for planning lessons that integrate science inquiry and language learning;	Going beyond promotion of vocabulary learning, and crossing community borders;
			Identification of learning objectives that integrate content and language;
		Co-planning, co-teaching, and collaborative reflection.	Identification of scientific and discourse practices in lesson planning.

(continued)

Table 1.3 (continued)

Chapter	Authors	Preparation model & activities	Aims & principles
17	Rasulo, De Meo, & De Santo (Italy)	Offers a CLIL-for Science methodology course and practicum;	Content and language integrated learning approach (CLIL) to **secondary** science;
		Workshops and online tutorials;	Emphasizes collaboration between science teachers and language specialists in school settings;
		Video delivered through Moodle;	
		Teachers conduct Action research projects (self-inquiry);	Focused 8 Functions of scientific English: classifying, comparing and contrasting, explaining and exemplifying, hypothesizing and predicting, questioning, sequencing, generalization, marked lexis, and paraphrasing.
		Teachers post lesson lessons and receive feedback through Moodle;	
		Provides observations and reflection tools.	
18	Wilmes (Lu- xembourg)	Inquiry-based science workshop;	Aimed at promoting expertise in science-literacy integration;
		Model science-literacy integration;	
		Provides teachers strategies to support learners (classroom investigations; journal writing).	Training **elementary** teachers to construct informal heteroglossic spaces (traslanguaging and multilingualism);
		Provides coaching and material support to teachers.	Scaffolding toward more challenging pedagogical modes (from context-embedded/cognitively undemanding to context-independent/cognitively demanding).
19	Cameron, O'Neill, & Quinn (UK)	Provides teachers with a glossary of technical words in sign language;	Promoting deaf awareness among science teachers (**K-12**);
		Provides teachers with online resources (images, videos, and texts);	Specialized scientific terms are translated into sign languages to make science accessible to deaf students;
		Workshops for teachers (handouts, PowerPoint), and science festivals.	Iconic and visual metaphors in the new signs can be students understand science concepts.

References

Andrews, S. (2010). Why do L2 teachers need to "know about language"? Teacher metalinguistic awareness and input for learning. *Language and Education, 13*(3), 161–177.

August, D., Artzi, L., & Mazrum, J. (2011). Science and vocabulary for English learners. *AccELLerate! 42*(3), 14.

Boroditsky, L. (2001). Does language shape thought? Mandarin and English speakers' conceptions of time. *Cognitive Psychology, 43*(1), 1–22.

Bowerman, M., & Levinson, L. (2001). *Language acquisition and conceptual development.* Cambridge: Cambridge University Press.

Bunch, G. C. (2013). Pedagogical language knowledge: Preparing mainstream teachers for English learners in the new standards era. *Review of Research in Education, 37*(1), 298–341.

Camarota, S. A. (2007). Immigrants in the United States, 2007. *Center for Immigration Studies.* Retrieved December 23, 2012, from www.cis.org/articles/2007/back1007.htlm

Cline, Z., & Nicochea, J. (2003). Specially designed academic instruction in English (SDAIE): More than just good instruction. *Multicultural Perspectives, 5*(1), 18–24.

Council of Chief State Schools. (2012). *Framework for English language proficiency: Development standards corresponding to the common core state standards and the next generation science standards.* Washington, DC: CCSSO.

Council of Europe. (2011). *Common European framework of reference for languages: Learning, teaching, assessment.* Retrieved from http://www.coe.int/t/dg4/linguistic/Cadre1_en.asp

Echevarria, J., Vogt, M., & Short, D. J. (2003). *Making content comprehensible for English language learners: The SIOP model* (2nd ed.). Boston: Pearson Allyn & Bacon.

Echevarria, J., Richards-Tutor, C., Canges, R., & Frances, D. (2011). Using the SIOP model to promote the acquisition of language and science concepts with English learners. *Bilingual Research Journal, 34*(3), 334–351.

Fathman, A. K., & Crowther, D. T. (2006). *Science for English language learners: K-12 classroom strategies.* Arlington: NSTA Press.

Freeman, Y. S., & Freeman, D. E. (2009). *Academic language for English language learners and struggling readers: How to help students succeed across content areas.* Portsmouth: Heinemann.

Fritzen, A. (2011). Teaching as sheltering: A metaphorical analysis of sheltered instruction for English language learners. *Curriculum Inquiry, 41*(2), 185–211.

Garcia, O., Kleifgen, J. A., & Falchi, L. (2008). *From English language learners to emergent bilinguals* (Equity matters: Research review no. 1). New York: Columbia University.

Gentner, D., & Boroditsky, L. (2001). Individuation, relativity, and early word learning. In M. Bowerman & L. Levinson (Eds.), *Language acquisition and conceptual development* (pp. 215–256). Cambridge: Cambridge University Press.

Gentner, D., Imai, M., & Boroditsky, L. (2002). As time goes by: Evidence for two systems in processing time → space metaphors. *Language and Cognitive Processes, 17*, 537–565.

Glass, R., & Oliveira, A. W. (2014). Science language accommodation in elementary school read-alouds. *International Journal of Science Education, 36*(4), 577–609.

Gunckel, K. L., Covitt, B. A., Salinas, I., & Anderson, C. W. (2012). A learning progression for water in socio-ecological systems. *Journal of Research in Science Teaching, 49*(7), 843–868.

Honigsfeld, A., & Dove, M. G. (2010). *Collaboration and co-teaching: Strategies for English learners.* Thousand Oaks: Corwin.

Interagency Language Roundtable. (2011). *Description of proficiency levels.* Retrieved from http://www.govtilr.org/Skills/ILRscale1.htm

Larsen-Freeman, D. (1997). Chaos/complexity science and second language acquisition. *Applied Linguistics, 18*(2), 141–165.

Lee, O., Quinn, H., & Valdes, G. (2013). Science and language for English language learners in relation to next generation science standards and with implications for common core state standards for English language arts and mathematics. *Educational Researcher, 42*(4), 223–233.

Luft, J. A. (2002). I do not understand. In D. J. Tippins, T. R. Koballa, & B. D. Payne (Eds.), *Learning from cases: Unraveling the complexities of elementary science teaching* (pp. 85–88). Boston: Allyn & Bacon.

Mackey, W. F. (1970). A typology of bilingual education. *Foreign Language Annals, 3*(4), 596–908.

Maton, K. (2013a). *Knowledge and knowers: Towards a realist sociology of education*. London: Routledge.

Maton, K. (2013b). Making semantic waves: A key to cumulative knowledge-building. *Linguistics and Education, 24*(1), 8–22.

McGlone, M. S., & Harding, J. L. (1998). Back (or forward?) to the future: The role of perspective in temporal language comprehension. *Journal of Experimental Psychology: Learning, Memory, and Cognition, 24*(5), 1211–1223.

National Council for Accreditation of Teacher Education [NCATE]. (2007). State requirements for pre-service teachers of ELLs. *Educating English language learners: Building teacher capacity* (Vol. III). Retrieved from http://www.ncela.us/files/uploads/3/EducatingELLsBuildingTeacherCapacityVol3.pdf

National Research Council [NRC]. (2012). *A framework for K-12 science education*. Washington, DC: National Academy Press.

NYSED. (2012a). *NYS bilingual common core initiative: Theoretical foundations*. Albany: Engage. Retrieved from https://www.engageny.org/file/135506/download/nysbcci-theoretical-foundations.pdf

NYSED. (2012b). *NYS bilingual common core Initiative: Teacher's guide to implement the bilingual common core progressions*. Albany: Engage. Retrieved from https://www.engageny.org/file/135511/download/teachers-guide-to-implement-the-bilingual-cc-progressions.pdf

Oliveira, A., Sadler, T., & Suslak, D. (2007). Analyzing language, interaction and outcomes in an inquiry-oriented classroom. *Cultural Studies of Science Education, 2*(1), 165–170.

Peregoy, S., & Boyle, O. (2013). *Reading, writing, and learning in ESL: A resource book for teaching K-12 English learners* (6th ed.). New York: Pearson.

Pilgreen, J. L. (2010). *English learners and the secret language of school: Unlocking the mysteries of content-area texts*. Portsmouth: Heinemann.

Pimm, D. (1987). *Speaking mathematically*. London: Routledge.

Plummer, J. D., & Maynard, L. (2014). Building a learning progression for celestial motion: An exploration of students' reasoning about the seasons. *Journal of Research in Science Teaching, 51*(7), 902–929.

Rosebery, A. S., & Warren, B. (2008). *Teaching science to English language learners: Building on students' strengths*. Arlington: NSTA Press.

Rush, L. S. (2002). I am a science teacher, not a reading teacher: An unfortunate categorization. In D. J. Tippins, T. R. Koballa, & B. D. Payne (Eds.), *Learning from cases: Unraveling the complexities of elementary science teaching* (pp. 71–72). Boston: Allyn & Bacon.

Shea, N., & Duncan, R. (2013). From theory to data: The process of refining learning progressions. *Journal of the Learning Sciences, 22*(1), 7–32.

Short, K. G. (2004). Researching intertextuality within collaborative learning environments. In N. Shuart-Faris & D. Bloome (Eds.), *Uses of intertextuality in classroom and educational research* (pp. 373–396). Greenwich: Information Age Publishing.

Silva, C., Weinburgh, M. H., & Smith, K. H. (2013). Not just good science teaching: Supporting academic language development. *Voices in the Middle, 20*(3), 34–42.

Smith, C. L., Wiser, M., Anderson, C. W., & Krajcik, J. (2006). Implications of research on children's learning for standards and assessment: A proposed learning progression for matter and the atomic molecular theory. *Measurement, 4*(1–2), 1–98.

Sutton, C. R. (1992). *Words, science and learning*. Buckingham: Open University Press.

Swales, J. (1990). *Genre analysis: English in academic and research settings*. Cambridge: Cambridge University Press.

Their, M. (2002). *The new science literacy: Using language skills to help students learn science*. Portsmouth: Heinemann.

U.S. Census. (2010). *Foreign born*. Retrieved December 23, 2011, from www.census.gov/population/foreign

Wei, L., & Garcia, O. (2013). *Translanguaging: Language, bilingualism and education*. New York: Palgrave Pivot.

Weinburgh, M. H., Silva, C., Malloy, R., Marshall, J., & Smith, K. (2012). A Science lesson or language lesson? Using the 5Rs. *Science & Children, 49*(9), 72–76.

Whorf, B. L. (1956). The relation of habitual thought and behavior to language. In J. B. Caroll (Ed.), *Language, thought and reality: Selected writings of Benjamin Lee Whorf* (pp. 134–159). Cambridge, MA: MIT Press.

WIDA. (2012). *2012 amplification of the English language development standards: Kindergarten-Grade 12*. Madison: Board of Regents of the University of Wisconsin System.

Wright, L. J. (2014). Academic conversations: Developing critical speaking and listening skills. Paper Presented at *National Professional Directors' Meeting* in Washington, DC. Retrieved from http://www.ncela.us/files/uploads/51/Academic%20Conversations.pdf

Part I
Pre-service Teacher Preparation

The Project

The Inheritances Books are a collaborative project between Ichabod Crane High School's Illustration and ELL students. It is made possible by a grant from the Berkshire Taconic Community Foundation.

The Student-Artist

Christian Szymanski, grade 12, 18 years old.

Chapter 2
Cultivating Teacher Knowledge of the Role of Language in Science: A Model of Elementary Grade Pre-Service Teacher Preparation

Marco A. Bravo

Introduction

A perfect storm forms for English Language Learners (ELLs) when the preservice teachers that will work with them have significant reservations about their efficacy to teach science *and* do not feel prepared to work effectively with this student population. Many teachers feel inadequately prepared to work with ELLs (Durgunoglu and Hughes 2010; Gándara and Santibañez 2016; Téllez and Manthey 2015) and this has an impact on the educational performance of ELLs (Turkan et al. 2014).

The challenge elementary teachers face of engaging an increasingly linguistically diverse population of students in learning about the full array of academic disciplines is particularly significant in science, a discipline that few elementary teachers are well-prepared to teach. In the 2013 Horizon Research National Survey of Science and Mathematics Education, only 39 % of elementary teachers reported that they felt well qualified to teach science, as compared to 77 % for mathematics and 81 % for language arts/reading (Banilower et al. 2013). In the same survey, only 2 % of teachers had undergraduate degrees in science or science education, and 47 % reported four semesters or less of college-level science coursework. Given these low levels of preparation, it is not surprising that many teachers feel inadequately prepared to teach science. This is troubling given that the more competent teachers feel, the more successfully they teach, and vice versa.

Reform-based approaches, accompanied by standards-based assessments of content area knowledge, require that deeper content be made accessible and comprehensible to all students, and that they be provided with a coherent view of how Science Technology Engineering and Mathematics (STEM) disciplines are investigated and applied (Duschl et al. 2007). These significant shifts require teachers to

M.A. Bravo (✉)
Santa Clara University, Santa Clara, CA, USA
e-mail: mbravo@scu.edu

A.W. Oliveira, M.H. Weinburgh (eds.), *Science Teacher Preparation in Content-Based Second Language Acquisition*, ASTE Series in Science Education,
DOI 10.1007/978-3-319-43516-9_2

build new attitudes, understandings, and knowledge about teaching, learning, and content. To foster learning in science and other domains, teachers need to have rich, flexible knowledge. "They must understand the central facts and concepts of the discipline, how these ideas are connected, and the processes used to establish new knowledge and determine the validity of claims" (Borko 2004, p. 5). Teachers are not able to teach what they do not know.

The trends discussed above are of great concern and have led to recommendations for significant national and state efforts aimed at increasing both teacher qualification and quality. Accordingly, the need to support preservice teachers in how to successfully address the science and language learning needs has never been greater. We cannot achieve educational excellence as a nation until we elevate the importance of and support teachers in providing ELLs with opportunities to learn deep science content and scientific ways of thinking while allowing space for them to sharpen their language skills.

This chapter explains a model for preparing preservice teachers to teach science and language in a synergistic manner. Two teacher education courses were revamped to take advantage of the 'curricular sweet spots' between the *Foundations in First and Second Language Acquisition* and *Elementary Science Methods* courses. The *Foundations in First and Second Language Acquisition* course used science content as the backdrop to consider English language development opportunities as well as unpacking approaches to scaffold the science learning of ELLs. In the *Elementary Science Methods* course, similarly language was used as a backdrop to explain how language plays a critical role in accessing and understanding science.

Cross-Curricular Integration and Coherence

In traditional teacher preparation models, language and science are segregated curricular enterprises. Coursework typically follows a sequence much like in Fig. 2.1 presented below.

In the model above, courses labeled as *Foundations* and those considered *Methods* often exist in silos. The science faculty address issues of science knowledge and processes while language educators typically focus on foundational theories and practices for acquiring a first and/or second language. The course content rarely seeps over to the other. The absence of cross-over is also felt in the *Clinical Experience* practicum where preservice teachers sharpen their pedagogy. Coherence between what happens in methods/foundations courses and field experiences does not always overlap.

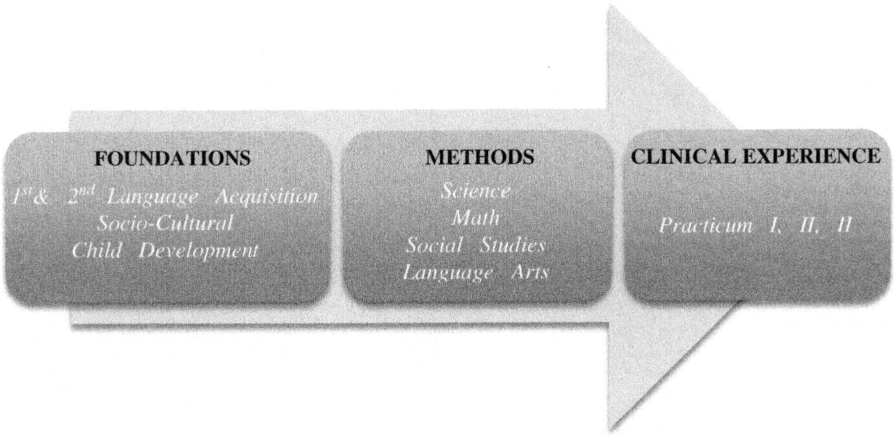

Fig. 2.1 Traditional teacher preparation model

Integrated and Coherent Model of Teacher Preparation

To reconsider the traditional teacher preparation model, faculty teaching in the teacher preparation program where I work considered the benefits to integration of coursework and coherence between the clinical experience and university course work. The model that we developed, capitalized on the natural convergences among these disciplines. Figure 2.2 illustrates the model being tested.

Science and Language Integration

The work began with a close view of how both science and language would be served if instead of trying to build teacher knowledge, skills, and dispositions about each discipline through an encapsulated approach, we integrated the two. The *Foundations in First and Second Language Acquisition* course for example, used science as the context in which to support preservice teachers' understanding of the necessary ingredients for optimizing language acquisition–meaningful language interactions (Halliday and Matthiessen 2004; Krashen 1982; Wong-Fillmore 2007)—which could be had in and were a part of science education. Similarly, we noted that science learning, particularly for ELLs, is facilitated when the language of science is demystified for students (August et al. 2009; Lee et al. 2013; Stoddart et al. 2002). Inquiry based science instruction was offered to the pre-service teachers as the meaningful context to acquiring the English language while explicit instructional attention to the language of science, including the modes of discussion, writing and reading science, was presented as essential for ELLs to better understand the nature of science.

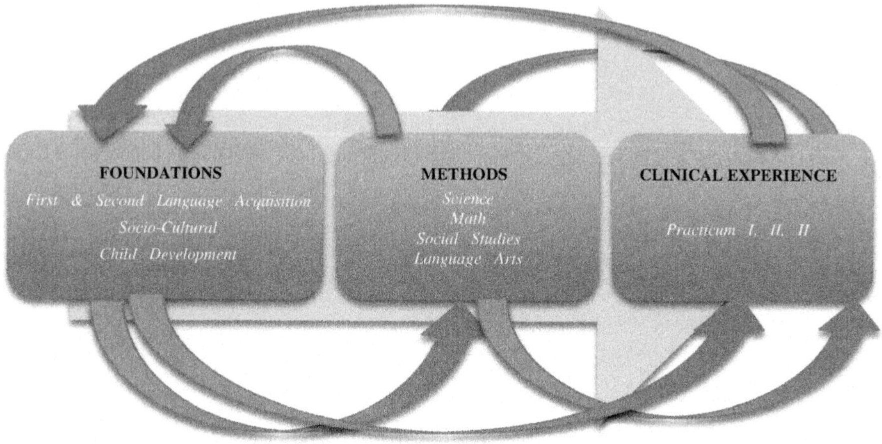

Fig. 2.2 Integrated teacher preparation coursework

The methods instructors also used a common curriculum that integrated science, language and literacy as material to exemplify the potential for integration. Moreover, both courses included readings about and activities exemplifying the natural convergences between science and language, including those listed above (see Appendix 1 for course plan). Delivery of these convergences to preservice teachers came in the form of meta-pedagogical discussions. These discussions are moments when the methods instructors step out of the role of elementary grade teacher and into the role of teacher educator to underscore a practice, detailing the theoretical origins, research support, connections to new standards, and how to enact a particular practice that illustrate convergence between the disciplines.

These natural convergences between language and science include the notion that robust vocabulary development, as viewed by language educators, is similar to what science educators refer to as conceptual development. Second, *Nature of Science* learning goals found in the Next Generation Science Standards (NGSS Lead States 2013) and expressive/receptive language goals in second language acquisition theory (Bialystok 2001) involve similar practices.

Vocabulary and Science Conceptual Development Synergy

In the *Elementary Science Methods* course, preservice teachers received opportunities to consider *words as concepts*. This consideration to vocabulary development moves beyond just knowing the definition of a word to more robust vocabulary knowledge, including seeing how key science terms relate to each other and how targeted vocabulary is used differently across different contexts. This type of contextual attention to vocabulary blurs the distinction with how language educators

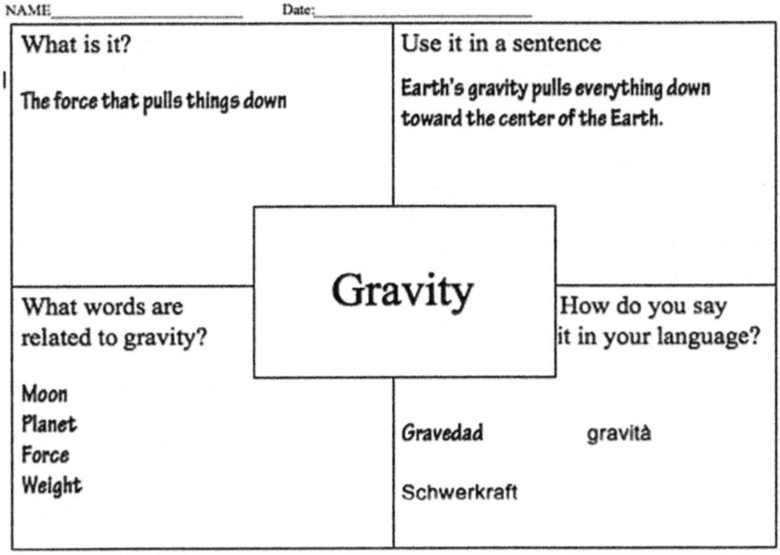

Fig. 2.3 Gravity semantic map

view vocabulary development and the way science educators consider science conceptual development.

To support a level of ownership of key science concepts, the science methods faculty discussed with preservice teachers the need to expose students to key terms repeatedly (Beck et al. 2002) in both first and second hand investigations (Bravo and Cervetti 2008), ensuring opportunities to both hear and use the key terms in science activities and to think about how words relate to each other. In the shared Space Science unit (below), the science faculty exemplified this by providing her preservice teachers with the semantic map in Fig. 2.3.

The science methods instructor introduced the concept/vocabulary *gravity* by definition, but then went on to explain other dimensions of 'knowing' the term *gravity*. This suggestion originated from the language educator's training when addressing vocabulary development in a second language. The meta-pedagogical discussion with pre-service teachers in the course is captured below (Table 2.1).

This level of attention to vocabulary provided support to preservice teachers' conceptual understanding and also served as a model for facilitating the teaching of science concept to ELLs in K-8 grade levels.

The science faculty also provided insights into the type of vocabulary the language educator could target in his class. The science methods faculty noted that many of her students have command of the everyday meaning to certain science words but do not recognize the nuance differences in meaning of the science word (e.g., dirt/soil; home/habitat; guess/predict). She reminded the language educator that part of the discourse of science is to use language with precision. The language educator in turn targeted vocabulary development in his course that would use what

Table 2.1 Vocabulary meta-pedagogical discussion

Participant	Contribution	Author commentary
Faculty	You noticed that before we got in too deep with the movement of planets, we discussed key vocabulary like gravity…so that students can use them in their discussions and writing and understand them when the teacher uses the vocabulary as well	Faculty in front of the classroom and pre-service teachers have models of planets at their desks
Pre-service Teacher 1:	So we frontload vocabulary but not just the definition	Pre-service teacher pointing at the gravity word chart. Frontloading vocabulary is a concept reviewed in the Language Methods course
Faculty	Yes. What else do you notice about this activity?	Calls on Pre-service teacher 2
Pre-service Teacher 2:	I like that we can use the students' native language	Faculty listens to student response
Faculty	What other categories can we include instead of Native language or words related? Think-Pair-Share and then we'll share some examples	Pre-service teachers in small groups write down other example and prepare to share
Faculty	What did this group find?	Group in the back is called on and other groups provide additional examples
Pre-service Teacher 3:	We have 1. How well do you know the word; 2. Is it a noun, verb, adjective; 3. How do you think we'll use this word in the activity?	Faculty listens to student response

students knew about a science meaning, but extended that to a more precise scientific understanding of the term.

Preservice teachers were informed that in science, as in other disciplines, there are particular ways of talking that help better explain science ideas. They were told scientists use words that can be similar but not identical in meaning. Examples were provided (e.g., study/investigate; look/observe; guess/predict) and explanations as to the science meaning were given (e.g., *observe* meaning the use of *all* of your senses not just sight; *predict* referring to a guess that is leveraged with strong and varied evidence).

An activity created by the language educators had preservice teachers use the Space Science curriculum to locate other words that have an *Everyday* meaning and a more precise *Science* meaning. Students were also given opportunities to see the relationship between the concepts on a chart paper where the language educator captured the word pairs identified by students. Figure 2.4 shows the word chart created.

The emphasis on robust vocabulary learning with this activity was not to focus on definitional aspects of word learning, rather to recognize that the terms are part of a conceptual network of concepts that define areas of science study.

Preservice teachers illustrated understanding of the potential of this activity. In the transcript below, preservice teachers discuss how this activity can be utilized with different types of vocabulary and other content areas as well (Table 2.2).

In the example above, preservice teachers appropriate the activity and find use for it across content areas.

Nature of Science and Expressive/Receptive Language Synergy

To help students see science as a process rather than a set of facts to be memorized, goals of NGSS Lead States (2013) include students learning about the practices of science (e.g., gather evidence, make and test models, organize and analyze data). Learning about how scientists go about their work provides students a view into the habits of mind involved in science. These include posing their own questions to investigate and to critically evaluate data gathered.

Fig. 2.4 Science/everyday word chart

Everyday Word	Science Word
See	Observe
Information	Data
Trip	Mission
Robot	Rover
Make	Design

Table 2.2 Everyday/academic word discussion

Participant	Contribution	Author commentary
PreService Teacher 5:	This is great for tier 1 and tier 2 words dirt and soil	Referencing discussion of Beck, McKeown and Kucan's (2002) categorization of words according to frequency of occurrence. (sitting with 4 classmates in small group)
PreService Teacher 6:	I think you could use this with other subjects too like in math or language arts	Students come up with other word pairs in other subject areas using vocabulary lists from teacher's curriculum guide and write them on chart paper
	...Do and evaluate and same and equivalent can work	
PreService Teacher 7:	Leave and Abandon I think works for language arts	Examples provided by an additional student sitting close by
Faculty	This is great work. Would this group mind sharing with the rest of the class?	Group shares with class their additional word pairs

The practices of science require students to understand language (receptive language) and produce language (expressive language). Expressive and receptive language are important concepts in the arena of second language acquisition because they speak to the need of native speakers of the language to serve as language models (receptive language) as well as opportunities to use the language (expressive language) with expectations of offering positive corrective feedback (Pérez and Torres-Guzmán 1992) with the ultimate goal of developing language proficiency. Preservice teachers are usually introduced to these concepts of receptive/expressive language in a decontextualized or inauthentic manner. Using science as the context to discuss expressive/receptive language helps preservice teachers see goals of the Common Core State Standards related to English Language Arts (CCSS-ELA) (National Governors Association Center for Best Practices and Council of Chief State School Officers 2010) in a more salient manner.

An example that illustrates this synergy also involved the Space Science unit (Session 2 on Appendix 1) which faculty in both science and language courses taught to develop preservice teachers' understanding of the nature of science and speaking/listening for their respective courses. In the science methods course preservice teachers experienced a science discourse circle to exemplify the practices of science. After reviewing and building their knowledge base about what makes a planet a planet, preservice teachers had a discussion regarding whether Pluto should be considered a planet or not. In the discourse circle activity below, teacher candidates constructed arguments for and against Pluto's planetary status and tried to convince others of either a "no" or "yes" planetary classification (Fig. 2.5).

Teacher candidates learned about the fact that scientists discuss their ideas with others, read the work of other scientists and leverage evidence to support their claims. This latter point allowed for a natural discussion about providing ELLs with receptive and expressive language learning opportunities. Of particular focus was ensuring to provide clear models of the expected outcomes (receptive language) and positive corrective feedback of ELLs' contributions (expressive language) during the discourse circle.

In the *Foundations for First and Second Language* course preservice teachers attended to the nature of science learning goals and receptive/expressive language-learning goals through a similar process and the use of an Earth science unit. In the unit, they read a book of a shoreline scientist interested in understanding why beaches are shrinking along shorelines that are populated by people. Using this backdrop such practices of science as asking questions, making predictions, making observations and using tools are provided. Each of these practices requires attention to either receptive or productive language. The instructor prompted students to categorize each science practice as involving receptive, productive language, or both. Such context to discuss this key second language learning goal, is optimized by using an authentic context as science rather than one that is contrived.

These examples of integration across the *Elementary Science Methods* course and *Foundations of First and Second Language Acquisition* course blurred the distinctions between the courses. Students often asked "what course am I in right now?" Preservice teachers across both courses were reminded that this kind of

NAME:____________________ DATE:________

Instructions. Discuss the question in the box below with your partner. Respond to **both** the (yes) and the (no) sides of the question. Give reasons why you answered *yes* and why you answered *no* on the lines provided.

Pluto should be considered a planet.

Yes, because....	No, because....
· Size is like a rocky planet	· Pluto is small.
· Pluto's surface is solid like a rocky planet.	· Pluto is not a sphere
· Pluto has moons.	· oval shape orbit.
	· tilted orbit.

Fig. 2.5 Science discourse circle

integration creates 'curricular economy', allowing preservice teachers to teach more in less time while increasing ELLs' science understandings and English language development.

Results

Two sets of findings illustrate the potential for the model proposed. First, a set of efficacy studies illustrated the potential to providing language and science lessons to K-8 grade-level students. The second source of evidence includes a series of quasi-experimental design studies with preservice teachers that tested participants' ability to teach science in an integrated manner.

ELL Learning Potential of Integrated Lessons

The science activities presented to the preservice teachers have been tested with K-8 ELLs through a series of efficacy studies (Bravo and Cervetti 2014; Cervetti et al. 2012, 2015). In one study involving 115 fourth-grade classrooms, half the teachers were randomly assigned to use a language and science integrated curriculum model

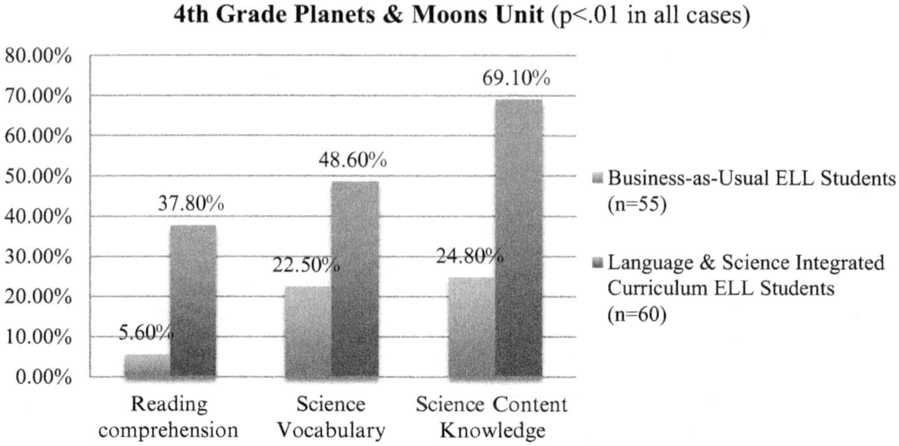

Graph 2.1 ELL percent growth

while the other half were asked to teach the district science curriculum with comparable science goals over an 8-week period for 45 min per day (Bravo and Cervetti 2014). Pre and post assessments of science vocabulary knowledge, science understanding, and reading comprehension were administered. The integrated curriculum provided students with opportunities to read, conduct firsthand investigations and discussions as well as writing tasks related to planets and moons. The comparison classrooms allowed students to conduct firsthand investigations and discuss their findings, but no explicit instruction on literacy was offered during science instruction. The researchers found that ELLs in the integrated classrooms significantly outperformed ELLs in the classrooms where the district curriculum was implemented in the area of science understanding, vocabulary and science knowledge. Graph 2.1 shows the comparison results for both groups of fourth grade students.

Considering students where the teacher used the district-adopted curriculum were involved in more first-hand investigations than the students in the integrated classrooms (time in these classrooms was balanced with reading, writing, vocabulary instructional focus as well as firsthand investigations), it is hypothesized that being able to do, read, write and talk about planets and moons is a more efficacious instructional approach.

Building Preservice Teacher Knowledge of Integrated Curriculum

Data from quasi-experimental design studies with elementary grade preservice teachers illustrate potential of the model for increasing preservice teacher efficacy with teaching science to English learners (Bravo et al. 2014; Teemant et al. 2011).

In one study (Bravo et al. 2014), teacher candidates (n=65) were provided with a science methods course that considered issues of language in science (i.e., vocabulary, reading, writing and talking about science). A comparison was made between this group of preservice teachers to preservice teachers (n=45) who received the more traditional science methods course. When controlling for various factors, including science background, language background, age, gender and other variables, treatment preservice teachers had stronger efficacy about science teaching and working with ELLs than did comparison teachers.

In a follow up study, researchers probed how much of a language focus was involved in the science instruction provided by preservice teachers when they taught science during their practicum experience (Bravo et al. 2015). The science methods course provided similar attention to language and scaffolds for ELLs during science teaching as the study referenced above. Preservice teachers in the treatment condition were more likely to scaffold for language development, promote more science talk and draw attention to the literacy needed to do science than preservice teachers in the more traditional teacher preparation model, where preservice teachers had to make the connection between their science methods and language foundations courses.

Conclusion

This chapter outlined an effort for science and language educators to consider the natural convergences between the two disciplines and the potential service each can have for the other when preservice teachers are prepared to teach to the 'sweet spots' between the two. Taking advantage of these natural convergences also excited preservice teachers, lessening apprehensions about science teaching. Moreover, supporting their training to optimize instructional time by addressing goals in two subject areas at the same time, leads to more productive learning opportunities in both disciplines. In the chapter, examples of the various ways in which science and language converge were offered as well as instructional approaches that can be used to prepare preservice teachers to address both the Common Core State Standards for Language Arts and the Next Generation Science Standards, both standards of which are proposing the type of integration that is suggested in this chapter.

Results from research interventions with children speak to the efficacy of such integration. ELLs are given the chance to sharpen their language skills in an authentic context that also serves developing their science understandings. A series of quasi-experimental design studies with preservice teachers illustrate the potential to learning to teach science and language in a synergistic manner. Preservice teachers whose teacher training program offered the integrated model for science and language felt better prepared to teach ELLs than those that did not receive the intervention methods courses and were more likely to address the language of science during their science teaching.

Given ELLs have 'double the work' (Short and Fitzsimmons 2007) in having to acquire the English language while learning content and under time constraints (meeting graduation requirements), instructional models geared toward ELLs must think more synergistically. Reaching ELLs by accomplishing more with less time and in an authentic manner is what was offered to preservice teachers in the methods courses described in this chapter. The methods faculty feel more confident that the preservice teachers would respond differently to surveys about their efficacy to teach science, work effectively with ELLs and specifically, to teach science to ELLs.

Appendix 1

Elementary science methods course plan		
Class session	Topics	Assignments
Session 1	Course overview	Introductory Science & ELL Teaching Efficacy Survey
	Visions of science teaching for ELLs	
	NGSS 101	
Session 2	Introduction of science teaching framework—Engage, Experience, & Explain with Evidence with embedded Structured English Immersion (SEI) strategies—leverage Ss' Native Language, Address Linguistic blindspots	Reading: Fathman & Crowther Science for English Language Learners: K-12 Classroom Strategies Ch.1
	Science Content: Space Science-*Movement of Planets & Moons*	Ready, Set, Science. Chapter 1, A New Vision of Science Education, pp1–16
Session 3	Unpacking NGSS and what this means for *all* learners, especially ELLs	Online Discussion—Post reflections of NGSS video and respond (questions, clarifications) to 2 posts of classmates
	View NGSS Videos	
	Science Content: Space Science-*Science & Technology, use of models*	Reading: Quinn, Lee & Valdés-Language Demands and Opportunities in Relation to Next Generation Science Standards for ELLs: What Teachers Need to Know

(continued)

Elementary science methods course plan

Class session	Topics	Assignments
Session 4	Planning Science Instruction I: Science Inquiry	Readings:
	Amplify Don't Simplify: SEI Strategies to scaffold abstract science concepts—Visual Representations	Fathman & Crowther Science for English Language Learners: K-12 Classroom Strategies Ch.3
	Science Content: Animal Diversity/ Skulls	BCSC 5E Instructional Model. In Ansberry & Morgan (2007) More Picture-Perfect Science Lessons
		Banchi, H. & Bell, R. (October 2008). The Many Levels of Inquiry. Science and Children, pp. 26–29.
Session 5	Planning Science Instruction II: Science Big Ideas	Readings:
	English Language Development Objectives in Science teaching—vocabulary are science concepts	Westervelt (2007) Schoolyard inquiry for English Language Learners. Science Teacher. March, pp. 47–51
	Share MT Interviews	Ready, Set, Science Chapter 4. Core Concepts
	Science Content: Animal Diversity/ Skulls	**Interview Master Teacher (MT)**: Goals for science learning & supports for ELLs during science
Session 6	Assessing Science & language Objectives	Readings:
	Formative Tools to Guide Instruction	Henriques Assessment in science: Practical experiences and education research, Chapter 2: Developing assessment items: A how-to guide (p. 15–30).
	Science Content: Life Science-*Plant & Animal Habitats*	Cox-Petersen, & Olson (Feb 2007). Alternate Assessments for English Language Learners. Science & Children 46–48.
Session 7	Science Literacy & Literacy in Science	Readings:
	Science Content: Life Science-*Plant & Animal Adaptations*	Cervetti, Pearson, Bravo & Barber (2006). Reading & Writing in the Service of Inquiry-based Science. In R. Douglas, M. Klentschy, and K. Worth (Eds.), *Linking Science & Literacy in the K-8 Classroom* (pp.221–244). Arlington, VA: NSTA.

(continued)

Elementary science methods course plan		
Class session	Topics	Assignments
Session 8	"Funds of Knowledge" in Science Teaching	Readings:
	Making Science relevant and contextualized in students' experiences in home & community	Stoddart, T. Tolbert, S., Solis, J. & Bravo, M. A. (2010). Effective Science Teaching with English Language Learners. In Gonzales, M (ED). Teaching Science with Hispanic ELLs in K-16 Classrooms. Charlotte, NC: Information Age Publishing.
	Science Content: Physical Science-*Light Energy (characteristics of light)*	
Session 9	Explicit Instruction of the Discourse of Science-*Science explanations & argumentation*	Readings:
	Video: Jonathan Osborne-*Argumentation in the Classroom*	Osborne, J.F. (2010) Arguing to Learn in science: The role of collaborative, critical discourse. *Science.*
	Science Content: Physical Science-*Light Energy (light as Energy Source)*	
Session 10	Share final Science Lesson Plans	Submit final Science Lesson Planning Projects.
	Next steps with inquiry based lessons	

References

August, D., Branum-Martin, L., Cardenas-Hagan, E., & Francis, D. J. (2009). The impact of an instructional intervention on the science and language learning of middle grade ELLs. *Journal of Research on Educational Effectiveness, 2*, 345–376.

Banilower, E. R., Smith, P. S., Weiss, I. R., Malzahn, K. A., Campbell, K. M., & Weis, A. M. (2013). *Report of the 2012 National Survey of Science and Mathematics Education*. Chapel Hill: Horizon Research, Inc.

Beck, I. L., McKeown, M. G., & Kucan, L. (2002). *Bringing words to life: Robust vocabulary instruction*. New York: The Guilford Press.

Bialystok, E. (2001). *Bilingualism in development: Language, literacy, & cognition*. New York: Cambridge University Press.

Borko, H. (2004). Professional development and teacher learning: Mapping the terrain. *Educational Researcher, 33*, 3–15.

Bravo, M. A., & Cervetti, G. N. (2008). Teaching vocabulary through text and experience. In A. E. Farstrup & S. Samuels (Eds.), *What research has to say about vocabulary instruction* (pp. 130–149). Newark: International Reading Association Inc.

Bravo, M. A., & Cervetti, G. N. (2014). Equitable science instruction for English language learners. *Equity and Excellence in Education, 47*, 230–245.

Bravo, M. A., Mosqueda, E., Solís, J. L., & Stoddart. (2014). Integrating science and diversity education: A model of pre-service elementary teacher preparation. *Journal of Science Teacher Education, 43*, 1–15.

Bravo, M. A., Mosqueda, E., & Solís, J. L. (2015). *Developing pre-service teacher expertise in integrating science and diversity education*. The American Educational Research Association annual meeting in Chicago, IL.

Cervetti, G. N., Barber, J., Dorph, R., Pearson, P. D., & Goldschmidt, P. G. (2012). The impact of an integrated approach to science and literacy in elementary school classrooms. *Journal of Research in Science Teaching, 49*, 631–658.

Cervetti, G. N., Kulikowich, J. M., & Bravo, M. A. (2015). The effects of educative curriculum materials on teachers' use of instructional strategies for English language learners in science and on student learning. *Contemporary Educational Psychology, 40*, 86–98.

Durgunoglu, A. Y., & Hughes, T. (2010). How prepared are the U.S. preservice teachers to teach English language learners? *International Journal of Teaching and Learning in Higher Education, 22*, 32–41.

Duschl, R. A., Schweingruber, H. A., & Shouse, A. W. (2007). *Taking science to school: Learning and teaching science in grades K-8*. Washington, DC: National Academies Press.

Gándara, P., & Santibañez, L. (2016). The teachers our English language learners need. *Educational Leadership, 73*, 32–37.

Halliday, M. A. K., & Matthiessen, C. M. I. M. (2004). *An introduction to functional grammar* (3rd ed.). London: Arnold.

Krashen, S. (1982). *Principles and practice in second language acquisition*. New York: Prentice- Hall.

Lee, O., Quinn, H., & Valdés, G. (2013). Science and language for English language learners in relation to next generation science standards and with implications for common core state standards for English language arts and mathematics. *Educational Researcher, 43*, 223–233.

National Governors Association Center for Best Practices & Council of Chief State School Officers. (2010). *Common core state standards for English language arts and literacy in history/social studies, science, and technical subjects*. Washington, DC: Authors.

NGSS Lead States. (2013). *Next generation science standards: For states, by states*. Washington, DC: The National Academies Press.

Pérez, B., & Torres-Guzmán, M. E. (1992). *Learning in two worlds: An integrated Spanish/English biliteracy approach*. New York: Longman Publishers.

Short, D., & Fitzsimmons, S. (2007). *Double the work: Challenges and solutions to acquiring language and academic literacy for adolescent English language learners – A report to Carnegie Corporation of New York*. Washington, DC: Alliance for Excellent Education.

Stoddart, T., Pinal, A., Latzke, M., & Canaday, D. (2002). Integrating inquiry science and language development for English language learners. *Journal of Research in Science Teaching, 39*, 664–687.

Teemant, A., Wink, J., & Tyra, S. (2011). Coaching on teacher use of sociocultural instructional practices. *Teaching and Teacher Education, 27*, 683–693.

Téllez, K., & Manthey, G. (2015). Teachers' perception of effective school-wide programs and strategies for English language learners. *Learning Environments Research, 18*, 111–127.

Turkan, S., de Oliveira, L., Lee, O., & Phelps, G. (2014). Proposing a knowledge base for teaching academic content to English language learners: Disciplinary linguistic knowledge. *Teachers College Record, 116*, 1–30.

Wong-Fillmore, L. (2007). English learners and mathematical learning: Language issues to consider. *Assessing Mathematical Proficiency, 53*, 333–344.

Chapter 3
A Language-Based Approach to Content Instruction (LACI) in Science for English Language Learners

Luciana C. de Oliveira

Introduction

Over the last half decade, the number of English language learners (ELLs) in the United States has increased dramatically. Given this increase, it is vital for teacher education programs to address the needs of ELLs in their courses. According to the National Clearinghouse for English Language Acquisition (2010), over 10 % of the K-12 student population is comprised of ELLs, which represents over five million students in schools. The largest number of these students is found in California, Florida, Illinois, New Mexico, New York, Puerto Rico, and Texas. However, states such as Arkansas, Alabama, Colorado, Delaware, Georgia, Indiana, Kentucky, Nebraska, North Carolina, South Carolina, Tennessee, Vermont, and Virginia have experienced more than 200 % growth in the numbers of ELLs in schools (NCELA 2010). The need to prepare teachers to work with these students in all U.S. states, then, is pressing. These rapid changes put pressure on teacher education programs to prepare teachers to work with ELLs (Athanases and de Oliveira 2011). This is a tall order in all content areas, but it is especially important in science.

Science presents disciplinary knowledge very differently from the ways in which meanings are constructed in students' everyday language (Fang 2006; Fang and Schleppegrell 2008). ELLs need to be able to understand the language of science, as language is an integral part of learning science content (Bunch 2013; Lee and Miller 2016; Lee et al. 2013. Therefore, science teachers need to be prepared for meeting ELLs' content and language needs. Because of the growing number of ELLs in mainstream classes, *all* teachers – not just bilingual or English-as-a-Second-Language (ESL) specialists – need to be prepared for working with ELLs (Lucas and Grinberg 2008).

L.C. de Oliveira (✉)
University of Miami, Coral Gables, FL 33146, USA
e-mail: ludeoliveira@miami.edu

A.W. Oliveira, M.H. Weinburgh (eds.), *Science Teacher Preparation in Content-Based Second Language Acquisition*, ASTE Series in Science Education,
DOI 10.1007/978-3-319-43516-9_3

The context of this chapter is a teacher preparation model that started in California and was modified over the years, being applied in Indiana, New York, and Florida. This model was implemented with both in-service teachers as part of their professional learning and pre-service teacher as part of their teacher education programs. Over the course of the past 10 years, elementary and secondary teachers learned about a language-based approach to content instruction (LACI) to address the content and language needs of ELLs in their classes. This chapter argues that LACI enables teachers to simultaneously focus on language and content in the content area of science. Using examples from classroom discourse in a fourth-grade classroom in Indiana, the chapter presents principles – the 6 Cs of support – for implementing LACI in science instruction.

Teacher Preparation Model

The teacher preparation model – a language-based approach to content instruction (LACI) – was developed over the past 10 years of research in content area classrooms with ELLs and based on recent scholarship on the language demands of schooling (e.g. Athanases and de Oliveira 2014; de Oliveira 2007, 2011; de Oliveira and Dodds 2010; de Oliveira et al. 2013; de Oliveira and Lan 2014; de Oliveira and Schleppegrell 2015; de Oliveira and Yough 2015; Fang 2006; Fang and Schleppegrell 2008; Schleppegrell 2004).

This approach attends to recent concerns about addressing the needs of ELLs in the era of standards-based education. The development of the Common Core State Standards (CCSS) (National Governors Association Center for Best Practices and Council of Chief State School Officers 2010a), starting in 2009, marked a new chapter in this era. The CCSS in English Language Arts, Mathematics, and Science, History/Social Studies and Technical Subjects were designed for a general student population and provide little guidance for teachers who have ELLs in their classrooms. The only direction given is a two-page document entitled "Application of Common Core State Standards for English Language Learners" (National Governors Association Center for Best Practices and Council of Chief State School Officers 2010b) that provides very general information about ELLs and their needs. This document does not provide any guidance for teachers in how to adapt and use the CCSS with ELLs, and nothing about how to address the demands and expectations of the standards with this student population. The Next Generation Science Standards (NGSS Lead States 2013) call for integration of language and science for all students, and especially ELLs. The NGSS highlight the need to develop strong literacy skills, as they are vital to building knowledge in science.

LACI clearly addresses the language and content demands and expectations of the CCSS and the NGSS for ELLs.

Understanding the Challenges of a Discipline in Linguistic Terms

Subject matter in schools is constructed in language that differs in key ways from the language we use to interact with each other in daily life (Schleppegrell 2004). This language of schooling (Schleppegrell 2004), often called *academic language*, is difficult for all students, but it is especially challenging for ELLs. Academic language is generally learned in school from teachers and textbooks, and only with proper instructional support (Schleppegrell 2004). Academic language contrasts with *everyday language*, the language used in everyday life. For students without opportunities to develop this language outside of school, the classroom needs to offer opportunities to learn language and content simultaneously. Much of the challenge of the content areas, science included, is linguistic.

Starting in elementary school, especially at the intermediate grades, to learn content, ELLs need to be able to see how language works in texts, read with comprehension, engage in discussion of complex issues, and critically evaluate the texts they encounter. Research focusing on the linguistic construction of secondary content areas (e.g. Fang 2006, 2012; Fang and Schleppegrell 2008, 2010; Schleppegrell 2004) has shown that authors present disciplinary knowledge very differently from the ways in which meanings are constructed in students' everyday language (Fang 2006; Fang and Schleppegrell 2008). My work in K-12 classrooms has demonstrated that, as ELLs progress in school, they need to understand how authors construct the discipline-specific discourse of the content areas. This research has shown that the key linguistic features found in the content areas at the secondary level are already present at the elementary school level. Therefore, identifying the linguistic demands of the content areas for ELLs at the elementary school level is important for teachers so these teachers can develop a better understanding of how authors construct disciplinary knowledge and address these demands in their teaching.

General strategies, such as creating collaborative groups, using visuals, and building on students' background knowledge, often are cited as strategies that work well for ELLs in the content areas (Keenan 2004; Hansen 2006). While these may be helpful strategies for ELLs at the beginning levels of English language proficiency, they are not enough for the science classroom because of the demands of the language of science. To make content accessible to ELLs, many content area teachers of ELLs draw on a variety of strategies and techniques to simplify language and to dilute content. While these strategies and techniques may be helpful for ELLs at the beginning levels of language proficiency, they are not appropriate for ELLs at intermediate to advanced levels, especially as they progress through the elementary grades. Under a watered-down curriculum, ELLs may not be taught academic language and they may never learn to read texts without modifications or adaptations (Gibbons 2006).

A Focus on the Content to Be Taught Through *Language*

A language-based approach to content instruction (LACI) places emphasis on language learning in the content classroom. Teachers must use language to teach content, rather than using content to teach language. Instead of finding relevant content to further language development goals, this approach focuses on enabling teachers to foreground the language as a way into the content. Talking about language *is* talking about content. LACI, with a focus on content *through* language rather than on language through content, can be a means through which instruction for ELLs can be accomplished in meaningful ways in a mainstream classroom.

LACI differs from content-based instruction (CBI). Content-based instruction (CBI) is an approach that focuses on the integration of language and content, common in many English as a second language (ESL) classes. CBI uses meaningful language to motivate students and enable content learning along with language learning (Davison and Williams 2001). In CBI classrooms, a focus on form and meaning should be balanced, indicating that form and meaning are seen as aspects of language that can be addressed separately. The content is considered a means of selecting appropriate, authentic, or motivating language instruction; therefore, teachers must find relevant content to further language development goals.

A key component of LACI is providing mainstreamed ELLs with *access* to the language of the different content areas, not by simplifying content but by enabling ELLs to manipulate language as it is written, without simplification. The notion of making content *accessible* is taken here to mean providing *access* to the academic language that constructs content knowledge. LACI draws on a functional theory of language, allowing for a simultaneous focus on the meanings that are made (the "content") and the language through which the meanings are expressed. This approach is a powerful tool for raising teachers' awareness about the challenges of learning content, and enables them to more effectively contribute to the language development of ELLs in their mainstream classes. The goal is to provide teachers with ways of talking about the language that enable them to focus on the content at the same time that they offer ELLs opportunities to develop academic language proficiency.

To develop this approach, I first observed classroom teaching to expand my understanding about the skills and knowledge teachers needed to help ELLs fully in the content area classroom. Teachers had a range of strategies for building background knowledge, for helping students predict from headings, layout, visuals and other features, and for using strategies such as collaborative work, graphic organizers and other techniques that have been shown to be helpful for ELLs. But teachers had few strategies to employ when it came to actually reading texts and getting meaning from texts. I also identified patterns in content area texts and developed language-focused strategies to help teachers understand the linguistic challenges presented in texts and tasks. Over time I discovered the ways of talking about language that teachers found most accessible, and identified the linguistic constructs that teachers most readily adopted in their own teaching. This chapter presents some of those constructs and describes the kind of focus on content that they promote. By putting the language focus into the content classroom, with a functional focus on the meanings expressed in the language, I enable teachers to simultaneously focus on language and content by seeing how language choices construct content. The chapter places emphasis on how teachers can select grade-level texts from textbooks and other materials, identify the potential challenges for ELLs they may present, and plan instruction to address those challenges.

Systemic-Functional Linguistics: A Meaning-Based Theory of Language

LACI draws on a meaning-based theory of language, *systemic-functional linguistics* (SFL) (Halliday and Matthiessen 2004). This theory does not separately address language and content, but instead sees language as the realization of meaning in context. This perspective puts the focus on content, helping teachers understand how the language works to construct knowledge in the discipline. It offers a way of getting meaning from the text itself, going beyond general reading strategies to provide a means of tackling a content area text, unpacking meanings clause by clause to examine how any content is presented in language. LACI enables a focus on language from each of these three angles: *presenting ideas, enacting a relationship with the reader or listener*, and *constructing a cohesive message* (de Oliveira and Schleppegrell 2015).

Presenting Ideas In terms of presenting ideas, we focus on the content of the message, looking at verbal and visual resources that construct the content presented in the nouns, verbs, prepositional phrases, and adverbs.

Enacting a Relationship with the Reader or Listener When we read, write, listen and speak, we draw on language resources that indicate the kind of relationship we are enacting; whether it is formal or informal, close or distant, and whether it includes attitudes of various kinds. We can explore the verbal and visual resources that construct the nature of relationships among speakers/listeners, writers/readers, and viewers, and what is viewed.

Constructing a Cohesive Message Some of the language choices we make are not about presenting content or enacting a relationship, but instead are in the service of constructing a message that holds together. For this we explore the verbal and visual resources that are concerned with the organization of the information and elements of texts and images used to present content in a cohesive way.

Application to Science

LACI was applied in pre-service and in-service teacher education to focus on the teaching of all content areas – English language arts, mathematics, science, and social studies. For this chapter, I use examples from classroom teaching (lesson plans that pre-service teachers developed and their reflections) to show how LACI was applied in the content area of science.

Implementation

Karla, a fourth-grade teacher, worked with me in Indiana for several years implementing LACI in her classroom. I use excerpts from her classroom teaching to exemplify implementation of LACI in the teaching of science. Karla teaches in a school district with 30 % culturally and linguistically diverse (CLD) students and 70 % White students. Many of the CLD students come from families which are associated with Purdue University, including children of international students and immigrants. Over the course of 5 years, the implementation of LACI went through three "phases". Phase 1 focused on reading science texts and developing lessons to address the challenges of science, and then moved to Phase 2 which addressed writing instruction about science experiments. Phase 3 focused on talking science, or the classroom discourse about science that supported and challenged ELLs (see de Oliveira and Lan 2014, for more details about each phase). Before designing language-based lessons for her classroom, Karla selected key texts to work on with students and used the following application framework to guide her planning.

LACI Application Framework for Teachers

The following analysis and application framework can guide teachers' application of LACI in their classrooms as they set goals based on key concepts, select texts to work with students, analyze these texts, and plan instruction for their ELLs.

Setting Goals, Based on Key Concepts The first step in this framework would be to set particular goals based on key concepts that students will need to develop. Specify the content knowledge that students need to develop.

Selecting a Text The second step is to identify a text – it could be two or three paragraphs that have significant content information related to the key concepts and the main points you want to make. Carefully read the text.

Table 3.1 Questions to guide text analysis

Presenting ideas	Enacting a relationship with the reader or listener	Constructing a cohesive message
Focus on content	Focus on relationships	Focus on organization
1. What is the text/image about?	3. What is the author's perspective?	5. How is the text/image organized?
2. What are the key concepts developed in the text/image?	4. How does the author of this text/image interact with the reader/viewer?	6. How does the text/image construct a cohesive message?

(a) What is most important for students to learn from the selection you have chosen? Write at least one ***guiding question*** that will guide your teaching of this content.
(b) What language challenges in the text may make it difficult for students to understand the content?

Analyzing a Text To explore different language patterns, follow the questions presented in Table 3.1 (adapted from de Oliveira and Schleppegrell 2015), looking at the language features of the text:

Planning Instruction with a Focus on Language and Content Here is when the teacher plans how to draw students' attention to the language as it is encountered in the text. Use these additional steps to guide your planning:

(a) Identify language features that will help students understand the content. Focus on those.
(b) Identify and discuss the main points necessary to understand the text with students.
(c) Write some discussion questions or a list of important questions/points that can be used to guide students to examine the language features and main points.

At this point in planning instruction, teachers use the following principles of implementation as the six Cs of support for ELLs.

Six Cs of Support

LACI builds on a variety of principles that have identified specific elements of instructional activities for CLD students. The principle of *connection* refers to the ways in which teachers can connect pedagogy and curriculum to students' backgrounds and experiences (Cochran-Smith 2004; Villegas and Lucas 2002). The principle of *code-breaking* involves explicitly teaching ways of doing school, academic literacy, and disciplinary, linguistic, and cultural codes of content learning (Fang 2006; Schleppegrell 2001, 2004). Code-breaking includes the integration of language and content as inseparable instructional components. Academic literacy is a process of making academic dimensions of subject matter transparent for ELLs. Bridges between everyday and academic language are essential for understanding of content (Gibbons 2006).

The principle of *community and collaboration* refers to joint productive activity in which students co-construct knowledge (Brown and Campione 1994; Lave and Wenger 1991). Teachers create communities of learners in their classrooms, where all students participate in activities to socially construct knowledge (Nieto 2000). The use of *culture* as a principle enables students to build on prior knowledge by accessing cultural and linguistic resources (Moll et al. 1992; Valenzuela 1999). Students' cultural and linguistic resources, or their "funds of knowledge" from home communities (Moll et al. 1992), are used to support academic learning as ELLs develop new resources to be able to participate in new situations, bridging home and school and enhancing opportunities for students to learn (Valenzuela 1999). *Challenge* relates to classroom goals and activity that explore disciplinary literacy and higher-order thinking and reasoning. High challenge and high academic standards and content are maintained for ELLs (Hammond 2006). The principle of *classroom interactions* focus on "interactional scaffolding," the use of oral discourse to prompt elaboration, build academic literacy, and move discourse and learning forward (Hammond and Gibbons 2005). Interactional scaffolding includes three main processes. (1) Linking to prior experience, pointing to new experiences, and recapping refers to teachers' ability to target a specific learning area to ELLs' current levels of knowledge and their English language abilities. (2) Appropriating and recasting students' contribution, typically during discussions or elicitations, involves the teacher's direction of students' contributions by means of recasting their words into more content, language, context appropriate discourse. (3) Using Initiation, Response, Feedback (IRF) sequence includes teachers offering strong verbal or gestural hints about expected responses, especially targeting specific students for specific purposes so students can say more and reflect on their understanding, that is, they ask for clarifications, probe a student's response, ask to explain a particular point in detail. The principles are summarized in Fig. 3.1.

Karla followed the implementation framework for lesson planning and delivery.

Application Framework Applied in Science: Karla's Animals Lesson

Setting Goals, Based on Key Concepts Karla identified as goals for her students to understand the different animals with backbones and where they lived and some of their key characteristics.

Selecting a Text Karla selected a text that she identified as being difficult for students but had all of the information that was necessary for students to understand the

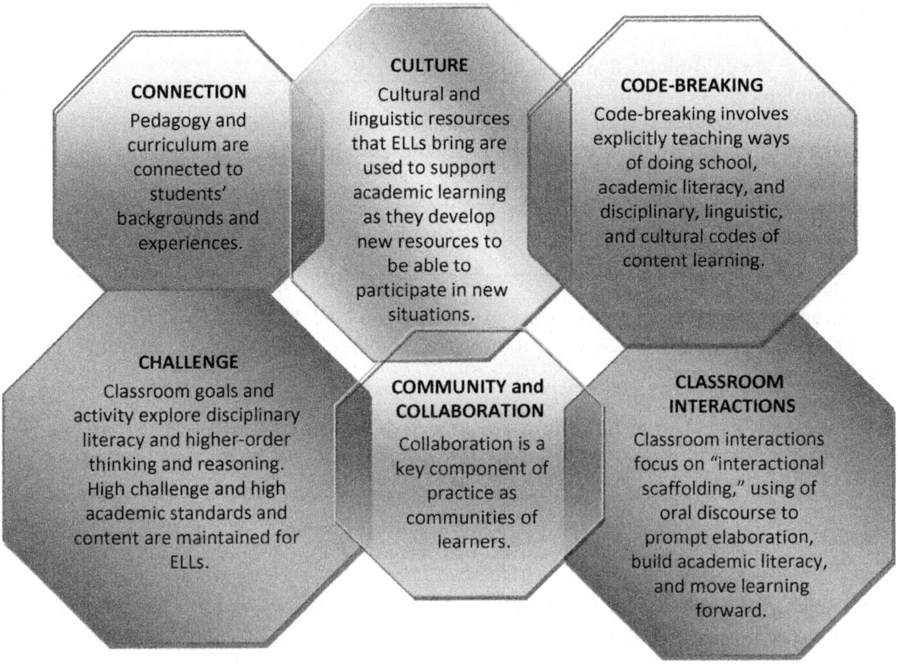

Fig. 3.1 Six Cs of support

key concepts. The selected text was from a textbook commonly used in Indiana and focused on how animals are classified (Harcourt 2005, p. 22).

(a) What is most important for students to learn from the selection you have chosen? Write at least one guiding question that will guide your teaching of this content.

Karla wrote the following guiding questions for this lesson: How are animals classified? What are their differences?

(b) What language challenges in the text may make it difficult for students to understand the content?

Karla explained in her lesson plan (quotes from the lesson plan are in quotation marks) that she selected this text because it presented key concepts about animals and their classification, including their bodies. In addition, she said that the format of the text – "set-up so that the definitions of each of the animals with backbones are separated into boxes" – may be challenging for ELLs.

The text that Karla used from the textbook is presented next (Fig. 3.2).

Karla conducted a text analysis following the concepts she learned about

Analyzing a Text Karla used the questions from Table 3.1 to guide her analysis. In terms of **presenting ideas**, teachers use two guiding questions: What is the text/image about? and What are the key concepts developed in the text/image? To address these questions, Karla identified the *participants, processes*, and

Fish Fish are usually covered with scales. They live only in water. Fish breathe mostly with gills. Fish are cold-blooded and most lay eggs.
Amphibians Amphibians are covered with skin. They can live both on land and in the water. They breathe with lungs or gills or both. They are cold-blooded. Amphibians hatch from eggs.
Reptiles Reptiles are covered with scales. Most reptiles live on land. Some can live in water. They breathe with lungs. Reptiles are cold-blooded. Reptiles usually lay eggs instead of having live births.
Birds Birds are covered with feathers. They usually live on land, but may birds spend much of their time in water. Birds use lungs to breathe. They are warm-blooded. All birds lay eggs.
Mammals All mammals have hair or fur. Most live on land, but few live in water. They breathe with lungs. Mammals are warm-blooded, they make their own heat. Most mammals have live births.

Fig. 3.2 How are animals classified? (Harcourt 2005)

the "scientific pattern" for each clause in the text. *Participants* are the who or what that are participating in the process, represented as a person(s) or thing(s). A *process* is a verb group that shows what is going on (the *doing, thinking, saying* or *being*). The *scientific pattern* is what came after the *doing* or *being processes* that described or explained something about the participants and processes.

This analysis helps to see the content that is presented and the key concepts such as *what animals have* or *what they are covered with*, *where they live*, and other important characteristics. The process column helps the teacher to understand information about the content of the text which she can then teach the students (Table 3.2).

For **enacting a relationship with the reader or listener**, teachers use two guiding questions: What is the author's perspective? and How does the author of this text/image interact with the reader/viewer? Karla noticed that this text presented information in declarative mood, that is, it represents actions or states as objective facts. She also noticed that the author did not interact with the reader by using interrogative sentences, that is, questions which sometimes are found in science texts, such as "Have you ever planted seeds in a garden?" to connect the content that will be presented to students' experiences.

In terms of **constructing a cohesive message**, teachers use two guiding questions: How is the text/image organized? and How does the text/image construct a cohesive message? Karla focused on the use of conjunctions such as *and* and *but* to link some clauses. The conjunction *and* showed addition and *but* showed contrast. She also noticed that many qualifiers such as *usually, mostly,* were used to build cohesion – keeping the text together as a message – and wanted to specifically address their use in the text (see Fig. 3.3). Karla planned a game to focus on these qualifiers (underlined in and in bold in Fig 3.3; see game description in the next section).

Table 3.2 Presenting ideas

Participants	Process	Scientific pattern
Fish	are usually covered	with scales.
They	live	only in water.
Fish	breathe	mostly with gills.
Fish	are	cold-blooded
(and) most	lay	eggs.
Amphibians	are covered	with skin.
They	can live	both on land and in the water.
They	breathe	with lungs or gills or both.
They	are	cold-blooded.
Amphibians	hatch	from eggs.
Reptiles	are covered	with scales.
Most reptiles	live	on land.
Some	can live	in water.
They	breathe	with lungs.
Reptiles	are	cold-blooded.
Reptiles	usually lay	eggs instead of having live births.
Birds	are covered	with feathers.
They	usually live	on land,
(but) many birds	spend	much of their time in water.
Birds	use	lungs to breathe.
They	are	warm-blooded.
All birds	lay	eggs.
All mammals	have	hair or fur.
Most	live	on land,
(but) few	live	in water.
They	breathe	with lungs.
Mammals	are	warm-blooded,
they	make	their own heat.
Most mammals	have	live births.

Planning Instruction with a Focus on Language and Content As Karla was conducting the text analysis, she already started to plan. To help students understand the text, Karla planned a "language dissection" activity, as she named these lessons. Students had been working on the concepts of *participants, process*, and *scientific description*, based on systemic functional linguistics and part of the implementation of LACI, introduced in previous lessons.

As Karla analyzed the text, she also found that the text had many "qualifiers" which we discussed as being words such as *some, many, usually* etc. She pulled the text out from the textbook and underlined and bolded key "chunks" of text. She introduced the text to students and asked them why certain chunks were highlighted. The students also reviewed the meaning of "qualifiers" to which they had been

Fish Fish **are usually covered** with scales. They live **only** in water. Fish breathe **mostly** with gills. Fish **are** cold-blooded and **most** lay eggs.
Amphibians Amphibians **are covered** with skin. They **can live** both on land and in the water. They breathe with lungs or gills or both. They **are cold-blooded**. Amphibians hatch from eggs.
Reptiles Reptiles **are covered** with scales. **Most** reptiles live on land. **Some can** live in water. They breathe with lungs. Reptiles **are** cold-blooded. Reptiles **usually** lay eggs instead of having live births.
Birds Birds **are covered** with feathers. They **usually** live on land, **but may** birds spend **much** of their time in water. Birds use lungs to breathe. They **are** warm-blooded. **All** birds lay eggs.
Mammals **All** mammals **have** hair or fur. **Most** live on land, **but few** live in water. They breathe with lungs. Mammals **are** warm-blooded, they make their own heat. **Most** mammals have live births.

Fig. 3.3 Qualifiers

introduced in the prior lesson. Karla also used the following to explain qualifiers and introduce students to the "poles on each end of the spectrum".

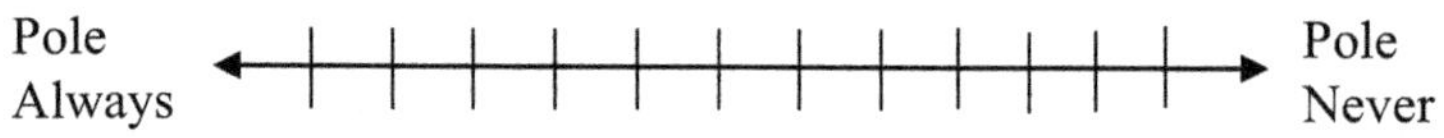

Karla explained to students that "there are different kinds of qualifiers in sentences, sometimes describing verbs, and sometimes describing nouns." Karla described "the polarity of the qualifiers" with the visual representation above. After the "language dissection," students played a game and continued to explore the language of the text.

Game directions:

1. The class will come up with words that can be used as qualifiers. The teacher will type these words or the students will write them on their own slips of paper. A list has been started below.
2. The students will work in groups, each receiving a different section of the text. They will cut out the sentences of the text, removing the qualifier completely from the text.
3. Students will replace and/or add qualifiers from the list that the class made together and discuss how the meaning of the sentence has changed. The amphibian example will need to add qualifiers. The class can discuss how adding qualifiers won't necessarily change the meaning.
4. The students will quiz other groups to find out who can remember the "correct" qualifiers.

To plan for providing the 6 Cs of support, Karla identified ways to *connect* her lesson to students' backgrounds and experiences. For *code-breaking*, she specifically planned to use the metalanguage of participants, process, and scientific description to help students identify patterns in text to promote content learning. For the principle of *community and collaboration*, Karla planned a joint productive activity in which students co-construct knowledge, a class discussion that promoted students' content knowledge development. For *culture*, Karla planned to build on prior knowledge by accessing cultural and linguistic resources. Karla also addressed *challenge* by focusing on disciplinary literacy and higher-order thinking and reasoning. She used *classroom interactions* to prompt elaboration, build academic literacy, and move discourse and learning forward.

Implementation

Karla developed a lesson that incorporated the challenges she identified through her text analysis and the key concepts essential for understanding of content. The following excerpt is a discussion about the text that Karla had analyzed and identified as presenting important content. At this point in the lesson, teacher and students are going through the text to identify the *participants, processes*, and *the "scientific pattern"* for each clause in the text, paying attention to specific sentence patterns.

K:	*Amphibians are covered with smooth skin.* What so far are we talking about with both of those sentences? Look at Alena's hand, she is writing up!
S (Alena):	Like what they have on their body.
K:	What they have on their bodies. Let's see if this continues. Carla, will you bring yours up, #1 go ahead and read it for me.
S:	*Reptiles are covered with scales.*
K:	*Reptiles are covered with scales.* Again are we talking about what's on the outside of the body? Edna, bring yours up and read it out loud for us.
Ss:	*Birds are covered with feathers.*
K:	*Birds are covered with feathers.* Does anybody notice a verb that keeps being repeated for the process? What keeps being repeated, Nora?
S:	*covered.*
K:	Yes, we have the word **covered** every time. Is the word **covered** in your process, too, Alena?
S (Alena):	Oh, no!
K:	We have a difference here! Is it still the same, are we still talking about the same process?
S:	Yeah, what it's covered in.

K: We're still talking about the same thing except that it didn't say covered with fur. What do you think that means, why do you think they didn't use the word covered for this one? -they used the word covered for every other one. Do you have an idea, Laura? Why didn't they use the word covered this time. Alena, you are up here why don't you tell us.

S (Alena): Because they aren't all covered, they could have hair or fur

K: Yeah, think about it, they aren't all covered. Are you covered, do we have some hair?

S: Yeah

K: But, we probably wouldn't say covered with hair. We don't look like bears! So they chose not to put covered for this very last one; kind of interesting. Let's see if our second sentence has a pattern, Laura. You guys have fish, so go ahead and read the fish sentence for me. You guys have been great listeners, by the way!

S: *They live only in water.*

K: *They live only in water.* What are we talking about this time? What is the topic of our sentence? What are we talking about, Carla?

S (Carla): Where they live.

K: Where they live. Let's see if the second group also talks about where they live. Brian. What you have?

S: *They can live on land and in water.*

K: *They can live on land and in water.* Do you think the authors purposely put these in the same order every time?

Ss: Yeah

K: Do you think you would have noticed right away that they put them in the same order?

Ss: No.

K: I wouldn't have noticed honestly but you know what, you may have noticed right away. But sometimes in science books and writing, you're going to see these types of patterns if you look at them carefully. Let's look at the next group to see if it also tells about where they live. Kathryn, go ahead.

This exchange exemplifies Karla's implementation of several Cs of support. First, Karla is using *classroom interactions* to focus on *code-breaking*. Karla and the students are discussing the use of the process *are covered* in the text, and Karla calls students' attention to this pattern and asks them to notice why *covered* is not used in other examples when the text discussed *hair or fur*. This part of the text was "All mammals have hair or fur." These examples show Karla's focus on disciplinary literacy. This is part of a collaborative activity in which students were engaged, exemplifying the principle of *collaboration and community*. The principle of *challenge* can also be seen in this example, as students are discussing challenging concepts. Karla is drawing on students' backgrounds and experiences to discuss the process covered, an example of the *connection* principle. Karla asks Alena, an ELL, a specific question, drawing on the ELL's linguistic resources, an example of *culture*.

Outcomes

The examples above show how teachers like Karla developed understanding of the role of language in science learning. Karla was introduced to ways of talking about language through the implementation of LACI. Karla used specific metalanguage – or a language to talk about language (de Oliveira and Schleppegrell 2015) – not only to discuss lesson design and conduct text analysis but also to implement lessons in her classroom. Over and over again, Karla reminded me of the difficulties with reading and writing the science textbook and the need for greater explicitness about how to engage students with science texts. Teachers like Karla can benefit from an approach such as LACI if it is introduced in teacher education programs systematically and applied by teachers in their classrooms. Despite repeated calls in the literature for greater attention to language in teacher education programs that prepare content area teachers, there are few empirical investigations of how professional development efforts support pre- and in-service teachers.

Teachers who have gone through pre- and in-service professional preparation with LACI have found that the text selection and the planning parts of the model are essential for them to implement LACI in the classroom. For example, teachers often ask about how to select the texts to present to ELLs and ways to focus on the language and content of the texts. We discuss how text selection is highly dependent on the key concepts students need to understand as well as the challenges the text may present to ELLs. In addition to LACI, pre-service teachers also receive an introduction to other models for teaching ELLs, including the Sheltered Instructional Observation Protocol (SIOP), Content-Based Instruction (CBI), and the Cognitive Academic Language Learning Approach (CALLA), which are contrasted with LACI. Every model has some benefits and some challenges, and we typically discuss what these are and how LACI involves a close look and discussion of texts that is not readily available and supported by these other models. We also discuss how LACI focuses on the content classroom and CBI and CALLA on the language classroom and SIOP in sheltered techniques. LACI, on the other hand, places a focus on content learning with a focus on language, more appropriate for mainstream content area teachers.

Summation

Choosing a particular text and deconstructing its language features provides more than an abstract focus on language. ELLs will have opportunities to explore the different patterns of language that construct different types of texts. By focusing on texts, we can show ELLs language patterns that present specific content, which encourages conversation in the classroom about which content is presented, who is represented and how, and how the text is organized.

In order for teachers to understand how language works in their discipline, they need practice in seeing how language expresses disciplinary knowledge. Since most teaching today is based on adopted standards, teachers should be able to work with texts that they choose to address the standards they must teach. In order to develop lessons, teachers begin with the selection of key concepts, often from their state standards. They then design units of study that incorporate language analysis to highlight the key concepts in their content curriculum. Teachers choose a related text and develop a guiding question that will focus their analysis and discussion. Then they engage in text deconstruction, first to learn more about it themselves, and then to design activities that could engage their ELLs in seeing the multiple meanings embedded in the text.

Mainstream, science teachers need knowledge and practical ideas about addressing the academic language needs of ELLs because they have the dual responsibility of facilitating ELLs' content learning while also supporting their ongoing English language development. LACI can accomplish that in significant ways. As the examples presented in this chapter show, teachers can develop ways to talk about both language and science in ways that help students access the language of the text so they can understand the scientific content better.

References

Athanases, S. Z., & de Oliveira, L. C. (2011). Toward program-wide coherence in preparing teachers to teach and advocate for English language learners. In T. Lucas (Ed.), *Teacher preparation for linguistically diverse classrooms: A resource for teacher educators* (pp. 195–215). New York: Routledge.

Athanases, S. Z., & de Oliveira, L. C. (2014). Scaffolding versus structured assistance for Latina/o youth in an urban school: Tensions in building toward disciplinary literacy. *Journal of Literacy Research, 46*(2), 263–299.

Brown, A. L., & Campione, J. C. (1994). Guided discovery in a community of learners. In K. McGilly (Ed.), *Classroom lessons: Integrating cognitive theory and classroom practice* (pp. 229–270). Cambridge, MA: MIT Press.

Bunch, G. C. (2013). Pedagogical language knowledge: Preparing mainstream teachers for English learners in the new standards era. *Review of Research in Education, 37*, 298–341.

Cochran-Smith, M. (2004). *Walking the road: Race, diversity, and social justice in teacher education*. New York: Teachers College Press.

Davison, C., & Williams, A. (2001). Integrating language and content: Unresolved issues. In B. Mohan, C. Leung, & C. Davison (Eds.), *English as a second language in the mainstream* (pp. 51–70). Harlow: Pearson Education Ltd.

de Oliveira, L. C. (2007). Academic language development in the content areas: Challenges for English language learners. *INTESOL Journal, 4*(1), 22–33.

de Oliveira, L. C. (2011). In their shoes: Teachers feel like English language learners through a math simulation. *Multicultural Education, 19*(1), 59–62.

de Oliveira, L. C., & Dodds, K. N. (2010). Beyond general strategies for English language learners: Language dissection in science. *The Electronic Journal of Literacy Through Science, 9*(1), 1–14. Available online at http://ejlts.ucdavis.edu/article/2010/9/1/beyond-general-strategies-english-language-learners-language-dissection-science de Oliveira, L. C., & Lan, S.-W. (2014). Writing science in an upper elementary classroom: A genre- based approach to teaching English language learners. *Journal of Second Language Writing, 25*(1), 23–39.

de Oliveira, L. C., & Schleppegrell, M. J. (2015). *Focus on grammar and meaning*. New York: Oxford University Press.

de Oliveira, L. C., & Yough, M. (Eds.). (2015) *Preparing teachers to work with English language learners in mainstream classrooms*. Charlotte: Information Age Publishing and TESOL Press.

de Oliveira, L. C., Lan, S.-W., & Dodds, K. (2013). Reading, writing, and talking science with English language learners. In J. Nagle (Ed.), *English learner instruction through collaboration and inquiry in teacher education* (pp. 3–23). Charlotte: Information Age Publishing. [100 abstract proposals were submitted for this book and 12 accepted]

Fang, Z. (2006). The language demands of science reading in middle school. *International Journal of Science Education, 28*(5), 491–520.

Fang, Z. (2012). Approaches to developing content area literacies: A synthesis and a critique. *Journal of Adolescent and Adult Literacy, 56*(2), 111–116.

Fang, Z., & Schleppegrell, M. J. (2008). *Reading in secondary content areas: A language-based pedagogy*. Ann Arbor: The University of Michigan Press.

Fang, Z., & Schleppegrell, M. J. (2010). Disciplinary literacies across content areas: Supporting secondary reading through functional language analysis. *Journal of Adolescent and Adult Literacy, 53*(7), 587–597.

Gibbons, P. (2006). *Bridging discourses in the ESL classroom: Students, teachers, and researchers*. London: Continuum.

Halliday, M. A. K., & Matthiessen, C. M. (2004). *An introduction to functional grammar* (3rd ed.). London: Arnold.

Hammond, J. (2006). High challenge, high support: Integrating language and content instruction for diverse learners in an English literature classroom. *Journal of English for Academic Purposes, 5*, 269–283.

Hammond, J., & Gibbons, P. (2005). Putting scaffolding to work: The contribution of scaffolding in articulating ESL education. *Prospect: An Australian Journal of TESOL, 20*(1), 6.

Hansen, L. (2006). Strategies for ELL success. *Science and Children, 43*(4), 22–25.

Harcourt. (2005). *Harcourt science: Indiana edition*. Orlando: Harcourt.

Keenan, S. (2004). Reaching English language learners: Strategies for teaching science in diverse classrooms. *Science and Children, 42*(2), 49–51.

Lave, J., & Wenger, E. (1991). *Situated learning: Legitimate peripheral participation*. New York: Cambridge University Press.

Lee, O., & Miller, E. (2016). Engaging in phenomena from project-based learning in a place-based context in science. In L. C. de Oliveira (Ed.), *The Common Core State Standards for literacy in history/social studies, science, and technical subjects for English language learners*. Alexandria: TESOL Press.

Lee, O., Quinn, H., & Valdes, G. (2013). Science and language for English language learners in relation to next generation science standards and with implications for Common Core State Standards for English language arts and mathematics. *Educational Researcher, 42*(4), 223–233.

Lucas, T., & Grinberg, J. (2008). Responding to the linguistic reality of mainstream classrooms: Preparing all teachers to teach English language learners. In M. Cochran-Smith, S. Feiman-Nemser, & J. McIntyre (Eds.), *Handbook of research on teacher education: Enduring issues in changing contexts* (3rd ed., pp. 606–636). Mahwah: Erlbaum.

Moll, L., Amanti, C., Neff, D., & Gonzalez, N. (1992). Funds of knowledge for teaching: Using a qualitative approach to connect homes and classrooms. *Theory Into Practice, 31*(2), 132–141.

National Clearinghouse for English Language Acquisition. (2010). *The growing numbers of limited English proficient students: 1999/2000–2009/10*. Retrieved from http://www.ncela.us/files/uploads/9/growing_EL_0910.pdf

National Governors Association Center for Best Practices & Council of Chief State School Officers. (2010a). *Common Core State Standards*. Washington, DC: Authors.
National Governors Association Center for Best Practices & Council of Chief State School Officers. (2010b). *Application of Common Core State Standards for English language learners*. Washington, DC: National Governors Association Center for Best Practices & Council of Chief State School Officers. Retrieved from http://www.corestandards.org/assets/application-for-english-learners.pdf
NGSS Lead States. (2013). *Next generation science standards: For states, by states*. Washington, DC: The National Academies Press.
Nieto, S. (2000). *Affirming diversity: The sociopolitical context of multicultural education* (3rd ed.). New York: Longman.
Schleppegrell, M. (2001). Linguistic features of the language of schooling. *Linguistics and Education, 12*, 431–459.
Schleppegrell, M. J. (2004). *The language of schooling: A functional linguistics perspective*. Mahwah: Erlbaum.
Valenzuela, A. (1999). *Subtractive schooling: U.S.-Mexican youth and the politics of caring*. Albany: SUNY Press.
Villegas, A. M., & Lucas, T. (2002). *Educating culturally responsive teachers: A coherent approach*. Albany: State University of New York Press.

Chapter 4
The Professional Development of Pre-service Teachers in an Integrated Science and Language Acquisition Curriculum with Third-Grade Students

Anita C. Hernández

Elementary schools today face an increasing number of diverse students, especially students with language and literacy needs. Therefore, there is a growing demand for more preservice and inservice teachers who are prepared to work with diverse students by accommodating their curricula to English language learners (Turkan et al. 2014). Knowing how to help students meet academic goals in reading and writing across subject matter is a challenge for any teacher, and knowing how to help second-language learners continues to be challenging for today's teachers (Gándara et al. 2005).

One of the ways to improve students' learning is through teacher professional development. There are a number of instructional models that focus on teaching both content and second-language development. Of these, I have selected the Guided Language Acquisition Design Project, or Project GLAD for short (OCDE 2011), because many bilingual and second-language classrooms use this model to develop students' general and specific reading, writing, and learning strategies in content areas such as science, social science, literature, and mathematics.

This chapter focuses on a classroom professional development program for preservice elementary-school teachers who are earning their teaching credential at a local university in New Mexico. During a 3-day session, ten preservice teachers met at "El Dorado Elementary School" in southern New Mexico to observe and work with bilingual learners in a third-grade Spanish-English classroom.[1] In the afternoons, the preservice teachers met at the university to reflect on what had and had not worked well in the classroom.

[1] Bilingual learners are students who are acquiring two languages academically in dual-language classrooms.

A.C. Hernández (✉)
New Mexico State University, Las Cruces, NM 88003, USA
e-mail: achernan@nmsu.edu

A.W. Oliveira, M.H. Weinburgh (eds.), *Science Teacher Preparation in Content-Based Second Language Acquisition*, ASTE Series in Science Education,
DOI 10.1007/978-3-319-43516-9_4

To support the preservice teachers during the 3-day session with third graders, I adapted a 6-day GLAD instructional unit for fourth graders and modeled language-acquisition strategies for a science unit taken from Boswell and Wattman-Turner (2012). The classroom teacher, "Mrs. Linda Lara," had requested lessons on energy, since she found this subject difficult to teach to her bilingual learners. In the science curriculum, the workshop was centered on luminous energy; in the language curriculum, the focus was on reading, writing, and academic oral discourse. In this chapter, I suggest ways that science educators can develop integrated science and language-acquisition lessons to model for preservice teachers in order to effectively accommodate both second language learners in English-only classrooms and bilingual learners in dual-language classrooms.[2]

Project GLAD (OCDE 2011) is a professional development program that was developed by expert teachers for educators to meet the academic language and literacy needs of second language learners. Project GLAD's professional development includes an instructional model that consists of oral language and literacy strategies that teachers can use to create content area units of instruction in language arts, science, social studies, and mathematics.

In the 3-day session, I only had time to model 20 strategies that were most appropriate for teaching scientific concepts to third graders. To support scientific inquiry learning, I added five experiments on light that the preservice teachers led with the third-grade students. In sum, while I modeled the GLAD strategies on all three days, the preservice teachers were mostly involved in the instruction on Days 2 and 3.

The Context

I am a teacher-educator who, in addition to my regular responsibilities, directs extracurricular workshops on language and content learning for TESOL/bilingual preservice teachers. I am also a former bilingual elementary-school teacher and a former consultant to school teachers who earned a 45-h certificate in working with second language learners.

The preservice teachers in the 3-day workshop were part of a national professional development project designed to improve instruction for bilingual students. All of these teachers were enrolled in a local university teacher preparation program to earn their elementary teaching credential and their TESOL or bilingual endorsement in order to be able to teach English language learners in public schools. Seven of the ten preservice teachers were Hispanic females, two were Hispanic males, and one was a Caucasian female.

The morning workshop for the preservice teachers was given in Mrs. Lara's third-grade dual-language classroom, in a rural school outside the city limits of a

[2] English language learners, or ELLs, are students who are learning English. Their English proficiency is identified through formal assessments of their listening, speaking, reading, and writing skills.

large city in New Mexico. Her students were 23 bilingual learners at various levels of English and Spanish language proficiency, ranging from beginning to advanced in their second language. Twenty of the students began the dual-language program in kindergarten, and by third grade all were reading in both of their languages.

There are a number of advantages to using a classroom as a learning lab for expanding preservice teachers' instructional repertoire. Modeling in classrooms provides preservice teachers ample opportunities to observe second-language and content area methods with bilingual learners and their learning in real time (Garcia and Wei 2013). Another advantage is that the preservice teachers gain practical teaching experience. In most teacher preparation programs, preservice teachers gain teaching knowledge through theory and research studies in university settings, with classrooms and students only imagined (Darling-Hammond 2005). In classrooms with real students, issues can arise that can then be analyzed and solved for future lessons (Lampert 2003). Thus, teaching becomes a form of problem-solving at which the preservice teachers gain experience.

Finally, because teaching often uses a "closed-door" approach, having all the preservice teachers present to observe the same students interacting in the same lessons creates a climate of open discussion of teaching in ways that are not available to many preservice teachers during their teacher preparation experience (Darling-Hammond 2005). In the present professional development program, the preservice teachers observed a condensed 3-day instructional sequence with various academic language and literacy instructional strategies and tools that would normally take 6 or 7 h to implement over 2 or 3 weeks.

Teacher Knowledge and Professional Development

The professional development described here is based on sociocultural theories of learning, language-acquisition theories, scientific inquiry methodologies, and professional development principles. The theoretical framework underpinning this professional development brings three areas together: second-language acquisition, science, and successful teaching. Researchers agree that successful teaching includes the following elements: content knowledge, disciplinary linguistic knowledge, pedagogical content knowledge, pedagogical learner knowledge, and analysis/reflection (e.g., Darling-Hammond 1999; Shulman 1987; Walqui and van Lier 2010. See Fig. 4.1).

Sociocultural Theories of Learning

The present professional development project is grounded in social learning theory (Lave and Wenger 1991; Vygotsky 1978, 1980). Participation is an integral part of the social learning process, as is dialogue and observation. Children learn through

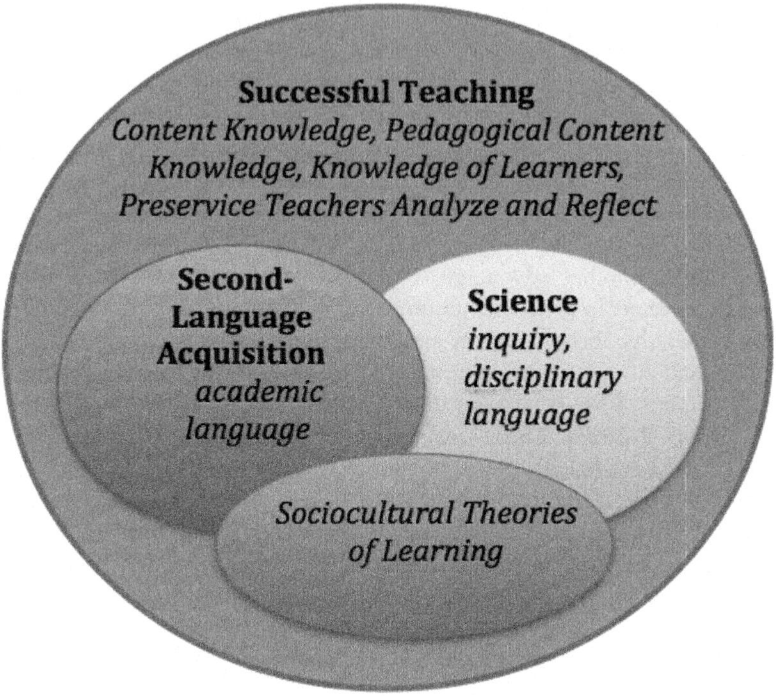

Fig. 4.1 Professional development for preservice teachers earning TESOL and bilingual education endorsements

participation and dialogue with others. Social learning is not unique to children, however, since it also applies to adults in trade apprenticeships, internships, student teaching, and business team and collaborative projects. Vygotsky (1978) defined what he called the "zone of proximal development" as "the distance between the actual developmental level as determined by independent problem-solving and the level of potential development as determined through problem-solving under adult guidance or in collaboration with more capable peers" (p. 86). In apprenticeships, individuals move from peripheral participation as they demonstrate learning that may be considered easy to full participation as they acquire competency (Lave and Wenger 1991).

For teachers, social learning means supporting second language learners, via cooperative learning with other students, to learn discipline-specific knowledge and discourse over time, even though the students' English may be at the beginning or intermediate level of proficiency. Preservice teachers are introduced to cooperative learning in the classroom with the third graders, since it is part of the small-group tasks, the jigsaw reading, and then followed by a discussion of the cooperative learning principles in the afternoon reflection sessions. Understanding how to organize a cooperative learning environment is key to student learning.

Cooperative learning introduces four important learning principles: (a) simultaneous interaction, (b) positive interdependence, (c) individual accountability, and

(d) equal participation (Aronson and Patnoe 1997; Cohen 2014; Johnson et al. 1985). *Simultaneous interaction* involves the teacher dividing the class into small groups of three to six students each, who then process information by sharing it among themselves—as the teacher observes them, to guide their thinking on the topic. *Positive interdependence* involves dividing subject matter into subcategories, about which each subgroup is responsible for reporting to the whole class. *Individual accountability* requires each student to make a presentation to the small group about his or her understanding of the subtopic, and later to make a written summary of his or her understanding of all the topics. *Equal participation* entails each member of a small group to contribute roughly as much effort as each of the other members.

Unlike cooperative learning classrooms, in traditional classrooms teachers on average do almost 80 % of the talking, with less than 20 % left for student talk (Goodlad 1984). Lingard et al. (2003) found that in classrooms with high numbers of students living in poverty, teachers talk more and students talk less. This results in superficial levels of understanding and critical thinking—in other words, less engagement with concepts. Traditional whole-class structures result in unequal participation because the high-achieving extroverted students usually raise their hands quickly to answer the teacher's questions (Johnson et al. 1985; Quinn et al. 2012).

Language-Acquisition Theories

The language-acquisition model used in this project is based on Krashen's (1988) concept of comprehensible input and Scarcella's (2003) concept of academic language. Comprehensible input refers to the understandable messages to which students must be exposed in order to acquire a language. Academic language includes the linguistic foundations in phonology, vocabulary, grammar, sociolinguistics, and discourse that are needed for second-language learners to complete assignments in various subjects. For example, academic language includes the use of sophisticated sentence structures and the use of multisyllabic Greek and Latin words. Furthermore, students who have receptive and productive academic language can extract meaning from increasingly difficult texts, show relationships among ideas, summarize concepts, infer meanings, and evaluate evidence and arguments (Scarcella 2003; Swain and Canale 1980). While this language distinction is often applied to English language learners, academic language is not a concern exclusive to second-language learners—social language and school language are concerns for many learners.

Professional Development and Successful Teaching

Gándara et al. (2005) recommended that school districts give a high priority to the professional development of their teachers because many mainstream instructors are not prepared to work with English language learners. In the United States, second language learners acquire varying levels of English competency, but not all attain the academic level necessary for reading and writing across subject matter such as mathematics, science, and social studies. Hence, teachers need to be prepared to teach English for a wide variety of contexts in order to meet the rigorous standards of their entire curricula.

Content Knowledge which is the foundation of teaching, includes knowing the core ideas of the various disciplines and relating those to each other. Teachers need to understand how inquiry is organized by the different disciplines (e.g., Darling-Hammond 1999; Shulman 1987). For example, scientists reason and support ideas differently from the way historians or mathematicians do. In science, introducing appropriate content knowledge, such as life, physical, and earth science to young children is key to their later foundational science knowledge. Observing, gathering evidence, constructing explanations are foundations in the development of students scientific understanding.

Pedagogical Content Knowledge includes understanding how specialists teach their various disciplines to students. In other words, teachers must be able to organize ideas in science, history, mathematics, and language arts for primary grade and intermediate grade students in ways that the children will understand (e.g., Darling-Hammond 1999; Shulman 1987). Insofar as science is concerned, the National Research Council (NRC 2012) notes that "students cannot comprehend scientific practices, nor fully appreciate the nature of scientific knowledge itself, without directly experiencing those practices for themselves" (p. 45). These scientific practices include asking questions, using models, testing ideas, graphing data, interpreting data, and drawing conclusions. Moreover, the National Science Teacher Association (NSTA 2004) provides the following rationale for scientific inquiry in the classroom:

> Scientific inquiry reflects how scientists come to understand the natural world, and it is at the heart of how students learn. From a very early age, children interact with their environment, ask questions, and seek ways to answer those questions. Understanding science content is significantly enhanced when ideas are anchored to inquiry experiences. (p. 1)

These statements by the NRC and the NSTA underscore the importance of understanding the nature of science through a scientific inquiry approach, instead of just reading scientific facts and concepts to students. Scientific inquiry is also emphasized by the Next Generation Science Standards (NGSS 2013). The scientific inquiry approach affords opportunities for language learners to use empirical evidence to answer questions they have come up with on their own and to negotiate meaning with their peers through empirical observation (Lee and Luykx 2006).

Disciplinary Linguistic Knowledge is understanding of the language needed to teach an academic discipline or content area. Turkan et al. (2014) argued that teachers of English language learners need a linguistic base for their disciplinary discourse in order to teach the discipline. Knowledge of the disciplinary discourse allows teachers to make transparent to second-language learners how to communicate appropriately, both orally and in writing. In terms of science, the discourse not only includes specialized vocabulary, but also includes particular ways of using language to describe observations, create hypotheses, and construct explanations.

Pedagogical Learner Knowledge comprises teachers' understanding of the wide range of students' linguistic and academic needs and designing instruction to which they can relate. Tapping into students' funds of knowledge and being familiar with differences in motivation, culture, language, family, and community help teachers to respond to students and adapt lessons for them (Darling-Hammond 1999; Moll et al. 1992; Shulman 1987). Walqui and van Lier (2010) noted that an important part of teachers' professional understanding must include knowledge about the language and cultural needs of second-language learners. Other dimensions of teacher knowledge include the depth and breadth of the curriculum, which are related to understanding how students learn at different age spans and grade levels, and how cultural and class differences impact learning. Specifically, understanding which concepts students can learn deeply and which concepts they can only learn superficially within the curricula for particular grade levels is key to teacher knowledge (Bransford et al. 1999; Darling-Hammond 1999). For second-language learners at different levels of language proficiency, teachers are challenged to scaffold lessons for greater understanding of concepts (Walqui and van Lier 2010).

Analysis/Reflection Effective teachers analyze and reflect on their own teaching practices and their students' work, continually asking themselves how they are doing at supporting students' understanding and at structuring their curricula for that purpose (Darling-Hammond 1999; Dewey 1933; Pine 2009). Lortie (1975) noted that when teachers fail to reflect on instructional decisions, their teaching becomes imitative rather than intentional.

Implementation

The preservice teachers met for three consecutive days, from 8:00 A.M. to 3:30 P.M. During the morning period at the school site, I modeled the luminous energy unit by using second-language acquisition principles and strategies for bilingual learners. The preservice teachers observed the lessons, took notes, monitored the activities during the small-group learning, guided the reading on light, and were responsible for the inquiries about light. Additionally, on the first day, I did most of the teaching and directing of activities; and on the second and third days, the preservice teachers led many of the activities (see Table 4.1). At appropriate moments, the classroom teacher took time to talk to the preservice teachers about her students, their reading proficiencies in both languages, her science curriculum, and her assessments of her students.

Table 4.1 Overview of the 3-day workshop on luminous energy

Day 1	Day 2	Day 3
Luminous energy: introduction	Luminous energy: release of responsibility to preservice teachers and third graders	Luminous energy: third graders and preservice teachers working together
Classroom standards: show respect, make good decisions, and solve problems	Preservice teachers **reviewed the 10/2 lecture through the chart on luminous energy.** Students had color-coded vocabulary cards based on the three kinds of luminous energy and placed their vocabulary on the chart	**Cooperative strip paragraph**. Teacher gave a topic sentence, and students in groups wrote a detailed sentence. Once all sentences were placed on the pocket chart, they were read and reordered for the paragraph to flow
T-Graph for social skills (focus was on Cooperation)		
Team points		
Small groups with **observations posters** & **exploration report**	Preservice teachers led chants on energy and "Lights Here Lights There"	Preservice teachers led chants on energy and "Lights Here Lights There"
Inquiry charts	**Cognitive content dictionary** (Energy)	**Sentence patterning**
		Students created sentences using adjective, noun, verb, and preposition phrases
10/2 lecture, **graphic organizer**, and **pictorial input chart on luminous energy**	**Home school connection**	Results of the home school connections. Preservice teachers praised students who returned their home-school connection paper with **literacy awards**
	"How does your family use luminous energy in your home?"	
Chanted "Energy Here Energy There"	Preservice teachers reviewed the definition and example of energy in the cognitive content dictionary	In pairs, preservice teachers led the five experiments on luminous energy
Focused reading: preservice teachers observed expert group A	Preservice teachers led the reading with expert groups D and E	Part 1: preservice teachers reviewed the readings with **expert groups** A, B, C, D, and E
Then preservice teachers led the reading with expert groups B and C		Part 2: **Jigsaw reading**
When students were not in their focused groups, they were working on **Team tasks**		Each jigsaw group had an A, B, C, D, and E member who discussed his or her aspect of luminous energy
Preservice teachers handed out **literacy awards** to the small groups	Preservice teachers handed out literacy awards to the small groups	Preservice teachers handed out literacy awards to the small groups

Day 1: Luminous Energy—Connecting Content with Students' Background Knowledge

The introductory lesson on energy for the third graders (and preservice teachers) used five different observation posters that incorporated different pictures of luminous energy. The purpose of these five observation posters was to activate students' prior knowledge and to anticipate their scientific understanding of luminous energy. The pictures for each of the five observation posters were created ahead of time (Fig. 4.2 illustrates two of the five observation posters).

This introductory activity is designed for students to discuss different sources of light and how light is used. Students are drawn to the content through the pictures and are guided in their discussions by the three questions on the report, which foster scientific thinking: observing, wondering, and predicting what can happen. The observation posters were placed around the room on chalkboards, file cabinets, and bulletin boards. In small groups of four, the third graders described what they saw, posed questions, and predicted what might happen next, as the preservice teachers listened in pairs. In several cases, the preservice teachers prompted the students, provided vocabulary, and encouraged shy students to participate.

To develop students' discourse in a second language, Gibbons (2003) recommended that student groups report orally to the rest of the class about their project; hence the posters should all be different to pique the interest of the class. The reporting not only informs the other students, but also allows the teachers to listen to the students' communication skills and gently guide them with precise tips about vocabulary and discourse. The third-grade students' efforts will be discussed here as the lessons were introduced.

Fig. 4.2 Two sample light observation posters and the exploration report, a guide for students to describe luminous energy

Table 4.2 Brainstorming of questions about what students wanted to learn

Students' responses	Translation
¿Como funciona la energía de luces?	How does light energy function?
¿Por qué se ve el arco iris como un arco, como el sol?	Why do we see the rainbow like an arc, like the sun?
¿Cómo las plantas usan energía?	How do plants use energy?
¿Cómo puede pasar la energía [luz] por toda la cuidad?	How does energy [light] travel throughout a city?
¿Como funciona las luces del los farolitos?	How does light energy function in the farolitos?[a]

[a] *Farolitos* are paper bags with candles burning inside

Once all five groups shared their descriptions of the pictures, their questions, and their predictions, I asked the class to brainstorm questions about luminous energy. Since the students had seen many pictures of luminous energy, several of the questions they posed were created from the pictures they had observed (see Table 4.2). In the brainstorming session, some students shared their questions in Spanish (translated here).

Pictorial Input Charts

I presented a 10/2 mini-lecture together with a pictorial input chart on light, using two large pieces of white paper to write down information about the properties, sources, and uses of luminous energy in everyday life. As I talked about sources of light, I drew pictures and wrote words on the pictorial input chart to illustrate my points. For example, as I described the sun as the primary source of light on Earth, I drew a picture of the sun on the paper. Figure 4.3 illustrates three aspects of luminous energy (properties, sources, and uses) that were presented in the mini-lecture on two picture charts:

Preservice Teachers' Reflections on the Lessons

During the whole-class activities, the preservice teachers listened as I drew the pictorial input chart and the labels of the concepts. After lunch, when the preservice teachers were back in the university classroom, I asked them to discuss what they had observed, how the observation posters could be used as pre-assessment data, and the type of learning they thought they had observed. A number of the preservice teachers noted that they were familiar with the use of visuals, but that they had not thought about organizing the visuals into categories of knowledge (e.g., properties, sources, and uses of light). Through inductive inquiry, I asked the preservice

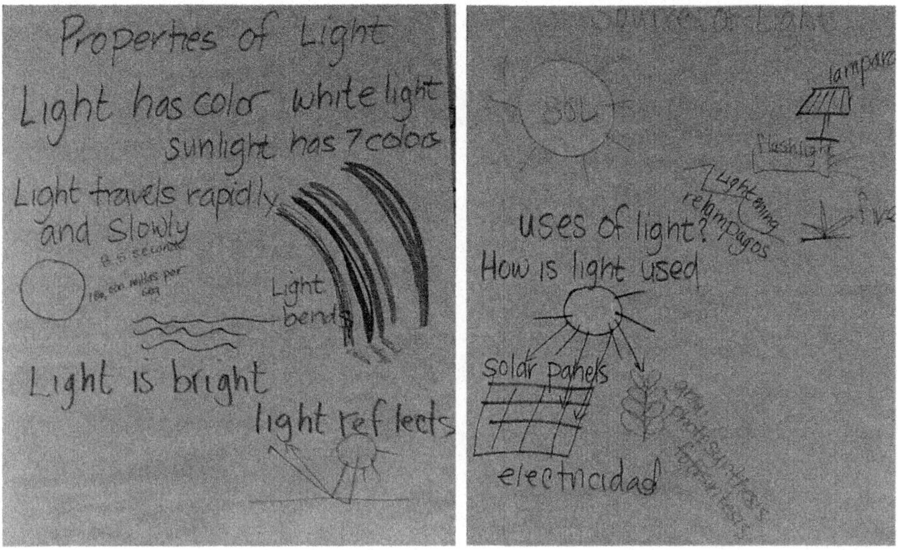

Fig. 4.3 Pictorial input chart — properties of light (*left*); sources and uses of light (*right*)

teachers how the pictorial input charts could be used for other activities. After some paired discussion, one of the preservice teachers suggested that the students could be asked to write three paragraphs about luminous energy, either as a whole-class assignment or an individual writing activity. The latter could be used to assess both content knowledge and writing fluency. Another preservice teacher noted that during independent work or when the teacher assigned center time activity, these posters could be used for students to practice their oral language with partners to demonstrate their understanding of the concepts. A third preservice teacher pointed out that the labels for the visuals could be phrases and sentences, in addition to individual words. In addition to the reflections, the preservice teachers helped prepared the materials for Day 2 that are described below.

Day 2: Working with Luminous Energy

On Day 2, the preservice teachers and I reviewed the light energy pictorial charts: properties, sources, and uses of light. To make the review interactive, we had written the key terms on pieces of sentence strips and distributed these to the students. As we reviewed the chart, children came up to it, read their vocabulary sentence strip, and placed it on the appropriate space on the chart (see Fig. 4.4).

To help students think about how science organizes information, the preservice teachers and I color-coded the charts and the word cards into the three aspects of luminous energy (properties, sources, and uses). Although the preservice teachers

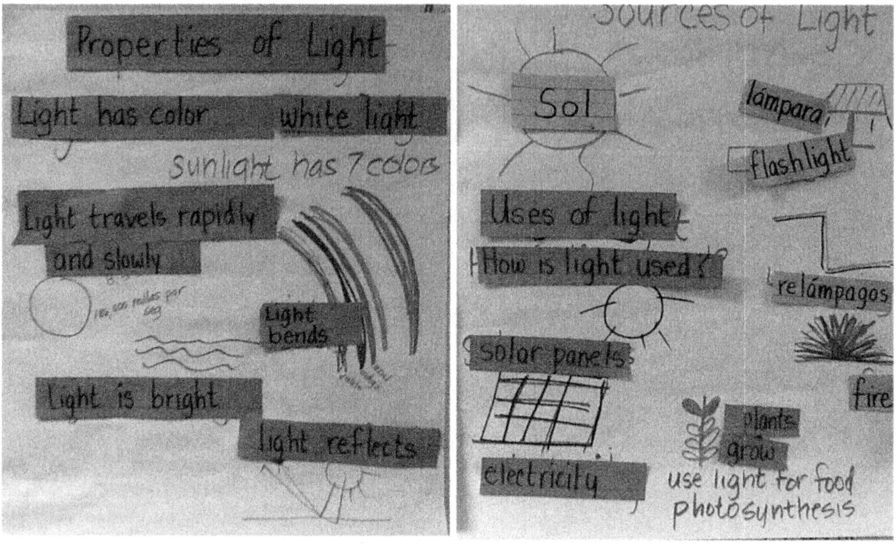

Fig. 4.4 Day 2 pictorial charts with science vocabulary cards

suggested that the students should write a three-paragraph essay about luminous energy at the end of the workshop, we ran out of time before the students could do this.

To help students understand scientific discourse, teachers can incorporate songs to which they can add facial and hand gestures. Children enjoy singing. I introduced the "Lights Here, Lights There" song by having all the students stand up and gather around the song chart (see Fig. 4.5).

A preservice teacher and I modeled the song, adding a few hand gestures to help with the participation. For example, with the word *here*, students pointed near themselves; with *there*, they pointed away from themselves; and with *everywhere*, they made a circle with their arms.

The science reading activity was organized into a jigsaw cooperative project that gave students an extra opportunity to report to their peers about information they had read (Aronson and Patnoe 1997; Gibbons 2003; Vygotsky 1978). Five expert groups, each one containing four students, read about light. The first group read about the primary source of light, the second group read about the secondary sources of light, the third group read about the properties of light, the fourth group read about how light travels, and the fifth group read about the human uses of light. During the small-group readings, the preservice teachers and I clarified the text, students' understandings of the text, and the scientific concepts. According to Vygotsky (1978, 1980), learning arises from the mediation in teacher-student and student-student interactions.

The GLAD Project (OCDE 2011) integrates three types of learning: whole-class work, small-group work, and individual work. In our three sessions, I first intro-

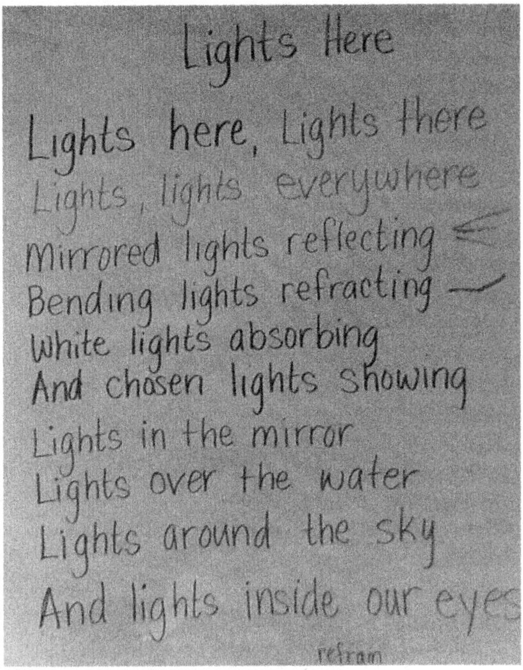

Fig. 4.5 "Lights here, lights there" song

duced information on light to the class as a whole. Then I directed the third graders to work in small groups that incorporated the information from the whole-class lessons. The preservice teachers, under my guidance, then led the small guided reading groups and the experiments. Finally, the students should have been able to work individually on their own learning, but we were not able to fit this in within the 3-day workshop. An additional day would provide time for individual work.

While visuals and hands-on inquiry activities are supported in the research literature, these activities by themselves are insufficient for second language learners to acquire the academic discourses. For example, in science, Gibbons (2003) recommended that teachers require language learners to report back on the findings of their hands-on inquiry to acquire the academic discourse needed for school.

The students had opportunities to work in small groups of three or four students on all of the activities introduced in the large group. For example, in the small groups, students could create their own pictorial chart about luminous energy, their own written exploration report, and their own questions about luminous energy. For students working in their second language, the small-group assignment provided an opportunity to participate with others in the planning and completion of a task, while also offering opportunities to ask questions about the task.

My afternoon session with the preservice teachers included a discussion of the instructional practices that they had observed and what they thought the students had learned. The preservice teachers discussed principles of cooperative learning

Table 4.3 The small-group centers

Experiments	Description
Experiment 1—Properties of Light	Used transparent, translucent, and opaque materials to demonstrate how light travels
Experiment 2—Potential Energy	Used a flashlight to demonstrate how energy is stored in batteries
Experiment 3—Refraction	Placed a pencil in a clear glass of water to demonstrate how light is bent
Experiment 4—Properties of Light	Used a metal ring and water and detergent to blow a bubble and place it on a glass slide to observe the colors in it
Experiment 5—Refraction (Rainbow)	Held a glass of water on white paper up to sunlight to observe the light refracting the rainbow colors on the paper

and how the team tasks were structured to foster simultaneous interaction, positive interdependence, equal participation, and individual accountability (Aronson and Patnoe 1997; Johnson et al. 1985). Organizing cooperative learning is a challenge for any teacher, especially preservice teachers because their experience with diverse learners is usually limited.

The preservice teachers analyzed (a) how the songs nurtured the students' academic language and knowledge of luminous energy, (b) the kinds of academic language the children used during the team tasks, and (c) the planning required for Day 2. During the reflection about Day 2, the preservice teachers noted how quickly the third graders began incorporating the scientific vocabulary and discourse introduced on Days 1 and 2. They also noticed the students' scientific discourse during the team tasks. Two of the preservice teachers pointed out that a couple of the English learners asked them what *reflection* and *refraction* meant. Instead of only learning pedagogical theories that can be applied to assist second language learners or bilingual learners, the preservice teachers were also asked to predict what the students would learn from the strategies and the scientific inquiries being put into practice. Most of them predicted that the experiments would enhance the students' scientific vocabulary. To be sure that the experiments would run smoothly, and that the preservice teachers understood the concepts involved, I monitored as they conducted the experiments (Table 4.3). This engagement in the inquiry allowed the preservice teachers to learn or re-learn luminous energy concepts.

Day 3: Eliciting Scientific Curiosity

On the morning of Day 3, the third graders were introduced to light energy experiments for which they had learned the theory on Days 1 and 2. The experiments were divided into five small-group centers (Table 4.3), which allowed for student-student and student-teacher co-construction of meaning about light (Lee and Luykx 2006).

Each of the pairs of preservice teachers briefly introduced the materials of the five experiments to the students and asked them to create questions about the experiments. For example, for the potential energy experiment, the preservice teachers showed the batteries and the different parts of the flashlight. We then divided the class into five groups of four students each, one group per center, each center monitored by two preservice teachers. We planned to give the students 20 min to conduct each experiment at their center and gather data about their inquiries. After that, the groups would rotate to the next experiment, until, after an hour, three of the experiments had been completed. However, in practice, each group only had time to complete two of the experiments. After recess, the students reported the findings of their last experiment to the whole class.

In the afternoon of Day 3, the preservice teachers and I dialogued about the lessons and strategies they had observed. These pedagogical discussions included analyses of the students' oral responses and written work. I asked the preservice teachers to share how the experiments progressed with the third graders. The preservice teachers who had been in charge of Experiment 1 (transparency) noted that the students enjoyed shining light through the three types of materials and seeing the differences. The preservice teachers who conducted Experiment 2 with the flashlights noted that their students were distracted by the ones who were doing Experiment 3 (refraction), so they were not able to hold the students' attention the entire time to explore the different parts of the flashlights and how they worked. The preservice teachers who had been in charge of Experiment 3 noted how excited the students got when they observed the image of the pencil being bent in water. The preservice teachers who conducted Experiment 4 noted that the students were awed by observing the rainbows in the bubbles. The preservice teachers who had been in charge of Experiment 5 noted that the students were delighted by the refraction of the white light through water to form a rainbow.

Outcomes

After examining the interviews with the preservice teachers conducted before the 3 days in the classroom, as well as the written questionnaires they filled out at the end of each day, I analyzed the preservice teachers' perceptions of what they had learned about classroom practice for language learners and about the integration of content with language acquisition principles. The pre-workshop interviews of the preservice teachers illustrated their theoretical knowledge of second-language acquisition and of science concepts that they had learned in their undergraduate courses. Prior to the workshop, eight of the ten participants had limited experience with language learners. They had only 36 classroom hours to fulfill the requirements for a credential. This is not surprising, because preservice teachers are busy taking undergraduate courses. Hence, their knowledge of teaching and learning was greatly enhanced during the professional development sessions.

The questions on the written questionnaires during the 3-day workshop asked the preservice teachers (a) to describe how the instructional strategies modeled helped the students to learn science through a second language, (b) what they thought the children learned in the lessons presented, (c) what problems they saw and how those might be solved, and (d) what they now knew about teaching content and language acquisition to second-language learners.

All ten preservice teachers chose the pictorial chart as the best instructional tool to help students learn through a second language. They felt that the visuals greatly helped the mini-lecture. Some of the responses included: "The pictorial chart helps students say what they already know while teaching new information"; "Great visuals for ELLs that help with vocabulary enrichment"; "The charts provide a framework when writing paragraphs"; and "The pictorial charts were fabulous because they were so simple, while providing a guide to organizing science information."

Several preservice teachers also praised the sentence strips used to process the science information on the second day. For example, one preservice teacher wrote: "The taping of the vocabulary cards on the pictorial input chart helped students review the information." Another preservice teacher wrote: "Students appropriate the academic language by going back to the vocabulary on the chart." A few preservice teachers chose to highlight other strategies such as the inquiry charts, the observation posters, and the cooperative reading of light energy.

When I asked the preservice teachers what they thought the children had learned about vocabulary and language, several noted that the children were able to work on public speaking. One observed that the students had been "guided on essentials like a clear and loud voice, standing straight, and keeping hands to their sides." When I asked the preservice teachers if they had learned any new scientific concepts, several stated that the luminous energy was a good review for them. One preservice teacher said "I learned a lot of light energy. I relearned the science and learn to teach it, while another noted "I learned a lot of light refraction and reflection. I have mainly worked in kindergarten and working with third graders has helped me to see how to organize science content." Another preservice teacher stated "I learned about different types of energy like kinetic energy, different types of light. I learned a lot just watching and participating in the hand-on activities about light."

When I asked the preservice teachers to identify problems they had observed and how they might be solved, several of them noted that not all the students participated 100 % of the time in their small groups. A few preservice teachers observed that some of the students were confused about the language in the chant (e.g., what *reflection* meant).

When I asked the preservice teachers which instructional strategies they already knew before the workshop, they mentioned the songs, the inquiry chart, the observation posters, and scientific experiments. When I asked them which of the remaining strategies they regarded as most effective, they mentioned cooperative learning, the 10/2 pictorial input charts, the small-group team tasks, and sentence patterning. Overall, they could see how a teacher can merge second-language acquisition strategies and content area pedagogy so that language learners can express themselves using scientific discourse.

Conclusions, Implications, and Recommendations

In the mornings during this 3-day professional development workshop, the synergy created among the third-grade students, the preservice teachers, and the workshop instructor (this writer) was unique. In the afternoon sessions, the preservice teachers worked together with a university professor (again, this writer) to observe and analyze how teaching and learning can be applied to particular contexts.

I modeled how a mix of learning in whole-class and small-group settings led to: (a) mediated reading, writing, and oral discourse within the activities, and (b) inquiry learning for second-language learners to become immersed in rich scientific discourse. The preservice teachers observed how the third-grade dual-language students expanded their scientific learning and their linguistic repertoire, which merged the theory of second-language acquisition with the scientific concepts learned in their undergraduate courses. Furthermore, the preservice teachers reflected on how the science and language arts were merged with language acquisition. They also saw how the curricular activities were built on each other to model information for students, and how the students enthusiastically engaged in those activities.

The classroom teacher, the preservice teachers, and I observed how the third graders built their own knowledge of science and were willing to speak in front of the class and in small groups about luminous energy. Helping me to plan lessons was also key to the preservice teachers' pedagogical content knowledge and disciplinary linguistic knowledge.

While this type of professional development—situated in a classroom—is considered ideal for classroom teachers, it may also be an opportunity for preservice teachers to learn more about the complexity of curriculum integration (i.e., science and language acquisition). Furthermore, it provides preservice teachers with specific experiences for accommodating their curricula to language learners. Finally, this school-university professional development program with dual language learners has implications for reimagining teacher preparation programs.

References

Aronson, E., & Patnoe, S. (1997). *The jigsaw classroom: Building cooperation in the classroom* (2nd ed.). New York: Addison Wesley Longman.

Boswell, D., & Wattman-Turner, M. (2012). *Energy forms, transfer, and economics*. Albuquerque: Project GLAD, Dual Language of New Mexico.

Bransford, J., Brown, A., & Cocking, R. (1999). *How people learn: Brain, mind, experience, and school*. Washington, DC: National Academy Press.

Cohen, E. G. (2014). *Designing groupwork: Strategies for heterogeneous classrooms* (3rd ed.). New York: Teachers College Press.

Darling-Hammond, L. (1999). Educating teachers for the next century: Rethinking practice and policy. In G. A. Griffith (Ed.), *The education of teachers: The ninety-eighth yearbook of the National Society for the Study of Education* (pp. 221–256). Chicago: University of Chicago.

Darling-Hammond, L. (2005). Developing professional development schools. In L. Darling Hammond (Ed.), *Professional development schools: Schools for developing a profession* (pp. 1–27). New York: Teachers College Press.
Dewey, J. (1933). *How we think. A restatement of the relation of reflective thinking to the educative process* (Rev. ed.). Boston: D. C. Heath.
Gándara, P., Maxwell-Jolly, J., & Driscoll, A. (2005). *Listening to teachers of English language learners: A survey of California teachers' challenges, experiences, and professional development needs*. Santa Cruz: The Center for the Future of Teaching and Learning.
Garcia, O., & Wei, L. (2013). *Translanguaging: Language, bilingualism, and education*. New York: Palgrave Pidot.
Gibbons, P. (2003). Mediating language learning: Teacher interactions with ESL students in a content-based classroom. *TESOL Quarterly, 37*(2), 247–273.
Goodlad, J. (1984). *A place called school: Prospects for the future*. New York: McGraw-Hill.
Johnson, D. W., Johnson, R., Roy, P., & Zaidman, B. (1985). Oral interaction in cooperative learning groups: Speaking, listening, and the nature of statements made by high-, medium-, and low-achieving students. *Journal of Psychology, 119*, 303–321.
Krashen, S. (1988). *Second language acquisition and second language learning*. Upper Saddle River: Prentice Hall International.
Lampert, M. (2003). *Teaching problems and the problems of teaching*. New Haven: Yale University Press.
Lave, J., & Wenger, E. (1991). *Situated learning: Legitimate peripheral participation*. Cambridge: Cambridge University Press.
Lee, O., & Luykx, A. (2006). *Science education and student diversity: Synthesis and research agenda*. Cambridge: Cambridge University Press.
Lingard, B., Hayes, D., & Mills, M. (2003). Teachers and productive pedagogies: Contextualising, conceptualizing, and utilising. *Pedagogy, Culture, and Society, 11*(3), 399–424.
Lortie, D. C. (1975). *Schoolteacher: A sociological study*. Chicago: University of Chicago Press.
Moll, L., Amanti, C., Neff, D., & Gonzalez, N. (1992). Funds of knowledge for teaching: Using a qualitative approach to connect homes and classrooms. *Theory Into Practice, 31*(2), 132–141.
National Research Council. (2012). *A framework for K-12 science education: Practices, crosscutting concepts, and core ideas*. Committee on a Conceptual Framework for New K-12 Science Education Standards. Board on Science Education, Division of Behavioral and Social Sciences and Education. Washington, DC: The National Academies Press.
National Science Teachers Association (NSTA). (2004). *National Science Teachers Association position statement*. Retrieved from http://www.nsta.org/docs/PositionStatement_ScientificInquiry.pdf/
Next Generation Science Standards: For States, By States (NGSS). (2013). *Next generation science standards*. Retrieved from http://www.nap.edu/catalog/18290/next-generation-science-standards-for-states-by-states/
Orange County Department of Education (OCDE). (2011). *OCDE project GLAD: Guided language acquisition design*. Retrieved from http://www.ocde.us/projectglad/Pages/default.aspx/
Pine, G. (2009). *Teacher action research: Building knowledge democracies*. Thousand Oaks: Sage.
Quinn, H., Lee, O., & Valdés, G. (2012). *Language demands and opportunities in relation to next generation science standards for English language learners: What teachers need to know*. Retrieved from http://ell.stanford.edu/sites/default/files/pdf/academic-papers/03-Quinn%20Lee%20Valdes%20Language%20and%20Opportunities%20in%20Science%20FINAL.pdf/
Scarcella, R. (2003). *Academic English: A conceptual framework*. Irvine: University of California Linguistic Minority Research Institute.
Shulman, L. (1987). Knowledge and teaching: Foundations of the new reform. *Harvard Education Review, 57*(1), 1–22.
Swain, M., & Canale, M. (1980). Theoretical bases of communicative approaches to second language teaching and testing. *Applied Linguistics, 1*(1), 1–47.

Turkan, S., Oliveira, L., Lee, O., & Phelps, G. (2014). Proposing a knowledge base for teaching academic content to English language learners: Disciplinary linguistic knowledge. *Teachers College Record, 116*, 1–30.

Vygotsky, L. (1978). *Mind and society*. Cambridge: MIT Press.

Vygotsky, L. (1980). *Thought and language*. Cambridge: MIT Press.

Walqui, A., & van Lier, L. (2010). *Scaffolding the academic success of adolescent English language learners: A pedagogy of promise*. San Francisco: WestEd.

Chapter 5
Developing an Adaptive Disposition for Supporting English Language Learners in Science: A Capstone Science Methods Course

Sarah A. Roberts, Julie A. Bianchini, Jin Sook Lee, Sarah Hough, and Stacey L. Carpenter

Introduction

High quality science teachers are crucial if all students, including English language learners (ELLs), are to acquire the science and engineering knowledge and practices needed to succeed in our twenty-first century, technologically complex, and globally connected world (California Council on Science and Technology 2007; National Research Council 2007, 2010). In California, 2.7 million students (or 43 % of the state's public school enrollment) speak a language other than English at home and 1.3 million (or 22 %) are designated as ELLs (California Department of Education 2014). California serves as the site of the National Science Foundation's Noyce preservice teacher scholarship program, STELLER (STEM Teachers for English Language Learners: Excellence and Retention), described in this chapter. Although California has been a somewhat novel context, as student demographics continue to change across the nation to reflect similarly the diversity in California, content courses focused specifically on ELLs will be a necessity in all teacher education programs.

Our model for effective science teacher preparation in content-based second language acquisition is situated in a 13-month, post-baccalaureate teacher education program in Central California. Originally, preservice science teachers completed two science methods courses, two courses focused on ELLs and English language

This research was supported by a grant from the National Science Foundation (DUE 1439923). However, any opinions, findings, and conclusions or recommendations expressed in this material are those of the authors and do not necessarily reflect the views of the National Science Foundation.

S.A. Roberts (✉) • J.A. Bianchini • J.S. Lee • S. Hough • S.L. Carpenter
University of California, Santa Barbara, CA 93106, USA
e-mail: sroberts@education.ucsb.edu; jbianchi@education.ucsb.edu;
jslee@education.ucsb.edu; sarahh@education.ucsb.edu; scarpenter@education.ucsb.edu

A.W. Oliveira, M.H. Weinburgh (eds.), *Science Teacher Preparation in Content-Based Second Language Acquisition*, ASTE Series in Science Education,
DOI 10.1007/978-3-319-43516-9_5

acquisition, and a yearlong course in professional issues that attended to their experiences in the field. Even with these courses, however, we found that there was inadequate space for preservice teachers to understand how to teach science to their ELLs, a diverse group of learners who come from different linguistic and cultural backgrounds and who have had myriad previous personal and educational experiences. As such, as part of STELLER, we developed a capstone science methods course to address this particular need – to better support preservice teachers in learning how to teach science in ways attentive to ELLs.

In our teacher education program, many preservice science teachers begin their courses expecting to receive a list of tools or a set of specific strategies that are deemed effective for teaching ELLs; however, we have found that such "tricks" and tools do not go far enough in supporting ELLs. Our goal is to foster a shift in our preservice science teachers from a naïve understanding that they only need pedagogical tools or strategies to support ELLs, to the recognition that they need to develop an adaptive disposition to engage in reflection on their teaching of ELLs. This shift includes knowing how to identify what resources all students bring to the classroom, how to build on students' understanding to promote deeper learning, and how to adequately assess what students have actually learned.

Our teacher education program also emphasizes the need to recognize the diversity present among ELLs – as a response to their traditional treatment as a homogeneous group of learners who have similar needs and who respond in the same ways to pedagogical strategies. Despite the overarching title "ELL," ELLs are heterogeneous, with differing levels of English proficiency, both socially and academically. ELLs bring "a variety of cultural and linguistic experiences" to learning science (Lee et al. 2013, p. 728). Concurrently, science teachers must use a variety of strategies to support ELLs in developing, engaging with, and using the "disciplinary language of science" in multiple ways (Lee et al. 2013, p. 231).

In this chapter, we provide the theoretical underpinnings for the intersection of ELLs, science, and instructional practice that form the foundation of our capstone science methods course. Our course embodies three principles: (1) building from students' funds of knowledge; (2) implementing cognitively demanding tasks; and (3) providing opportunities for rich language and literacy exposure and practice. We discuss preliminary outcomes from our first cohort of preservice science teachers and close with recommendations on the future design of our course.

Overview of Teacher Education Program and Capstone Science Methods Course

The model for our capstone science methods course is grounded in the conviction that preservice science teachers require a coordinated, multi-faceted approach to teacher development within a safe context to learn about and reflect on the teaching and learning of science to ELLs. The need for a reflective process in teacher education is long standing (e.g., Dewey 1904). We extend this reflection using an inquiry model (Tom 1985) that is nested in the situated nature of preservice teachers' work – that acknowledges the preservice science teachers' physical and social contexts as well as their activities as integral to their learning (Putnam and Borko 2000). Because preservice teachers are student teaching in local secondary science classrooms while enrolled in this capstone course, the course emphasizes the mis/connections between theoretical constructs and classroom practice as part of learning about and reflecting on teaching.

More specifically, to promote reflection, we engage preservice science teachers in multiple cycles of a reflective process: (a) to learn about attending to ELLs in science; (b) to try out ideas presented in the course in their own classrooms; (c) to come back to the capstone course to share and reflect upon those experiences; and (d) to devise future plans based on those experiences, reflections, and feedback from their peers and instructor. We also attempt to foster in our preservice science teachers an adaptive disposition, so that these cycles of implementation, data gathering, and reflection become an integral part of who they are as teachers of ELLs. In addition, our approach moves away from equating effective instruction for ELLs with a singular focus on teaching vocabulary terms or other categories of what may constitute science-related academic language (Bunch 2013). Rather, we present language as an essential mediator of the teaching and learning process (Dutro and Moran 2003) – by supporting students' development of sense-making talk across everyday and disciplinary-specific languages, as well as students' ability to communicate effectively their understanding of core ideas, crosscutting concepts, and practices orally and in writing to others (Lee et al. 2013).

Below, we first introduce our larger model of teacher preparation, which includes three course components. We do so to provide readers a sense of how the teacher education program is organized to develop preservice science teachers' knowledge and practices over time. We then discuss in detail the three fundamental principles of our capstone science methods course, the focus of this chapter.

Teacher Education Course Components

Our 13-month teacher education program involves three sets of courses and experiences, which we represent as overlapping circles in Fig. 5.1.

Component 1: Field Experiences and Professional Issues Course Throughout their entire year, our preservice science teachers take a Professional Issues in Teaching Science course. They also engage in student teaching in grades 7–12 science classrooms, moving from observer in the early months of the school year to classroom teacher in the second semester. These course and teaching opportunities allow preservice science teachers to try out and reflect on their own practice.

Component 2: ELL Courses Preservice science teachers take two courses focused specifically on ELLs. They complete the first course, Foundations of Academic Language, during the summer, and then the second course, English Language Development (ELD)/Specially Designed Academic Instruction in English (SDAIE) Methods and Procedures, in the fall. These two courses provide preservice science

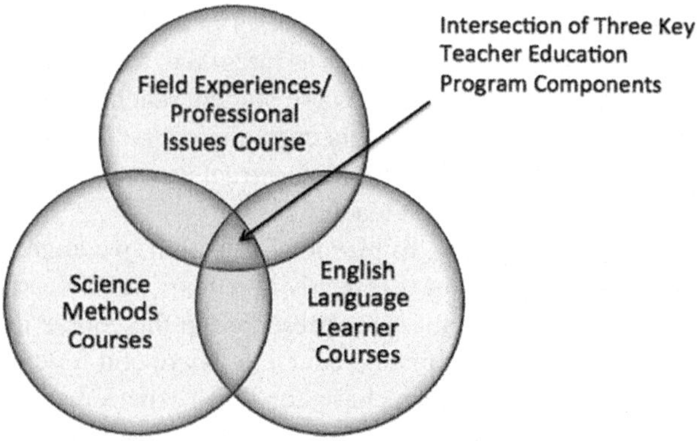

Fig. 5.1 Visual model of the three key components of the teacher education program for preservice science teachers

teachers with foundational information about academic language in addition to general practices to support reading comprehension and fluency, English language development, and disciplinary content for students learning English.

Component 3: Science Methods Courses Preservice science teachers take three secondary science methods courses. Two science education faculty members teach these methods courses: a full-time lecturer and a tenured faculty member. Both instructors have prior secondary teaching experience and are PI/Co-PIs of the STELLER grant. The first two methods courses are offered in the summer and fall, respectively. These courses provide preservice science teachers with experiences that support their development of pedagogical content knowledge, methods, curriculum design, research, and theory related to the teaching and learning of science at the secondary level. The third methods course, offered in the spring and the focus of this book chapter, serves to deepen and broaden knowledge and skills acquired in the earlier courses. This capstone science methods course is found at the intersection of the three domains of field experiences, ELL instruction, and science methods (see again Fig. 5.1). In this course, preservice science teachers apply the pedagogical theories, principles, and practices of second language acquisition to the discipline of science. In particular, they examine ways to effectively teach the disciplinary-specific language, core ideas, and practices outlined in the *Next Generation Science Standards* (NGSS; NGSS Lead States 2013) to students in their classroom.

Capstone Science Methods Course Model

Our capstone science methods course is organized around three key principles that we expect preservice science teachers to attend to in their instruction. We use these ideas to ground the work we do and the conversations we have about ELLs in the teaching and learning of science. As stated above, these broad ideas go beyond strategies and tools to provide the foundation for an adaptive disposition for teaching ELLs in science classrooms.

Funds of Knowledge Our first principle is *building on and using students' funds of knowledge and resources* (Lee et al. 2008; Moll et al. 1992; Moschkovich 2002) in the teaching and learning of science. Preservice science teachers are encouraged to attend to this principle by examining their ELLs' cumulative files to identify English language competency; having conversations with or giving surveys to students to identify ELLs' prior schooling experiences, life experiences, and community resources; appreciating the diversity of ELLs by identifying the multiple languages and different dialects students speak, and/or the varying levels of literacy they produce and display; using community resources to make science relevant and meaningful; and recognizing and utilizing their ELLs' native languages and cultural knowledge as resources for learning. One way science teachers can accentuate students' funds of knowledge in their classrooms is to connect their science instruction to local and indigenous cultures and communities. For example, Chinn (2007) encouraged science teachers in Hawaii to study sustainability and biodiversity, to learn from local elders about traditional practices, to monitor and restore local ecosystems, and then to integrate local knowledge into their science lessons.

Cognitively Demanding Work The second principle in our model is *providing students with cognitively demanding work* (Lee and Fradd 1998; Tekkumru-Kisa et al. 2015; Tobin and Kahle 1990; Understanding Language 2013). We draw from Lee et al.'s (2013) discussion of the NGSS to inform our recommendations for the kinds of analytical tasks students should complete: These analytical tasks should require students to move away from "detailed facts or loosely defined inquiry" (p. 223) to focus instead on the practices (including those practices deemed language-intensive), crosscutting concepts, and core ideas identified in the NGSS. Teachers must ensure activities and assignments are made accessible to ELLs without reducing the cognitive demand of such tasks. They can incorporate this principle into their instruction by including visuals, introducing complex language structures gradually, identifying which contexts ELLs need help understanding, determining what realia facilitate understanding of tasks, actively engaging their students in NGSS science and engineering practices, and asking students to connect science concepts to everyday life, their sense of place, and/or socioscientific issues. In Rosebery et al.'s (2010) example of cognitively demanding work in science classrooms, a teacher and her 9–11 year-old students investigated heat transfer and the particulate nature of matter, topics students usually learn in middle or high school. From their engagement in a number of investigations and simulations, students were able to develop an understanding of and an ability to explain these scientific concepts. Cognitively demanding work in science entails engaging students in the work of science.

Opportunities for Rich Language and Literacy Exposure and Practice Our third principle is *providing students opportunities for rich language and literacy exposure and practice* (Bleicher et al. 2003; Khisty and Chval 2002; Khisty et al. 1990; Lee and Buxton 2013; Lee et al. 2013; Moschkovich 2007). Science teachers must offer ELLs multiple pathways to learn the "disciplinary language of science" (Lee et al. 2013, p. 231) by creating opportunities to receive comprehensible input, to produce comprehensible output, and to engage in negotiations of meaning that will advance their English language acquisition, their mastery of the disciplinary language of science, and thus, their science learning. Preservice science teachers can address this principle by orchestrating discourse and engaging their ELLs in using multiple modes of communication: speaking, listening, reading, and writing. In other words, teachers can model target language (e.g., what science argumentation looks like); use sentence frames; encourage students' use of their home languages; promote any language register/discourse with which ELLs are comfortable; and use gestures, models, and graphics/visuals. To encourage ELLs' use of disciplinary language aligned with the NGSS, teachers can ask ELLs to prepare written or verbal arguments, construct explanations or design solutions, develop questions and critiques to evaluate scientific information, and/or provide summaries of scientific information for specific audiences or purposes (Lee et al. 2013).

Capstone Science Methods Course for ELLs: Implementation and Materials

As explained above, the capstone science methods course was developed pedagogically around the ideas that preservice teachers should consistently be engaged in cycles of reflection and that they should attend to the three key principles of ELL instruction. The course, held in the spring, is 10 weeks long and meets once a week for 3 hours. The instructor and preservice science teachers spend approximately 3 weeks focusing on each of the course's three key principles. Students read about each principle, conduct observations and write reflections, and discuss their thoughts in class. When possible, we include examination of video records of teaching from the preservice teachers themselves or from other sources available online. Further, while completing this course, preservice teachers also teach in a secondary science classroom; they plan and carry out instruction as the teacher of record in one class period.

Below, we discuss two specific examples from our capstone course that focus on the intersection of the three circles shown in Fig. 5.1: (1) video records and (2) observations and reflections. These activities and assignments provide our preservice science teachers with the space and structure to enact the reflective process in their work with ELLs and to begin to develop an adaptive disposition.

Video Records

Examining classroom video records is our first example of an activity we include in our capstone methods course. This activity is implemented several times across the course. It allows preservice teachers to view a permanent record of their own or others' practice from multiple perspectives (van Es and Sherin 2010) and helps foster conversations investigating teaching and learning (Borko et al. 2008). As stated above, video records are selected from preservice teachers' own classroom practice or from video collections of teaching available online. In the former instances, preservice teachers are asked to share a video segment from their edTPA (www.edtpa.com), a portfolio of materials from their clinical student teaching experiences submitted as a performance assessment. During and after viewing the video, the instructor and preservice teachers engage in content-specific discussions that attend to academic language and ELLs. Watching these videos creates opportunities for preservice science teachers to develop skill sets around asking generative questions about their instruction that can positively affect their future growth and development. It also supports them in learning how to reflect on their teaching – how to practice enacting a reflective process during and after they leave the teacher education program.

Observations and Reflections

The second example from our capstone course is a set of three assignments in which we ask our preservice science teachers to complete classroom observations and written reflections related to ELLs. Each observation-reflection assignment aligns with one of our three fundamental principles discussed above.

- **Observation of and Reflection on an ELL (Due Week 3):** Preservice teachers are asked to observe a science lesson in another teacher's class that includes several ELLs. They are to select one ELL to follow during the lesson and to take notes on what this ELL says and does. After their observation, they are to write a one-to-two paragraph reflection where they respond to the following questions: What successes and struggles did the ELL encounter? How was the ELL supported (or not) by his or her teacher and/or peers? What implications do the ELL's words and actions have for future instruction? This assignment and the subsequent discussion foreground ELLs' funds of knowledge.
- **Observation of and Reflection on Language Demands (Listening, Reading, Writing, and Speaking) for ELLs (Due Week 5):** Preservice teachers are asked to observe a second science lesson in a class that again includes several ELLs. During the lesson, they are expected to take notes on what language demands these ELLs encounter. After the observation, they craft a one-to-two paragraph reflection that responds to the following questions: What were the listening, reading, writing, and/or speaking demands placed on ELLs? How did the ELLs respond to these language demands? What implications do these language demands have for future instruction? This assignment and the subsequent discussion focus on the opportunities for rich language in a science classroom.
- **Observation of and Reflection on the Teaching of Science for ELLs (Due Week 7):** Preservice teachers are asked to observe a third science lesson in a class that again includes several ELLs. During this lesson, they are expected to

take notes on the various ways the teacher supports ELLs' participation and learning. Their one-to-two paragraph reflection responds to the following questions: Which strategies did the teacher use? Which ones specifically addressed the needs of ELLs? How might you strengthen this teacher's implementation of strategies for ELLs in future lessons? This assignment and the subsequent discussion highlight how a teacher can provide cognitively demanding work in science classrooms for her or his ELLs.

Our preservice science teachers complete one of these observations approximately every 3 weeks during the course in science classrooms at their student teaching sites. On the day the assignment is due, we spend approximately 45 min in class discussing the observations and reflections and considering how they can inform subsequent instruction. These three assignments allow our preservice science teachers to examine multiple perspectives on and for supporting ELLs explicitly tied to our three key principles for ELL instruction.

While our three principles to support ELLs are not novel, our course is unique in several respects. It provides preservice science teachers the opportunity to examine ELL concepts and strategies grounded both in research and theory and in the context of science student teaching. The use of media and technology, collective discussion, and the reflective cycle of teaching are emphasized. The course also offers sustained study of ELLs' learning of the discipline of science over an entire term, rather than as a single "ELL lesson." Further, our course supports preservice science teachers in developing an adaptive disposition toward working with ELLs in their science classrooms.

Research on Our Capstone Science Methods Course: Successes and Challenges

Below, we present findings from our first cohort of preservice science teachers who completed the capstone science methods course in Spring 2015. We discuss patterns that emerged from our qualitative analysis of these preservice teachers' individual interviews. Our methods and analysis were guided by three research questions: How did preservice science teacher participants understand the three principles of our capstone course? What successes and struggles did they identify when attempting to use these three principles to inform their classroom practice? What suggestions did they have for ways to improve the course?

Research Methods

We conducted and audio recorded individual semi-structured interviews with all 10 of the preservice science teachers enrolled in the capstone course. Members of the research team who were not involved in the course's design or implementation conducted these interviews at the conclusion of the course. We transcribed each interview, coded each first individually and then identified any differences across coders, and finally met as a team to discuss and to reconcile any differences in coding. We used five a priori codes: funds of knowledge, cognitively demanding work, language rich opportunities, self-perceived needs of preservice teachers to improve, and preservice teachers' suggestions for course improvement. We were interested in how and to what extent these preservice science teachers understood and enacted the big ideas of the course.

Findings

Our interview findings are organized into three major sections. In the first section, we discuss the strengths and limitations of preservice science teachers' understanding of the three principles of effective science instruction for ELLs. These principles are presented in the order of most to least often discussed by the preservice teachers; this order also mirrors the depth and complexity of their understanding of these principles. The second section identifies additional theoretical and practical guidance preservice science teachers stated that they needed to teach ELLs in their classrooms. The final section outlines preservice science teachers' recommendations for improving the capstone science methods course.

Three Principles of ELL Instruction

Language Rich Opportunities In their interviews, our preservice science teachers routinely emphasized the importance of providing ELLs with language rich opportunities to engage in the academic language and practices of science. These language rich opportunities most often began with helping ELLs understand science vocabulary terms. Attention to science terms included posting word banks, identifying the Greek or Latin roots of terms, rephrasing definitions of science terms in multiple ways or asking students to devise their own definitions, providing students opportunities to observe and/or investigate a phenomenon before introducing a vocabulary term, and asking students to practice using the vocabulary terms in appropriate contexts. For example, one preservice science teacher participant, Bryan, not only reviewed the definitions of science terms with his students, but he also provided them with practice in using these terms in meaningful contexts as well.

> [In discussing native, non-native, and invasive species, we would identify] plants [and] animals, what effects that organism has [on the ecosystem], and then we would discuss is that non-native or is it invasive, and then we'd do the whole thumbs up, thumbs down sort of discussion with that. So I've been trying to focus a lot more on having them actually use the vocabulary and practice the vocabulary in that way, as opposed to me just breaking down the definition.

Most preservice science teachers also described specific ways they supported their ELLs in listening, reading, writing, and speaking the academic language of science. They were aware that their ELLs, at times, struggled to understand parts of lessons, experienced frustration when completing tasks, and/or were reluctant to express their ideas. As such, our preservice science teachers employed various scaffolds to address academic language demands. These included using visuals along with text in worksheets, assessments, and PowerPoint presentations; providing peer support, for example, implementing think-pair-share and small group activities; providing writing support, for example, using scaffolded notes, sentence starters, templates for well-argued paragraphs, and/or a series of questions that students first answered and then used to construct an argument; and providing students the space and opportunity to speak in both their home language and in English. For example, Nancy listed the varied ways she and her cooperating teacher attempted to support ELLs' understanding and participation in her science class: "So we scaffold worksheets a lot. We give sentence starters. We do pair share. We also have a lot of different ways of presenting information. So we have a lot of visuals."

Less common, however, were discussions of ways preservice science teachers went beyond science terms and scaffolds to engage their students in more complex, language rich opportunities. A few preservice teachers discussed writing assignments and/or debates they implemented to help students make connections between the science concepts they were learning and socioscientific issues relevant to their everyday lives. As one example, Kim asked students in her conceptual physics course to write letters to a cell phone company.

> [In these letters, students were to discuss] if the waves from cell phones cause cancer or not. And so we were getting at ionizing versus nonionizing waves and longer versus shorter wavelengths [through this letter writing assignment]. We provided them with a couple articles and then also encouraged them to [conduct] research their own.

As a second example, Monty asked his students to construct an evidence-based argument after they read a narrative from a girl who lived along the Gulf of Mexico – "about how eutrophication is affecting her dad's business, and the investigations that she did to find out what the cause of the eutrophication was." Students peer edited the initial drafts of others, learning about constructive critique in the process.

Thus, although the preservice teachers seemed to have attained a good grasp of the various pedagogical strategies that are employed to support vocabulary learning among ELLs, their breadth and depth of understanding about how to support ELLs' language development beyond the lexical level was limited. Perhaps expertise in the design and implementation of language rich opportunities cannot be attained in a single year of teacher education. Still, a more explicit focus on how to support language development at the sentential and discourse levels during the capstone science methods course appears to be needed.

Cognitively Demanding Work We found preservice science teacher participants talked less about cognitively demanding work than they did about rich language opportunities. In their descriptions of classroom practice, they identified a number of common elements among activities and assignments that made these tasks cognitively demanding yet accessible to ELLs. These elements included hands-on work, collective classroom experiences, connections among ideas, connections to everyday life, and opportunities for students to construct their own arguments or solutions. Most commonly, the preservice teachers discussed using "hands-on" activities to provide tangible, concrete examples and experiences for students to better understand and remember concepts. For example, Riley included dissections in her biology instruction: "So we try to do a lot of hands-on stuff.... We did a lot of dissections, just something to make it really, really concrete. It seemed to help. They were grossed out, but you know, they will remember that."

Some preservice teachers went beyond hands-on activities and described more substantive, powerful ways of engaging students in cognitively demanding work. They created opportunities for students to draw from a collective classroom experience, required students to make connections among concepts, and/or contextualized science concepts in real-life situations. For example, to introduce the topic of electromagnetism, Kim first had her students figure out how to build an electromagnet, a cognitively demanding task. Building electromagnets oriented her students to the topic and served as a reference for students as they further developed their understanding of the concept of electromagnetism. As Kim described:

> Before we knew anything about electromagnets, for example, I had them build them to get them to work. Then we could talk about all the parts and what was going on. Not just for ELLs, but for everyone, they can have something tangible to relate it to.

Only a few preservice science teachers discussed cognitively demanding tasks that provided their students opportunities to construct their own arguments or solutions, that is, cognitively demanding tasks that were also language rich opportunities. Kim's cell phone letter and Monty's evidence-based argument assignments, discussed above under language rich opportunities, are two such examples. Kim emphasized this point of intersection when recounting a second writing assignment where students constructed their own solutions to a hypothetical energy crisis: "Because it was their thoughts, they could write more." In short, in the capstone science methods course, a more concerted effort to examine tasks that are both cognitively demanding and language rich appears fruitful and necessary.

Funds of Knowledge In contrast to the two principles discussed above, we found that preservice teachers did not explicitly discuss how they used students' funds of knowledge to inform their instruction of ELLs. Most preservice teachers acknowledged the importance of "knowing" their students, including having an awareness of their students' first languages, English language competencies, backgrounds, and interests. Although this as a valuable starting point for building from and using students' funds of knowledge, movement from knowing students' interests to enacting lessons that explicitly build from their resources is needed.

We did identify two preservice teachers who went beyond simple acknowledgement to take concrete steps to better understand their students; although, they, too, did not fully address how they built from students' funds of knowledge in their instruction as a result. As one example, Riley interviewed several of her students about their science experiences outside of school for her Master's thesis. She found that a number of students did science outside of school, including at home and in the Boy Scouts. She learned that her students enjoyed "making things," which she thought fit well with the engineering emphasis in the NGSS. However, she did not clearly discuss how she used this knowledge of her students in her teaching. As a second example, Alison surveyed her students and found that they liked "creative activities." She thought that was why her students enjoyed an activity where they had to model a DNA strand using beads. However, she did not specify that she implemented this activity because she knew her students liked creative activities. Thus, instilling a better understanding of how students' funds of knowledge can be integrated and utilized as a resource in the teaching and learning process is another aspect of the course that requires improvement.

Preservice Science Teachers' Self-Perceived Needs

In this section, we address self-perceived challenges and needs that the preservice science teachers thought remained after the completion of the course. Nine of our 10 preservice teachers commented on perceived needs that included knowing how to differentiate instruction for ELLs, to motivate students, to increase opportunities for ELLs to participate in class, to develop effective lessons, and to address gaps in their own knowledge of science content. Of these categories, the most commonly identified challenges were knowing how to identify the needs of ELLs, closely followed by how to better differentiate instruction for students and how to motivate students.

At the end of the course, all preservice science teachers continued to struggle with knowing how to identify the needs of their ELLs. Moreover, this very skill was closely associated with their other perceived challenges of differentiating instruction and motivating students. That is, in order for preservice teachers to know how to differentiate instruction appropriately and effectively for students and to motivate them so that they were engaged, actively participating, and asking questions, they needed to begin with a good understanding of who their students were and what their needs were. In his response, Ken identified not only the need to better understand "developmentally where high schools students are," but also commented that "the lesson planning aspect of making sure that the lessons are going to accommodate ELLs is…an ongoing thing that I still need to practice." In short, preservice teachers recognized the value and importance of attending to the needs of their students; however, they required more practice with identifying these needs.

As introduced above, the lack of practice in identifying the needs of students also seemed to be the basis for the difficulties preservice science teachers perceived in differentiating instruction, in particular for ELLs. Hugh commented:

> I think differentiation is one of my biggest challenges. There's just – especially in this last placement – there's such a varying degree of performance in different areas even. So there's, you know, varying levels of ELs. And then, there's varying levels of math ability. And then

> there's varying levels of content mastery as well....I had over 30 students in a conceptual physics class and there's just – it's so hard to meet the needs of every individual student....I think there's a lot of students whose needs weren't met 100 percent. And, I struggle with that.

Hugh continued:

> I think I still struggle with differentiating between an ELL struggling with content and struggling with language needs. I think it can be really easy to misdiagnose a struggling ELL as not understanding the content when it could be both. I mean, it could be [that] they're not understanding the content because their language needs aren't being met.

Hugh's comments identify a critical gap in the field's understanding of how to separate language issues from content comprehension issues and whether distinguishing the two completely is even possible because of the way that language and content are so tightly intertwined. Again, the preservice teachers recognized and understood that there was a need to differentiate instruction, however, the actual practice of differentiating instruction, especially for ELLs, appeared beyond their expertise.

A third commonly perceived need of the preservice science teachers was knowing how to motivate students to learn both science content and academic English. For example, Monty commented:

> I definitely think the readings we did and the types of discussions and activities that we were doing in the class just made me question engagement in students, and how I was saying, how students can all be engaged in an activity or be participating in an activity, and how those two things are different. Also, how can I ensure that one student doesn't feel... excluded from an activity because they don't have the reading comprehension skills to participate in it.

Working with ELLs was a significant challenge, because in order to support ELLs' language development, the preservice teachers recognized the importance of creating continued opportunities for ELLs to be exposed to and able to use their English. Therefore, preservice teachers saw the need to motivate students in two ways: to keep students engaged and interested in learning science and to support their language learning by fostering opportunities to develop their language skills. In sum, the preservice science teachers' perceived challenges centered around their lack of opportunities to further develop their actual teaching skills in terms of identification of student needs, differentiation of instruction based on those identified needs, and sustaining and increasing the motivation of students.

Preservice Science Teachers' Suggestions for Course Improvement

Finally, we identified recommendations from the preservice teachers about how to improve the course. Their recommendations ranged from suggestions about what to include in the course to when to offer it. Strongly aligned with the preservice teachers' perceived needs, the most common suggestion was to provide more time and space for actual teaching practice coupled with feedback about their teaching. For example, Bryan talked about how he enjoyed opportunities in class to reflect on what had worked in his teaching of a lesson in terms of supporting ELLs and why. He wanted more of those opportunities:

> It's not like we can necessarily ask what strategy works for this and that. It depends on the student. It depends on the situation. So I don't feel like there's too much that could really enhance it. I like the activities that we've done [in the course]. Maybe there should be just a little bit more introspection. Actually, we've done quite a bit of it and it's been really nice, but reflecting on my own work is something I try to do as much as I can, but being forced to do it is helpful, too.

Bryan and other preservice teachers emphasized that scaffolded reflection, where they could receive guidance and structure in their own reflections of their teaching practices, as well as where they could receive explicit feedback about their pedagogical ideas from others, would be highly beneficial.

As their second most mentioned request, preservice teachers also asked for more examples of teaching activities appropriate for ELLs from the disciplines of chemistry, physics, and earth sciences, in addition to biology. As novice teachers, it was difficult for them to digest examples given in class and to translate them into their own specific teaching contexts. For example, Riley, a candidate earning her biology credential, commented:

> I like picking apart the lessons that people present and I'm sorry that there isn't the diversity [of science disciplines] among topics. But the majority of us are life science teachers, which means that the majority of the things [covered in this class] will be life science topics, which is unfortunate for the physics and chemistry candidates.

Alison's request for more variety was connected to her own perceived need to better understand her content: "I would have liked it to be a physics class with ELLs, but that's just partially because I would like a better background in physics." The suggestions for more varied examples might be satisfied if preservice teachers were provided more time and opportunities in class to implement their pedagogical ideas and strategies specific to their individual teaching contexts.

Half of the preservice teachers suggested adding even more opportunities for practice in writing and planning lessons with subsequent peer review and/or self-reflection. For instance, Alison stated: "It would be cool to bring in an idea of something we're planning and then plan it out with our peer support, particularly with that eye on English learners and how we're going to support them in that lesson." Similarly, Bethany stated that they should be encouraged:

> to write a little bit more curriculum, with an emphasis on supporting ELLs….I think if we were to go back to those first units we saw [in our placements] and rewrite to have an ELL focus for a lesson that you saw your CT [cooperating teacher] do, I think that could be super beneficial.

Finally, several preservice teachers' requests for improvement were connected to the timing of the course; preservice teachers thought the course should be scheduled earlier in the program. However, revising the timing of the course for future offerings may not be easily accommodated due to the constraints of other programmatic needs.

Implications

The findings presented above make clear that learning to teach science to ELLs is a complex and uncertain process. While our course provided opportunities for preservice science teachers to better understand their ELLs and how to support them in science instruction, we were disappointed by the reach of our course. Here, we provide suggestions for moving forward with the next iteration. As one example, we found that our preservice teachers did not discuss our three key ELL principles in the depth and detail we had hoped for at the conclusion of the course. In particular, one of several aspects of the capstone course that could be improved is closer attention to ELLs' funds of knowledge and instructional needs. We found no examples of our preservice science teachers discussing their students' funds of knowledge as resources they could elicitly build from during their instruction. More clearly defining and explaining the three key principles of effective ELL instruction to our preservice science teachers, as well as providing additional concrete examples of each principle in action will be important in future iterations of this course.

As a second example, preservice teachers identified meeting the needs of ELLs, in terms both of knowing what those needs were and how to differentiate instruction to meet those needs, as gaps in their understanding that remained even after completion of the course. One suggestion preservice teachers made for course improvement could help address this challenge. Creating, practicing, and reflecting on lessons with their peers could help preservice teachers better identify ways to tailor their instruction to their ELLs' strengths and needs. Although we had incorporated several opportunities for preservice teachers to do so in the course, apparently they were not enough. Preservice teachers wanted frequent and regular practice with and feedback on lessons to effectively reflect on their practices. A second suggestion would be to create additional in-class activities and assignments that focus specifically and repeatedly on learning about their ELLs, recognizing the heterogeneity within this group of students, and learning to design lessons that identify and build from their diverse strengths. Finally, a third suggestion would be to extend our preservice teachers' conceptions of supporting language development from the lexical to the discourse level, perhaps through the use of rehearsals of discourse practices (Lampert et al. 2013) that preservice science teachers might use with their ELLs.

Conclusion

We know that there is no one-size-fits-all approach to supporting ELLs. We also know that, to teach ELLs well, preservice teachers need a strong understanding of ELL issues, including theories of second language acquisition, science content and methods, and classroom context. In the context of our teacher education program, then, our capstone science methods course attempts to create a designated time and space for preservice teachers to gain hands-on experiences and to experiment with and reflect on strategies with peers and instructors – to learn to support ELLs' learning in the context of their actual classroom practice. We attempt to prepare beginning science teachers to become reflective practitioners who are able to recognize the needs of individual ELLs and to formulate and adapt strategies that best support individual ELLs to attain their learning goals. Findings from our first year of implementation make clear that we can certainly improve this capstone course. Still, our goal remains to foster the development of an adaptive disposition so that preservice science teachers can effectively work with their unique and diverse group of ELLs, not only as student teachers in a temporary placement but also as beginning teachers in their own science classrooms.

References

Bleicher, R. E., Tobin, K. G., & McRobbie, C. J. (2003). Opportunities to talk science in a high school chemistry classroom. *Research in Science Education, 33*(3), 319–339.

Borko, H., Jacobs, J., Eiteljorg, E., & Pittman, M. E. (2008). Video as a tool for fostering productive discussions in mathematics professional development. *Teaching and Teacher Education, 24*(2), 417–436.

Bunch, G. C. (2013). Pedagogical language knowledge: Preparing mainstream teachers for English learners in the new standards era. *Review of Research in Education, 37*(1), 298–341. doi:10.3102/0091732X12461772.

California Council on Science and Technology. (2007). *Critical path analysis of California's science and mathematics education system*. Sacramento: Author.

California Department of Education. (2014). *CalEdFacts*. Retrieved from http://www.cde.ca.gov/re/pn/fb/index.asp

Chinn, P. W. (2007). Decolonizing methodologies and indigenous knowledge: The role of culture, place and personal experience in professional development. *Journal of Research in Science Teaching, 44*(9), 1247–1268.

Dewey, J. (1904). The relation of theory to practice in education. In C. A. Murray (Ed.), *The relation of theory to practice in the education of teachers* ((Third yearbook of the National Society of the Scientific Study of Education, Part I), pp. 9–30). Chicago: University of Chicago Press.

Dutro, S., & Moran, C. (2003). Rethinking English language instruction: An architectural approach. In G. Garcia (Ed.), *English learners: Reaching the highest level of English literacy* (pp. 227–258). Newark: International Reading Association.

Khisty, L. L., & Chval, K. B. (2002). Pedagogic discourse and equity in mathematics: When teachers' talk matters. *Mathematics Education Research Journal, 14*(3), 154–168.

Khisty, L. L., McLeod, D., & Bertilson, K. (1990). Speaking mathematically in bilingual classrooms: An exploratory study of teacher discourse. In G. Booker, P. Cobb, & T. Mendicutti (Eds.), *Proceedings of the fourteenth international conference for the psychology of mathematics education* (Vol. 3, pp. 105–112). Mexico City: CONACYT.

Lampert, M., Franke, M., Kazemi, E., Ghousseini, H., Turrou, A. C., Beasley, H., Cunard, A., & Crowe, K. (2013). Keeping it complex: Using rehearsals to support novice teacher learning of ambitious teaching in elementary mathematics. *Journal of Teacher Education, 64*, 226–243.

Lee, O., & Buxton, C. A. (2013). Teacher professional development to improve science and literacy achievement of English language learners. *Theory Into Practice, 52*(2), 110–117.

Lee, O., & Fradd, S. H. (1998). Science for all, including students from non-English-language backgrounds. *Educational Researcher, 27*(4), 12–21.

Lee, O., Deaktor, R., Enders, C., & Lambert, J. (2008). Impact of a multiyear professional development intervention on science achievement of culturally and linguistically diverse elementary students. *Journal of Research in Science Teaching, 45*(6), 726–747.

Lee, O., Quinn, H., & Valdés, G. (2013). Science and language for English language learners in relation to next generation science standards and with implications for common core state standards for English language arts and mathematics. *Educational Researcher, 42*(4), 223–233. doi:10.3102/0013189X13480524.

Moll, L. C., Amanti, C., Neff, D., & Gonzalez, N. (1992). Funds of knowledge for teaching: Using a qualitative approach to connect homes and schools. *Theory Into Practice, 31*(2), 132–141.

Moschkovich, J. (2002). A situated and sociocultural perspective on bilingual mathematics learners. *Mathematical Thinking and Learning, 4*(2&3), 189–212.

Moschkovich, J. (2007). Using two languages when learning mathematics. *Educational Studies in Mathematics, 64*(2), 121–144.

National Research Council. (2007). *Rising above the gathering storm: Energizing and employing America for a brighter economic future*. Washington, DC: The National Academies Press.

National Research Council. (2010). *Rising above the gathering storm, revisited: Rapidly approaching category 5*. Washington, DC: The National Academies Press.

NGSS Lead States. (2013). *Next generation science standards: For states, by states*. Retrieved from http://www.nextgenscience.org/next-generation-science-standards

Putnam, R. T., & Borko, H. (2000). What do new views of knowledge and thinking have to say about research of teacher learning? *Educational Researcher, 29*, 4–15.

Rosebery, A. S., Ogonowski, M., DiSchino, M., & Warren, B. (2010). "The coat traps all your body heat": Heterogeneity as fundamental to learning. *The Journal of the Learning Sciences, 19*(3), 322–357.

Tekkumru-Kisa, M., Stein, M. K., & Schunn, C. (2015). A framework for analyzing cognitive demand and content-practices integration: Task analysis guide in science. *Journal of Research in Science Teaching, 52*(5), 659–685.

Tobin, K. G., & Kahle, J. B. (1990). *Windows into science classrooms: Problems associated with higher-level cognitive learning*. Bristol: The Falmer Press/Taylor & Francis Group.

Tom, A. R. (1985). Inquiring into inquiry-oriented teacher education. *Journal of Teacher Education, 36*(5), 35–44.

Understanding Language. (2013). *Six key principles for ELL instruction*. Retrieved from Stanford University, Graduate School of Education, Understanding Language website: http://ell.stanford.edu/content/six-key-principles-ell-instruction

van Es, E. A., & Sherin, M. G. (2010). The influence of video clubs on teachers' thinking and practice. *Journal of Mathematics Teacher Education, 13*(2), 155–176.

Chapter 6
Preparing Pre-service Secondary Teachers to Teach Science to English Learners: Theory into Practice

Trish Stoddart, Jorge Solis, Edward G. Lyon, and Sara Tolbert

Introduction

This chapter focuses on the preparation of pre-service teachers to teach science to English Learners (ELs) and is based on the SSTELLA (Secondary Science Education with English Language and Literacy Acquisition) project that has been implemented in four pre-service teacher education programs in Arizona, California and Texas, all states with large populations of ELs. The project addresses two national priorities: (1) improving the preparation of teachers to work with the rapidly expanding population of ELs; and (2) implementation of the Next Generation Science Standards (NGSS) (NRC 2012) and Common Core State Standards for English Language Arts (CCSS ELA) (CCSI 2010). These two agendas are interrelated: research on effective teaching practices for ELs has demonstrated that integrating English language and literacy development in contextualized instruction in science increases ELs achievement in both science, and academic language and literacy (Echevarria et al. 2012; Lee et al. 2008; Rivet and Krajcik 2008; Rosebery and Warren 2008; Stoddart et al. 2002) and the frameworks for the NGSS and CCSS

T. Stoddart (✉)
The University of California, Santa Cruz, CA 95064, USA
e-mail: stoddart@ucsc.edu

J. Solis
University of Texas at San Antonio, San Antonio, TX 78249, USA
e-mail: Jorge.Solis@utsa.edu

E.G. Lyon
Sonoma State University, Rohnert Park, CA 94928, USA
e-mail: lyone@sonoma.edu

S. Tolbert
University of Arizona, Tucson, AZ 85721, USA
e-mail: saratolbert@email.arizona.edu

A.W. Oliveira, M.H. Weinburgh (eds.), *Science Teacher Preparation in Content-Based Second Language Acquisition*, ASTE Series in Science Education,
DOI 10.1007/978-3-319-43516-9_6

explicitly recognize the fundamental relationship between the teaching of academic language and literacy and the learning of school subjects (NRC 2012). Currently, however, few novice or experienced teachers feel prepared to teach ELs and the majority of pre-service teacher education programs do not include explicit preparation on integrating the development of academic language and literacy into subject matter teaching (Ballantyne et al. 2008; California Legislative Analyst's Office [LAO] 2007–2008; Darling-Hammond 2006; Gándara et al. 2005; NCES 2011; Villegas and Lucas 2002). As expected there are differences across teacher education programs that relate to statewide differences in teacher accreditation and as well as socio-historical differences with respect to student and teacher backgrounds and statewide educational policies. Texas and California for example represent the two states with the largest number of second language learners nationally, yet they differ greatly in terms of bilingual programs available for students in elementary school. Moreover, teacher education candidates in California are mostly graduate students as opposed to Texas where the majority teacher candidates are undergraduates. However, novice teachers in all three states must be prepared to teach ELs. The goal of SSTELLA project is to provide an explicit practical model of teacher preparation that can be implemented by the majority of pre-service teacher education programs.

English Learners, NGSS and CCSS ELA

The relationship between science learning and English language and literacy development is reciprocal and synergistic. Through the contextualized and authentic use of language in scientific, mathematical and engineering practices, students develop and practice complex language forms and functions while simultaneously, through the use of language functions such as explanations and arguments in STEM investigations, students make sense of abstract ideas (Baquedano-Lopez et al. 2005; Cervetti et al. 2007; Lee et al. 2008; Ovando and Combs 2012; Rivet and Krajcik 2008; Rosebery and Warren 2008; Short et al. 2011; Stoddart 2005; Stoddart et al. 2002). This approach to teaching science and English language and literacy to second language learners is consonant with the NGSS and the Common Core State Standards (CCSS) for English Language Arts (ELA). Four of the eight NGSS science and engineering practices are particularly language intensive: developing and using models; constructing explanations (science) and designing solutions (engineering); arguing from evidence; and obtaining, evaluating, and communicating information. Science content and language intersect as students, for example, construct oral and written explanations and engage in arguments from evidence (Cheuk 2012; Lee et al. 2013), two practices that echo CCSS for English Language Arts. Concurrently, the ELA reading and writing standards for literacy in science and technical subjects require that students engage with technical (e.g., lab reports, scientific research articles) and non-technical (e.g., newspaper articles, letters to the editor) texts that are discipline specific by writing arguments, translating written information into visual forms (e.g., tables, graphs), and comparing/contrasting findings presented in various sources. Moreover, the pedagogical exemplars described in this paper highlight how teachers in California, Arizona, and Texas attempt to address both science and language learning for their students. While broader statewide language policies can influence local classroom contexts for teaching and learning science (McEneaney et al. 2014), we also take the position that science teachers can institute sound pedagogical policies in their classroom for second language learners (Langman 2014). Secondary science teachers can transform their classroom contexts when they put into action science-driven language and literacy practices for scientific sense-making (Lyon et al. 2016).

SSTELLA Teacher Preparation Model: Establishing Program Coherence

The SSTELLA project uses a practice-focused model of teacher education based on research demonstrating that the development of expertise in novice teachers is facilitated by engaging them in observation, analysis and experience with explicit models of the instructional approaches they are being prepared to teach (Abell and Cennamo 2004; Goldman et al. 2007; Hewson and Hewson 1988; Roth et al. 2011; Schwartz and Hartman 2007; Sherin 2004; Thompson et al. 2013; Wilson et al. 2001) and providing them with opportunities to practice these instructional approaches with the student population they are being prepared to teach with intensive feedback, coaching and support (Joyce and Showers 1995; Speck and Knipe 2001). However, for this approach to be effective conceptual and practical coherence must be established across the pre-service teacher education program by articulating the integrated instructional model throughout the coursework and field practicum. In many teacher education programs there is no explicit model for teaching diverse learners articulated across the program (Wilson et al. 2001). Courses on subject matter teaching typically give little attention to the importance of valuing and incorporating the linguistic needs and cultural experiences of the students being served (Cochran-Smith et al. 2008; Godley et al. 2006; Lee and Luykx 2006; Stoddart et al. 2011). Issues relating to cultural and linguistic diversity, when taught, are presented in separate courses that often focus on social conditions and not pedagogy (Ball and Tyson 2011; Trent et al. 2008; Zeichner 2003). In addition, there is often discontinuity between the pedagogical model presented in the university courses and the teaching practices modeled in field practicum (Stoddart 1993; Wilson et al. 2001).

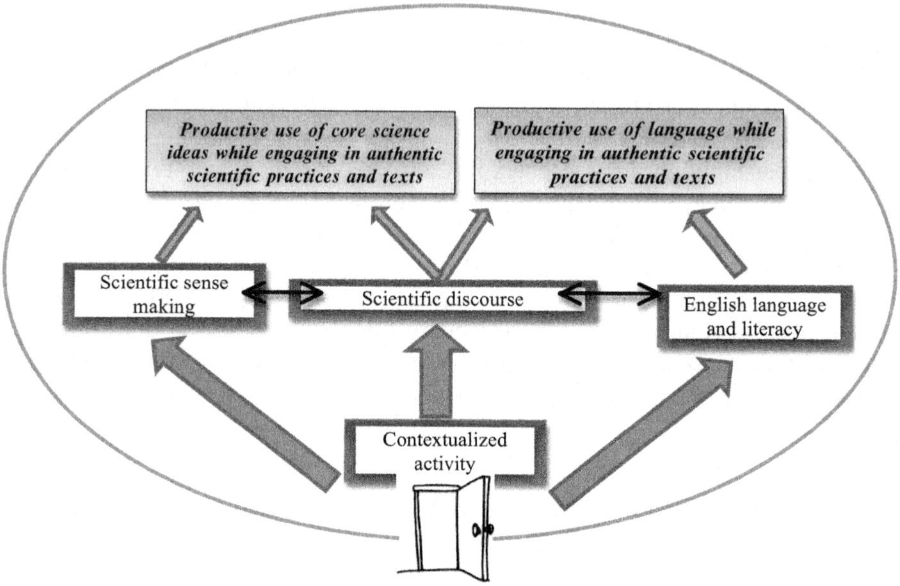

Fig. 6.1 SSTELLA instructional practices

Video and Instructional Cases

The first step in developing SSTELLA program coherence was to define the core conceptual and practical framework for the SSTELLA program and develop practical representations of these practices in instructional cases and videos. The four SSTELLA practices – Scientific Sense-making, Scientific Discourse, English Language and Literacy Development and Contextualized Science Activity – are based on the intersection between the NGSS and the CCSS ELA practices (see Fig. 6.1). The instructional cases describe in detail the teaching of a secondary science lesson (e.g. how a secondary science teacher could integrate the SSTELLA practices into a biology lesson). The vignette is followed by commentary on how the lesson exemplifies specific elements of SSTELLA instructional practices. Two general approaches have typically been used to supporting second language learners' content and language learning in K-12 classrooms: (1) the use of the primary language for instruction, often referred to as bilingual education and (2) sheltered English instruction strategies (sometimes referred to as Specially Designed Academic Instruction in English, or SDAIE). The goal of both approaches is to support second language learners in the content areas through rigorous academic content and language skills. SSTELLA examples shown here illustrate how teachers incorporate English learners (ELs) into focal science lesson activities and draw from their sociocultural experiences by using appropriate SDAIE strategies (i.e. graphic organizers, grouping structures, paralinguistic cues, etc.) to augment science content comprehension while also addressing English language development skills[1]. SSTELLA practices provide opportunities for students and teachers to negotiate and appropriate emerging understandings of scientific concepts and practices through translanguaging and discussions of familiar/home contexts (Canagarajah 2011; García and Wei 2013; Hammond 2001; Langman 2014; Rodriguez 2013).

Language and Literacy Practices

A central premise behind the SSTELLA framework is that language is part and parcel of all science learning activity and that particular attention to how language and literacy are used in the service of science learning supports greater student learning of both language and science especially for ELs. While SSTELLA practices represent an integrated and overlapping approach to science education, three specific language and literacy practices promote opportunities for English language development and access to science content by (1) attending to opportunities for student-student and teacher-student interaction, (2) supporting relevant science vocabulary and general academic terms, and (3) activating disciplinary literacy tasks. All science students benefit from support and assistance in using the language and literacy tasks that surrounds science activities such as reading, creating, and using scientific models, writing observational records, and presenting and sharing data. This attention is especially important for EL students from a range of English language proficiencies (i.e., beginning, intermediate, advanced). SSTELLA practices focus on scientific and engineering practices as the central context that requires the use of specific receptive and productive language and literacy functions. Therefore, understanding and teaching the features of the language of science activities is not the goal of language and literacy SSTELLA practices but rather serve as potential sources for providing feedback and instruction to students when appropriate.

Deconstructing or even teaching the language and literacy functions in science (i.e., often referred to as explicit teaching of academic language), without a focus on scientific and engineering practices, often misses the ultimate goal of addressing both language and content learning. Science teachers can become familiar with the language and literacy features of science learning as a way to promote greater access to science when using SDAIE strategies (e.g., graphic organizer, jigsaw reading groups, quick-writes). The language of science texts for example require making sense of dense clauses, hierarchically structured information, and a mixture of general academic vocabulary and highly specialized terms. Zwiers (2008) lists several language features in science where science teachers can address, plan for, and even distinguish from other disciplinary uses. The following language and literacy

[1] These strategies however need to be considered carefully to sustain nature of science activities being addressed.

features of science texts describe the varied nature of language use in science classrooms (Zwiers 2008):

- Describe procedures explicitly with procedural language, such as measures, observe, calculate, graph, record, watch, place, make, seal, hold, predicts, remove, examine, prevent, dissolve, attach, connect, mark, insert, and align. These are used mostly in lab directions and lab reports.
- Use many new and big words with new meanings, many of which are nominalizations, such as condensation, refraction, induction, resonance, reaction, radiation, fusion, erosion, and most other –ation words
- Describe relationships of taxonomy, comparison, cause and effect, hypothesis and interpretation. Unlike language arts and history, science texts have few stories or narratives. The text structure is dense and hierarchical (topic, subtopics, details).
- Connect abstract ideas illustrated by various media. Photos, diagrams, graphs, charts, math and chemistry symbols, lab experiences, and text all overlap to communicate concepts.

These language and literacy features in science classrooms are not exhaustive but more importantly, should be understood as generic language and literacy features that will differ and change depending on the scientific and engineering practices implicated and the accompanying materials used. However, understanding the language and literacy features and related tasks in any given science lesson can help teachers support students who may have more difficulty with certain language practices.

9th grade example: Mrs. Bird and the Chronology of the universe

The following classroom example of a 9th grade teacher (Mrs. Bird) using SDAIE strategies involves a science lesson in California. The ELs[2] in this class are considered mostly intermediate English language learners as tested by the California English Language Development Test (CELDT). The lesson focuses on the chronology of the universe since the Big Bang. Albeit with some challenges, the following example illustrates some of the ways that language and literacy activities in science serve to move forward complex and authentic understandings of scientific concepts and practices. SSTELLA pre-service teacher (PSTs) used examples like this to come to learn how to notice and analyze their own teaching practices as well as those of their mentor teachers. In this case, PSTs were introduced to the overall context of the lesson and asked to view four selected video-clips approximately

[2] EL or English Learner/Limited English Proficient terms are used in many official state and federal policies to refer to students acquiring a second language in school. We acknowledge that these terms are problematic terms because these students are very diverse from age of arrival to the U.S., to proficiency in the primary language, and prior academic preparation among other significant differences. ELs can't be defined as a homogeneous or absolute group; moreover, state assessments used to classify EL designations and language levels differ from state to state. The term "EL" moreover is a restrictive identifier that fails to acknowledge non-English primary languages and the emergence of bilingual abilities.

2–3 min each before coming to class. They were asked to note instances where the teacher engaged students in understanding the tasks related to the universe timeline activity and how the teacher supported students' understanding of key science content and literacy. During the class session, PSTs were engaged in a discussion where they had to identify specific aspects of the lesson related to language and literacy development. The following are some guiding questions used in the noticing protocol used with this video example:

- What kind of visuals, gestures, questions, or other strategy does the teacher use to explain key words and phrases? What about figurative language references like "soup of things" or analogies?
- When the teacher asks a pair of students "how are we going to fix it" pointing to a student illustration representing their universe timeline epoch from the reading, what kind of writing feedback does the teacher provide? How does she do this?
- Does the lesson provide students an opportunity to *develop and use models.* Did you agree or disagree? If you agree, how are students in any way developing, revising, and/or using models to illustrate or predict the relationship between two or more parts of a concept or system? If you disagree, how could this lesson example connect classroom activities more closely to developing and using models in science?

The lesson also illustrates areas where the teacher could have extending further the development of language and literacy without diluting the scientific practices:

Mrs. Bird lists two lesson objectives for this lesson where students will be able to describe the timeline of the universe by completing a collective class timeline and will able to explain the theory of evolution through different forms of evidence. Every pair of students received a reading section of a larger document representing an era of the universe since the Big Bang. The focal activity during the lesson is for students to work in pairs interpreting a jigsaw reading and then sharing index card summaries from the reading to make a collective classroom timeline. Once students complete their reading and notes on their index card they are asked to hang their index card on a yellow string hanging in the back of the classroom. Of note here is that before completing the classroom timeline on a string, the teacher worked closely for the majority of the class period with students in pairs and one-on-one to help students summarize their understanding of the reading. The teacher rotated through each pair helping students with understanding the reading and the writing task. Students were asked to draw an image on the other of their index card describing their era. During several instances Mrs. Bird provided help by simply reading aloud sections of the reading to pairs and thereby, helping students hear the pronunciation of unfamiliar terms. She often asked questions like "how can you put that into different words?" and "what does transition mean?" as students summarized their reading. When a pair of students had difficulty creating an image of their epoch, the teacher engages them by asking a series of questions such as "what would it look like", "what do you see in your mind", "is it a giant soup?", "what is it really hot or really cold?", "what do the particles look like?" and "what's in the center of a nucleus?" In another example, she asks if their image is going to look "like a flat pancake or something else?" as she gestures with her arms in an up and down motion and then opening her palms up. In another instance, Mrs. Bird repeatedly goes back to the same pair noticing that a student's illustration did not appear to be complete. Here the teacher reminds the student pair what it means to have charge differences in atoms. She does this by noting that atoms can collide much like what happens in a soccer match where players can collide.

Throughout this lesson, Mrs. Bird directly addresses the lesson objectives by assisting students' comprehension of the reading, which is challenging for most students in the class. Most of her support of language and literacy is focused on key vocabulary, but it also focuses helping students gain access to the science content by:

- Using visuals, contextual cues, graphic representations, paraphrases, and some definitions to help students comprehend new vocabulary.
- Recognizing students' developing scientific understandings using "everyday" words but encourages students to use key terms as appropriate.
- Providing opportunities for student to read or write focusing on scientific/engineering practices (i.e., asking questions and defining problems, developing and using models, planning and carrying out investigations, analyzing and interpreting data, etc.).
- Using strategies to support discipline-specific writing (e.g., constructing models, drawing graphic representations of data, etc.).

This lesson addresses some aspects of developing and using models (Scientific and Engineering Practice #2) (NRC 2012) where students work on revising and using a model of the Big Bang "to illustrate and/or predict the relationships between systems or between components of a system" (Appendix F) as students illustrate their impression of particular epochs in relation to the Big Bang. There is an implicit reference to creating and using models throughout the lesson. The explicit focus remains on reading and writing about the chronology of the universe and in particular, summarizing the jigsaw reading and illustrating an image of that reading. The science lesson culminates with students co- constructing a chronology of the universe through the hanging of their index cards on the string. This example is an instructive example because it also shows the types of support teachers can provide students in accessing the language and content of a science lesson through a variety of SDAIE strategies. But this lesson also highlights the challenge for ELs often observed in secondary school science classrooms where scientific and engineering practices and reading and writing tasks are not explicitly addressed by the teacher.

Experiencing, Analyzing, and Approximating SSTELLA Practices in a Secondary Science Method Course

The next stage in developing program coherence was to bring the teacher education and collaborating school faculty in collaborative professional development to assist them in redesigning teacher education courses and field practicum. The science methods instructors (SMI), teacher supervisors and cooperating teachers were engaged in analysis of the SSTELLA practices using NGSS and CCSS ELA standards and the SSTELLA instructional scenarios and video cases. The SMI across the four sites spent 6 months working in a team to infuse the SSTELLA practices and materials into their science methods courses. Each instructor developed a model SSTELLA science lesson to be enacted in their science methods course and developed course syllabi to engage secondary science pre-service teachers in analysis and implementation of SSTELLA practices through (a) an introduction to the theory behind language-science integration, (b) participation in SSTELLA model-lesson activities, (c) study of the SSTELLA framework through analysis of video and instructional cases and (d) development and teaching of a SSTELLA model lesson. Teacher supervisors and cooperating teachers also engaged in professional development that included analysis of the SSTELLA practices and training in coaching and mentoring using the SSTELLA rubric for assessing student teachers' performance and providing them with coaching and feedback.

A key feature of the project intervention was to use the SSTELLA Framework (Fig. 6.1) to enhance the content and structure of secondary science method courses at partner institutions. Four science method instructors met virtually and face-to-face over the course of a year to improve self understanding about effective science teaching for ELs and how a science method course might best prepare science teacher candidates to teach science to English learners while still being responsive to the institutional and geographic context.

Teaching Practices At the Core At the core of the method courses was an emphasis on discipline-specific teaching practices that best leverage learning opportunities for a diverse group of students (Ball and Forzani 2009; Grossman and McDonald 2008). According to Windschitl et al. (2012) these practices need to be: (1) accessible to novice teachers; (2) applicable to everyday work of teaching; and (3) function synergistically to form a coherent model of teaching and learning so that instructional approaches are grounded in theory of how students learn science. For the project, these practices were grounded in the SSTELLA Framework. To prepare teacher candidates to implement core practices, method instructors along with project PIs developed materials to engage candidates in three primary types of activities, represented in Fig. 6.2: experiencing the SSTELLA practices, noticing and analyzing the SSTELLA practices, and then approximating the SSTELLA practices (Abell and Cennamo 2004; Roth et al. 2011; Sherin 2004). Figure 6.2 shows how the four SSTELLA practices were addressed as part of professional development activities with novice teachers. This cycle of learning the SSTELLA practiced involved modeling and approximating SSTELLA practices, experiencing these practices in action, and engaging in activities where teachers had to learn how to notice and analyze these practices for their development and as a way to provide feedback to others.

Experiencing SSTELLA Practices Through Model Learning Segments The method instructors developed four learning segments (multi-day lessons) that focused on a range of grade levels and subjects. Each learning segment modeled various ways to contextualize science activity and integrate language, literacy, and

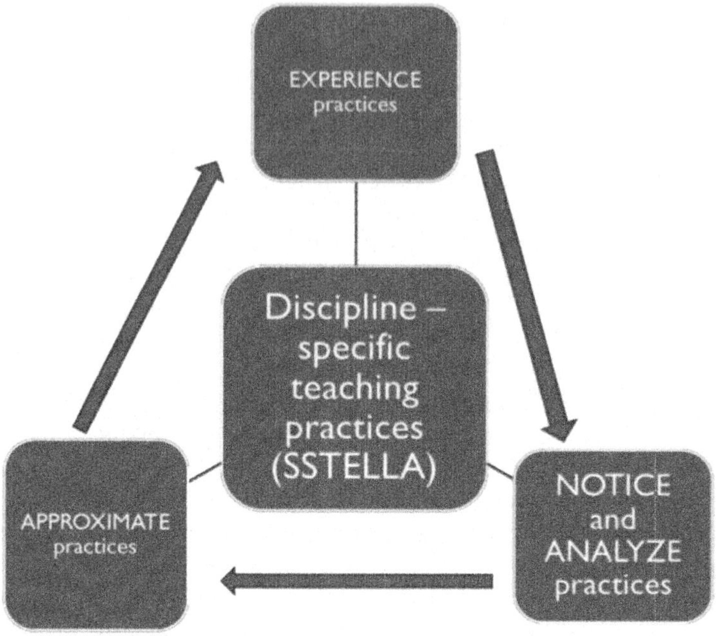

Fig. 6.2 Cycle of teacher as learner

science, but through varied emphases. For instance, the model learning segment, "It's about Time," designed for middle school space science, took candidates through the development and use of scientific models with strategies to support the sense-making process (e.g., visual representations, graphic organizers, hands-on materials, etc.) while the lesson, "Antibiotic Resistance of Bacteria," designed for a high school biology class, focused on helping students use their understanding of natural selection to engage in authentic science literacy tasks, including close readings of various texts and a written evidence-based explanations. The learning segments were documented with a similar template that included annotations for how it reflected SSTELLA practices. Each learning segment differed in terms of the focal SSTELLA practices that it was meant to model. For example, the lesson "The Antibiotic Resistance of MRSA," (Lyon 2016) focused on the core idea of how species change over time due the process of natural selection, the scientific practice of constructing evidence-based explanations, and the SSTELLA practice of English Language and Disciplinary Literacy Development. The vignette below describes part of this learning segment as it was experienced in the method course.

Candidates were first shown an anticipatory question:

Recall an experience with hospitals, such as when you (1) were injured, (2) waited for your brother, sister, or cousin being born, or (3) visited a sick family member or friend. Also think about your own knowledge of hospitals. Do you think someone could be harmed from bacteria while staying in a local hospital? Write your response with a reason in your science notebook.

After writing a response, candidates shared responses with a partner, and then the instructor invited two students to share with the class: a student who responded with "yes" and another student who responded with "no." A student shared a personal story about a friend who came to the hospital to get better, only to get sick with something different. The instructor asked follow up questions to the class, such as "do you agree with..." and "what kind of evidence would support our response?" The instructor informed candidates that they will not come to a consensus answer yet, since this real world problem, like most, is complex. However, everything we will be learning in the "upcoming weeks" [if a real high school class] will help us address this question.

At this point, the instructor stopped to do a quick check-in with candidates, asking about the purpose of these first minutes. Candidates had experienced, discussed, and practiced themselves framing instruction through contextualized activities that could connect authentic science to student lives and provide opportunities for student contributions. Thus, candidates readily noticed that what they experienced was intended to bring relevance to the lesson as well as communicate big ideas.

The instructor proceeded to show a short video clip from the url https://www.youtube.com/watch?v=bevhCDOoYeE#t=30 that depicted a newscaster from 2005 reporting on the increased presence of a "superbug" called Methicillin resistant Staphylococcus aureus (or MRSA). The clip ended with a reporter asking an expert: "What causes these so called Superbugs?" The instructor stopped the clip there and stressed to the class that we will exploring this phenomenon "What causes Superbugs," which will help them understand the big idea: "How do species change over time?" The teacher then asked the candidates to use a graphic organizer given to try making sense of some key words, such as antibiotic that were presented in the multimedia text. Candidates completed the worksheet and terms were addressed as a class, with the instructor noting that they will need to use this precise language when explaining what causes superbugs. The instructor then showed an abbreviated timeline that would be revisited later in the unit indicating four key points related to the MRSA "superbug." The instructor then asked candidates in small groups to create an outline (via a timeline, bulleted list, storyboard, etc.) to provide a tentative explanation about how the species Staphylococcus aureus changed (from 1880s to Present) so that over 60 % of the species is methicillin resistant (visuals or charts could help represent "0 %" and "60 %").

The instructor modeled the process first on a document camera: drawing a picture of a colony of bacteria on a petri dish to represent Staphylococcus aureus and then probes them to consider how we could represent this new "variation" of the species that was identified in 1941 and what words/phrases we could use to describe what happened in between. The instructor posted all students' initial models on the walls so that students can engage in a "gallery walk" where they view each other's models. The teacher closed the experience by pointing out the variation in students' models (both the content and how they decided to represent).

Analyzing SSTELLA Practices The part of learning segment described in the vignette would have represented the first day of a high school biology unit on natural selection. The instructor informed teacher candidates to put back on their "teacher's hat" and gave instruction for them to deconstruct this first day of the learning segment from the lens of both scientific practices and potential standards from the Common Core State Standards for English Language Arts (CCSS ELA): Literacy in the Social Sciences, Sciences, and Technical Subjects.

Candidates had read about and experienced scientific practices and even incorporated them into their own mini lessons taught in the method course. Thus, they were quickly able to identify "developing and using models" as the primary scientific practice, and also predicted that "constructing explanations" would become a central focus later in the segment. However, this was their first foray into looking at CCSS for ELA in the context of science teaching. The advantage of this particular deconstruction activity was that candidates explored the standards after experiencing a lesson in which the doing of science (e.g., scientific practices) was intimately tied to reading and writing in the discipline of science, so they could directly see overlaps between scientific practices and CCSS for ELA. After discussing what would happen in the upcoming week for the unit, candidates then added another layer of deconstruction by examining an outline of the California ELD Framework (see http://www.cde.ca.gov/ci/rl/cf/elaeldfrmwrksbeadopted.asp), specifically (1) standards related to interacting in meaning ways through collaboration, interpretation, and productive use of language, (2) a document developed by WestEd to connect the ELA/ELD Framework with science content and activities (http://www.cde.ca.gov/sp/el/er/eldstandards.asp) and (3) descriptions of the new categories of EL proficiency (emerging, expanding, bridging). Finally, the SSTELLA Practices Progression was used to discuss the development of particular instructional moves supporting student mastery of NGSS, CCSS ELA, and ELD standards. This instrument was used to code lessons in our research and also as a coaching tools for mentor teachers in professional development sessions. In the case of science literacy tasks, for example, teachers might attend to some forms of language and literacy support by introducing specific science vocabulary or having students copy down laboratory procedures; this form of language and literacy support is considered less developed (introducing level) compared to examples where the teacher links back language and literacy support in the context and service of doing scientific investigations and practices.

Analyzing Video Cases Preparing novices to become reflective practitioners requires that teacher educators help them "hone in on what is important in a very complex situation" (Van Es and Sherin 2002, p. 573). Videos can be effective as "cases" to develop this ability to notice and analyze teaching practices (Abell and Cennamo 2004; Ash 2007; Roth et al. 2011; Sherin 2004). The advantage of videos is that novices can observe teaching in a real-life context (more authentic than a lesson modeled in the method class) with opportunities to "replay" events for further noticing and analysis.

After candidates experienced the last part of the Antibiotic Resistance lesson, they participated in activities to distinguish between two ways of using vocabulary during science instruction. First, they read aloud dialogue from the article by Bruna et al. (2007) which is a case study of how one high school EL science teacher focused on narrow aspects of academic language instruction (i.e. vocabulary fluency, isolated definitions associated with the rock cycle) and supplanting opportunities for students to use vocabulary to make sense of the rock cycle itself. Candidates then watched a video clip related to the vignette of Ms. Bird that was described earlier in the chapter. The video was carried out in a similar schooling context as the one in Bruna et al. (2007), but in contrast the teacher supported students' use of vocabulary through multiple ways as students read texts and eventually developed a class model to depict the chronology of the universe. The video in particular allowed candidates to see and hear interactions between teacher and individual students, which reflected several strategies they experienced in the Antibiotic Resistance Lesson. They were then able to go back to the Antibiotic Resistance lesson and identify various supports and how those supports might be augmented depending on the EL proficiency of students.

Approximating SSTELLA Practices Beyond experiencing and analyzing, candidates need opportunities to practice instructional approaches with effective mentoring and support (Joyce and Showers 1995; Loucks-Horsley et al. 1998; Speck and Knipe 2001). The field experience component to a teacher education program provides the most meaningful and authentic experiences, supported through in SSTELLA by the mentor PD, but opportunities to approximate what they would do in a real classroom can also happen in a science method course. Candidates were assigned multiple "approximation" assignments. In each approximation, candidates wrote out and then carried out in real time a particular practice indicated in the SSTELLA practices progression: introducing a contextualized big idea, debriefing students' initial scientific models, using talk moves to facilitate an instructional conversation, and scaffolding an authentic literacy task. For example, one students' description of the approximation she carried out in class was as follows:

Alright, class we were able to brainstorm and share some of our own ideas of climate change and as a visual for us to use to remember where we began this unit I have assembled a Wordle for us to see the concepts we came up with.

Today we are going to start our exploration of the science behind climate change by doing a jigsaw reading activity. I have found 4 articles that I believe will help us begin to understand the factors that are contributing to the phenomenon of climate change. These articles will introduce us to new key terms and concepts that will help us eventually answer our big question for this unit. "How do human activities influence climate change and in what ways will climate change affect the biosphere?"

I have given each of you a worksheet titled Climate Change Jigsaw with several articles attached to it numbered 1–4. Please write your name and the date at the top of your worksheet (hold up worksheet and indicate where to write this information)

I will first demonstrate how to annotate an article and record key terms and concepts on your worksheet. Please find article #1, titled Global Warming: News, Facts, Causes & Effects, and annotate along with me marking key words and concepts as I go. Article projected onto board. Teacher reads article aloud, circling key words and underlining key concepts.

Candidates received both written as well as oral feedback from the instructor and peers directly after approximating the practice. These approximations scaffolded the candidate's implementation of SSTELLA practices, which then led directly to their culminating product: to develop their own 3-day learning segment and teaching 30 min from it in class to their peers. During the first half of the course, each group of four subject-specific candidates (e.g., biology, earth science) was given a specific NGSS and given the task to outline a 10 day unit plan with (1) a central "big idea", (2) a culminating performance task, and (3) daily learning objectives in addition to the daily classroom activities. Candidates then developed their learning segment from this unit outline.

As evidence by the exemplars and descriptions above, the method course aligned with SSTELLA practices as candidates experienced the practices, analyzed them, and approximated them. Model lessons and videos were key materials to communicate nuances with the SSTELLA practices, how they relate to science and literacy standards, and to depict varying levels of implementing the practices.

As part of the SSTELLA study, pre-service teachers participating in both the unrestructured (control) and restructured (treatment) method course and mentoring were observed twice during student teaching or clinical practicum. Trained SSTELLA observers conducted and scored science lessons taught by teacher candidates using the SSTELLA Classroom Observation Rubric (SCOR). Table 6.1 provides a description of the SSTELLA practice addressing language and literacy in science used in the rubric. The full SCOR instrument addresses the four SSTELLA practices including (1) Scientific Sense-making, (2) Scientific Discourse, (3) English Language and Literacy Development and (4) Contextualizing Science Activity. Each SSTELLA practice moreover was measured through related sub-practices comprising each SSTELLA practice. Table 6.1 identifies the sub-practices corresponding with each SSTELLA practice.

Observations were videotaped and accompanied by an audio-recorded post-observation debrief interview with the teacher candidate. SCOR results allowed us to capture the pedagogical development of SSTELLA practices of each participant and determine the impact of the intervention on teacher implementation of effective science teaching practices for ELs (Lyon et al. 2016). Observations were scored on a scale of 0–3 (0 = Not present, 1 = Introducing, 2 = Implementing, 3 = Elaborating). An Analysis of the SCOR results reveal that teacher candidates who participated in the SSTELLA redesigned method course and SSTELLA-informed mentoring implemented four sub-practices at a statistically significant higher score than those receiving the unrestructured method course and mentorship. The sub-practices that scored higher in the SCOR for the SSTELLA participants included *Framing* Contextualizing Science Activity, *Adapting and Applying* Contextualizing Science

Table 6.1 SSTELLA practices and sub-practices

SSTELLA practices	Sub-practices
1. Scientific sense-making	Communicating the "Big Idea"
	Pressing for model- or problem-based scientific/ engineering practices
2. Scientific discourse through scientific/ engineering practices	Facilitating productive student talk
	Pressing for scientific explanation & argumentation
3. English language & disciplinary literacy development	Promoting opportunities for English language development for ELs through student interaction
	Promoting opportunities for English language development for ELs through vocabulary
	Pressing for authentic science literacy tasks
4. Contextualizing ccience activity	Framing (Planned)
	Adapting (Unplanned) & applying

Activity, Promoting Opportunities for English Language Development for ELs through *Student Interaction*, and Facilitating Productive science *Student Talk*. Analyses also reveal differences across sub-practices in the percentage of teachers reaching higher levels of implementation (Scores of 2 or 3) of the reform pedagogy. For the student interaction sub-practice, 64 % of treatment teachers reached on average the implementation level compared to 47 % of control teachers. For productive science talk, 47 % of treatment teachers reached on average the implementation level as compared to 24 % of control teachers. Finally, adapting and applying contextualizing activities and student contributions, 23 % of treatment teachers reached implementation level compared to 8 % of control teachers. It is important to note that there are a range of mediating factors that contribute to the appropriation of SSTELLA practices by teacher candidates. Future analysis of teacher background variables, student-teaching contexts, and relationship with student achievement outcomes will be addressed. This analysis of the SCOR results, however, indicates that the SSTELLA intervention shows promise for better preparing teacher candidates to address sociocultural dimensions of science learning compared to teacher candidates not receiving the intervention.

Pre-service teachers were also asked to rate their confidence for teaching science to ELs on a five point scale at the beginning and end of the SSTELLA pre-service teacher education program. An analysis of pre- post ratings showed that a significant improvement in novice teachers efficacy for teaching ELs ($F (1.251) = 27.562$, $p < 0.001$)

Conclusion

In this chapter, we describe the SSTELLA framework for integrating language and literacy development into secondary science teaching and describe the development and use of instructional exemplars and video cases in four pre-service secondary science teacher education programs in Arizona, California and Texas. We emphasize the importance of developing coherence across the coursework and field practicum components of teacher preparation programs by engaging science methods instructors, teacher supervisors and cooperating teachers in collaborative professional development on the SSTELLA model. In model building it is important to examine the efficacy of the intervention and analyze the impact on the participants – in this case pre-service teachers. This chapter includes the results of an analysis of classroom observations of pre-service teachers in their student teaching practicum which demonstrate that teacher candidates who participated in the SSTELLA intervention were more likely to use practices that promote science and language learning for ELs than novice teachers who were not part of the intervention. These findings are particularly timely because recent reviews of research teacher education reveal very little empirical evidence of the positive impact of programs of preservice teacher preparation (Sleeter 2015).

The development, field testing and dissemination of practical models to improve the preparation novice teachers to work with ELs is particularly important in an environment where few novice or experienced teachers feel prepared to teach ELs and the majority of pre-service teacher education programs do not include explicit preparation in teaching subject matter to ELs (Ballantyne et al. 2008; Darling-Hammond 2006; Gándara et al. 2005; NCES 2011; Villegas and Lucas 2002). As a consequence, each year thousands of new teachers enter the profession unprepared to teach this vulnerable student population. Our findings show that the SSTELLA program not only significantly improved their confidence in their preparedness to teach ELs. Our primary goal with the SSTELLA project, therefore, is to develop a model to scaffold science teacher educators application of theory to practice to prepare student teachers to more effectively teach science to the rapidly growing population of ELs across the United States. Prior research has demonstrated that professional development with experienced teachers can support the implementation of integrated practice and improve the achievement of ELs (see for example, Lee et al. 2008). This research demonstrates that the model can also be applied to pre-service teachers.

Acknowledgment This material is based on work supported by the National Science Foundation (NSF) Discovery Research K–12 Program under Grant No. DRL-1316834. Any opinions, findings, conclusions, or recommendations expressed in this material are those of the authors and do not necessarily reflect the views of NSF.

References

Abell, S. K., & Cennamo, K. S. (2004). Videocases in elementary science teacher preparation. In J. Brophy (Ed.), *Advances in research on teaching: Using video in teacher education* (Vol. 10, pp. 103–130). New York: Elsevier JAI.

Ash, D. (2007). Using video data to capture discontinuous science meaning making in non- school settings. In R. Goldman, R. Pea, B. Barron, & S. J. Derry (Eds.), *Video research in the learning sciences* (pp. 207–226). Mahwah: Erlbaum.

Ball, D. L., & Forzani, F. M. (2009). The work of teaching and the challenge for teacher education. *Journal of Teacher Education, 60*(5), 497–511.
Ball, A., & Tyson, C. (Eds.). (2011). *Studying diversity in teacher education*. New York: Roman.
Ballantyne, K. G., Sanderman, A. R., & Levy, J. (2008). *Educating English language learners: Building teacher capacity Roundtable Report*. National Clearinghouse for English Language Acquisition & Language Instruction Educational Program. George Washington University, Graduate School of Education and Human Development.
Baquedano-Lopez, P., Solis, J. L., & Kattan, S. (2005). Adaptation: The language of classroom learning. *Linguistics and Education, 6*, 1–26.
Bruna, K. R., Vann, R., & Perales Escudero, M. (2007). What's language got to do with it?: A case study of academic language instruction in a high school "English Learner Science" class. *Journal of English for Academic Purposes, 6*, 36–54.
California Legislative Analysts Office (LAO). (2007–2008). Budget book – Education analyses, February 21, 2007. California Legislative Analyst's Office Report, 2007–08.
Canagarajah, S. (2011). Codemeshing in academic writing: Identifying teachable strategies of translanguaging. *The Modern Language Journal, 95*(3), 401–417.
Cervetti, G. N., Pearson, P. D., Barber, J., Hiebert, E., & Bravo, M. A. (2007). Integrating literacy and science: The research we have, the research we need. In M. Pressley, A. K. Billman, K. Perry, K. Refitt, & J. Reynolds (Eds.), *Shaping literacy achievement*. New York: Guilford.
Cheuk, T. (2012). Relationships and convergences found in the common core state standards in mathematics (practices), common core state standards in ELA/Literacy*(student portraits), and a framework for K-12 science education (science & engineering practices). Unpublished work. Stanford University, Understanding Language Initiative.
Cochran-Smith, M., Feiman-Nemser, S., McIntyre, D. J., & Demers, K. (Eds.). (2008). *Handbook of research on teacher education: Enduring questions in changing contexts*. New York: Routledge Taylor & Francis and the Association of Teacher Educators.
Common Core State Standards for English Language Arts & Literacy in History/Social Studies, Science, and Technical Subjects (Common Core State Standards Initiative, 2010).
Darling-Hammond, L. (2006). Constructing 21st-century teacher education. *Journal of Teacher Education, 57*(3), 300–314.
Echevarria, J. J., Vogt, M. E., & Short, D. J. (2012). *Making content comprehensible for English learners* (4th ed.). Boston: Pearson.
Gándara, P., Maxwell-Jolly, J., & Driscoll, A. (2005). *Listening to teachers of English language learners: A survey of California's teachers' challenges, experiences, and professional development needs*. Santa Cruz: Center for the Future of Teaching and Learning.
García, O., & Wei, L. (2013). *Translanguaging: Language, bilingualism and education*. New York: Palgrave Macmillan.
Godley, A. J., Sweetland, J., Wheeler, R. S., Minnici, A., & Carpenter, B. D. (2006). Preparing teachers for dialectally diverse classrooms. *Educational Researcher, 35*(8), 30–37.
Goldman, R., Pea, R., Barron, B., & Derry, S. J. (Eds.). (2007). *Video research in the learning sciences*. Mahwah: Erlbaum.
Grossman, P., & McDonald, M. (2008). Back to the future: Directions for research in teaching and teacher education. *American Educational Research Journal, 45*(1), 184–205.
Hammond, L. (2001). Notes from California: An anthropological approach to urban science education for language minority families. *Journal of Research in Science Teaching, 38*(9), 983–999.
Hewson, P. W., & Hewson, P. W. (1988). An appropriate conception of teaching science: A view from studies of science learning. *Science Education, 72*(5), 597–614.
Joyce, B., & Showers, B. (1995). *Student achievement through staff development*. White Plains: Longman.
Langman, J. (2014). Translanguaging, identity, and learning: Science teachers as engaged language planners. *Language Policy, 13*(2), 183–200.

Lee, O., & Luykx, A. (2006). *Science education and student diversity: Synthesis and research agenda*. New York: Cambridge University Press.

Lee, O., Maerten-Rivera, J., Penfield, R., LeRoy, K., & Secada, W. (2008). Science achievement of English language learners in urban elementary schools: Results of a first-year professional development intervention. *Journal of Research in Science Teaching, 45*(1), 31–52.

Lee, O., Quinn, H., & Valdés, G. (2013). Science and language for English language learners in relation to next generation science standards and with implications for common core state standards for English language arts and mathematics. *Educational Researcher, 42*(4), 223–233.

Loucks-Horsley, S., Hewson, P. W., Love, N., & Stiles, K. E. (1998). *Designing professional development for teachers of science and mathematics*. Thousand Oaks: Corwin Press.

Lyon, E. (2016). The antibiotic resistance of MRSA: Teaching natural selection with literacy development for English learners. *Science Activities: Classroom Projects and Curriculum Ideas, 53*(1), 33–47.

Lyon, E., Tolbert, S., Solis, J. L., Stoddart, T., Salinas, I., Knox, C., Roth, B., & Butler, M. (2016, April). *Teaching English learners through science-language integration: Linking a conceptual framework to secondary teacher preparation.* Strand 7 Symposium at the annual conference of the National Association of Research in Science Teaching, Baltimore, MD.

Lyon, E. G., Tolbert, S., Solis, J. L., Stoddart, T., & Bunch, G. C. (2016). *Secondary science teaching for English learners: Developing supportive and responsive learning contexts for sense-making and language development*. New York: Rowman & Littlefield Publishers.

McEneaney, E. H., López, F., & Nieswandt, M. (2014). Instructional models for the acquisition of English as bridges into school science: Effects on the science achievement of US Hispanic English language learners. *Learning Environments Research, 17*(3), 305–318.

National Center for Education Statistics. (2011). (NCES) The Nation's Report Card: Science 2009. Institute of Education Sciences.

National Research Council. (2012). *A framework for K-12 science education: Practices, crosscutting concepts, and core ideas*. Washington, DC: National Academies Press.

Ovando, C. J., & Combs, M. C. (2012). *Bilingual and ESL classrooms* (5th ed.). Boston: McGraw-Hill.

Rivet, A., & Krajcik, J. (2008). Contextualizing instruction: Leveraging students' prior knowledge and experiences to foster understanding of middle school science. *Journal of Research in Science Teaching, 45*, 79–100.

Rodriguez, G. M. (2013). Power and agency in education exploring the pedagogical dimensions of funds of knowledge. *Review of Research in Education, 37*(1), 87–120.

Rosebery, A. S., & Warren, B. (2008). *Teaching science to English language learners*. Alexandria: National Science Teachers Association.

Roth, K. J., Garnier, H. E., Chen, C., Lemmens, M., Schwille, K., & Wickler, N. I. (2011). Videobased lesson analysis: Effective science PD for teacher and student learning. *Journal of Research in Science Teaching, 48*(2), 117–148.

Schwartz, D. L., & Hartman, K. (2007). It is not television anymore: Designing digital video for learning and assessment. *Video Research in the Learning Sciences*, 335–348.

Sherin, M. G. (2004). New perspectives on the role of video in teacher education. In J. Brophy (Ed.), *Advances in research on teaching: Using video in teacher education* (Vol. 10, pp. 1–27). New York: Elsevier JAI.

Short, D. J., Vogt, M., & Echevarría, J. (2011). *The SIOP model for teaching science to English learners*. Boston: Pearson.

Sleeter, C. E. (2015). Towards teacher education research that informs policy. *Educational Researcher, 43*(3), 146–153.

Speck, M., & Knipe, C. (2001). *Why can't we get it right? Professional development in our schools*. Thousand Oaks: Corwin Press.

Stoddart, T. (1993). The professional development school: Building bridges between cultures. *Educational Policy, 7*(1), 5–23.

Stoddart, T. (2005). *Improving student achievement with the CREDE five standards pedagogy* (Technical Report No. J1). Santa Cruz: University of California, Center for Research on Education, Diversity and Excellence.

Stoddart, T., Pinal, A., Latzke, M., & Canaday, D. (2002). Integrating inquiry science and language development for English language learners. *Journal of Research in Science Teaching, 39*(8), 664–687.

Stoddart, T., Bravo, M., Solis, J., & Mosqueda, E. (2011). Preparing pre-service elementary teachers to teach science to English language learners. *Proceedings of the international conference on education*, Honolulu, HI.

Thompson, J., Windschitl, M., & Braaten, M. (2013). Developing a theory of ambitious early-career teacher practice. *American Educational Research Journal, 50*(3), 574–615.

Trent, S. C., Kea, C. D., & Oh, K. (2008). Preparing pre-service educators for cultural diversity: How far have we come? *Exceptional Children, 74*, 328–350.

Van Es, E. A., & Sherin, M. G. (2002). Learning to notice: Scaffolding new teachers' interpretations of classroom interactions. *Journal of Technology and Teacher Education, 10*(4), 571–595.

Villegas, A. M., & Lucas, T. (2002). *Educating culturally responsive teachers: A coherent approach*. Albany: State University of New York Press.

Wilson, S. M., Floden, R. E., & Ferrini-Mundy, J. (2001). Teacher preparation research: Current knowledge, gaps, and recommendations. Document R-1-3. 1–83. Center for the Study of Teaching and Policy.

Windschitl, M., Thompson, J., Braaten, M., & Stroupe, D. (2012). Proposing a core set of instructional practices and tools for teachers of science. *Science Education, 96*(5), 878–903.

Zeichner, K. M. (2003). The adequacies and inadequacies of three current strategies to recruit, prepare, and retain the best teachers for all students. *Teachers College Record, 105*(3), 490–519.

Zwiers, J. (2008). *Building academic language*. London: Newark International Reading Association.

Chapter 7
Online Professional Learning for Science Teachers of Multilingual Learners

Kara Mitchell Viesca, Elizabeth Mahon, Christopher D. Carson, and The eCALLMS Team

In its 2009 position statement *Science for English Language Learners*, the National Science Teachers Association (NSTA) recommended "that teacher preparation and professional development programs for teachers, regardless of area of certification, focus on science content and pedagogy for English language learners" (p. 2). Since that time, widespread adoption of both English language developments standards such as WIDA(https://www.wida.us) and comprehensive, rigorous science standards such as NGSS(http://www.nextgenscience.org) have provided extensive support in describing what bilingual students can and should be doing in science. While most science teachers have access to professional development to support the teaching practices described in either NGSS or WIDA resources, there are few opportunities to support the integration of both language and science standards.

Without specific support for integration of language and science, teachers may perceive rigorous science standards as beyond the capabilities of bilingual students with emerging English proficiency (Cho and McDonnough 2009; Lee et al. 2013; Verplaetse 1998). In crafting the Next Generation Science Standards (NGSS Lead States 2013), the National Academy of Sciences made it clear that the standards apply to all learners, including "students who have traditionally struggled to demonstrate mastery" (v 25, 25). Language and literacy instruction is embedded into the NGSS, and the shift toward greater emphasis on science and engineering practices allows for even greater opportunity for language acquisition. (Lee et al. 2013).

Since 2011, the e-Learning Communities for Academic Language Learning in Mathematics and Science (eCALLMS)[1] project has been working to craft

[1] eCALLMS is a National Professional Development grant from the federal Office of English Language Acquisition, award #T365Z110177.

K.M. Viesca (✉) • E. Mahon • C.D. Carson • The eCALLMS Team
University of Colorado Denver, Denver, CO 80204, USA
e-mail: kara.viesca@unl.edu; elizabeth.mahon@ucdenver.edu; christopher.carson@ucdenver.edu

A.W. Oliveira, M.H. Weinburgh (eds.), *Science Teacher Preparation in Content-Based Second Language Acquisition*, ASTE Series in Science Education,
DOI 10.1007/978-3-319-43516-9_7

professional learning opportunities that support the integration of language instruction in science as is called for by the NGSS standards and the NSTA recommendations. By creating innovative online resources that support professional learning communities of teachers to explore various aspects of language development in relationship to content teaching, eCALLMS (see http://ecallms.ucdsehd.net/) is offering rigorous opportunities for science teachers to meaningfully integrate both language and science content development.

eCALLMS eWorkshop Format and Guiding Principles

Our approach to this professional development reflects our philosophy about the assets of multilingualism and language as a sociocultural practice. We have designed our work grounded in the linguistically responsive teaching framework (Lucas and Villegas 2010, 2011; Lucas et al. 2008) that suggests the orientations, knowledge, and skills content teachers of multilingual learners should have. We also emphasize the value of bilingualism and multilingualism by using the term "bilingual learners" or "multilingual learners" rather than "English language learners" as an effort to help teachers recognize the children they are working with for their assets and linguistic abilities rather than simply their perceived or real deficiencies in English (Brisk 2006; García et al. 2008; Mitchell 2013). Our eWorkshops also promote critical sociocultural instructional practices as operationalized by the Standards of Effective Pedagogy (Teemant and Hausman 2013; Teemant et al. 2014) and are grounded in the WIDA standards for English language development. Increasingly, our work is also grounded in the literature focused on translanguaging and the social turn in second language acquisition (i.e., García 2009; García and Wei 2014; Valdés et al. 2015).

Informed by these guiding principles and frameworks, we have designed an eWorkshop format that assists in-service science teachers to further their professional expertise around supporting language and content development. Our eWorkshops take an asset-based approach to our audience of practicing teachers and offer differentiated activity choices that ensure the learning in the eWorkshops is applicable and relevant. They are designed for collaborative use by professional learning communities of teachers, rather than by single participants in isolation. Additionally, the eWorkshops were created so that they would not need to be moderated at the university level, rather could be used flexibly by instructional and teacher leaders within schools and districts to further local professional learning goals.

We strive to strike a balance between competing goals: (1) facilitating learning about specific aspects of linguistically responsive teaching, as informed by the literature and what we know about language acquisition, and (2) giving our professional audience control over their own learning and how they apply the learning in their specific context. To accomplish this, each of our eWorkshops has an essential question that provides an overarching framework for the learning that occurs in the eWorkshop. Then, guiding questions that lead the inquiry and learning for each unit

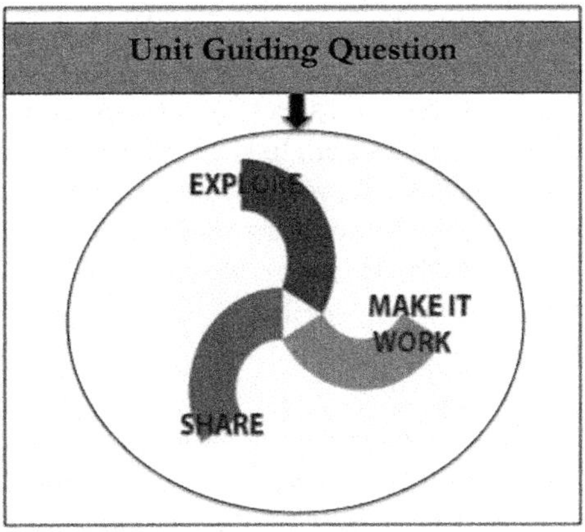

Fig. 7.1 Visual representation of the eCALLMS model for professional learning

of the eWorkshop (there are six in total) offer meaningful opportunities for in-service teachers to grow as professionals. We then offer an *Explore* section where teachers examine new ideas and content through self-selected differentiated learning. Next, there is a *Make it Work* section where we have created multiple options for teachers to apply the ideas and content from the *Explore* section to their practice. Our effort here is to ground theory and research into relevant and applied learning opportunities that are also inquiry oriented. Finally, we have a *Share* section where teachers have the opportunity to collaborate online and *Share* their ideas, successes, failures, questions, etc. after having done the *Explore* and *Make it Work* sections. Each eWorkshop has been developed with this format repeating in each unit over a six-unit learning period, requiring roughly 2 h of a teacher's time per unit. A visual representation of this model of professional learning that our eWorkshops are developed around is provided in Fig. 7.1. We will briefly describe these four components below.

1. Essential/Guiding Questions: An Asset-based Perspective of our Science Teacher Audience
   Just as K-12 instruction should build on the wealth of cultural and linguistic knowledge our bilinguals bring to the classroom, professional learning opportunities for teachers of bilingual students should build on the strengths of teachers' existing practice. For example, by asking the question "How can engaging STEM activities lead to a rich writing and revising practice for bilinguals?," we ask teachers to reflect on the most meaningful learning experiences used in their existing practice. While teachers have the option of exploring example science activities that enable bilinguals to use science in highly contextual ways, we start with the assumption that teachers already do great work, and need time to explore ways to integrate language instruction into the science lesson design.
2. Explore: Differentiated Learning
   Just as bilingual learners can be supported through differentiated learning opportunities and multiple ways to show proficiency, we believe professional learning for teachers should provide freedom for teachers to be in control of their own

learning. Each unit of our eWorkshops provides multiple options for teachers to explore content, research and theory related to the essential question and guiding questions. In respect for teachers' busy lives and demanding schedules, teachers are offered a path to keep their work in this section to around 30 min. Our project strives to offer manageable pieces of information that teachers will find relevant and easy to apply to their local context without feeling overburdened or overwhelmed. However, teachers are also offered extended resources to continue exploring the ideas and research that is most interesting and relevant to them as they choose.

3. Make it Work: Relevant and Applied Learning
   Both science content and language learning are facilitated when the learning is highly contextualized to the students' lived experience. We take the same approach to professional learning, by putting the application of learning at the heart of each eWorkshop. Each unit, during the *Make it Work* phase, teachers put the learning into action in a way that makes sense in their own science teaching environment. While the guiding question for the unit is the same for all learners, we provide a variety of options for how teachers may apply new knowledge or ideas garnered from the *Explore* section. Teachers may adjust an activity to intentionally plan for language acquisition, collaborate with language teachers, integrate a new practice into their teaching, or plan for future lessons.
4. Share: Collaboration
   Language learners benefit from making meaning in a collaborative space. In this same vein, we designed the eWorkshops to create an online learning community for teachers. While it is possible for a teacher to explore the resources alone, the learning experience is made more powerful through collaboration. Each unit, teachers discuss their learning online and share the result of their own attempts to apply content from *Explore* and *Make it Work* into their new learning.

The eCALLMS Model of Professional Learning: Related Literature

In developing the eCALLMS eWorkshops, we drew heavily on the findings from Desimone et al. (2002) longitudinal study suggesting the characteristics of professional development programs that were most likely to impact change in teacher practices (i.e. sustained, collaborative, active learning orientation, etc.). We also drew on the work of Cochran-Smith and Lytle (1999, 2009) and their focus on "Inquiry as Stance." They suggest that professional learning communities of teachers should evolve around "knowledge-*of*-practice" where "the knowledge teachers need to teach well is generated when teachers treat their own classrooms and schools as sites for intentional investigation at the same time that they treat the knowledge and theory produced by others as generative material for interrogation and interpretation" (1999, p. 205). The structure of eCALLMS eWorkshops described above creates meaningful learning opportunities for professional learning communities of teachers to treat their own classrooms and schools as sites of intentional investigation where "knowledge-*of*-practice" can be generated and that knowledge can thoughtfully impact the ongoing pedagogical development of science teachers working at the intersection of language and content development.

Additionally, we drew on the literature related to online professional learning that suggests online approaches can be at least as effective as face-to-face coursework (Carr 2010; Fishman et al. 2013) and that it can have positive effects on teachers' instructional practices and content knowledge (Borko 2004; Cady and Reardon 2009; Cavanaugh and Dawson 2010; O'Dwyer et al. 2010; Russell et al. 2009). Research suggests that quality online professional learning environments should offer ways for participants to get to know one another and build a sense of trust with their online peers (Carr and Chambers 2006; Carter 2004; Smith 2014; Sung 2009). The same researchers suggest that to be successful, participants need to be comfortable with the online discussion tools as well as have a strong sense of the expectations for when, where and how to respond to prompts. Further, online work can suffer from low participation and completion rates (Reeves and Pedulla 2011), but is most successful when materials are offered in a variety of multimedia formats (Carter 2004) and there is a consistency in the format and content of the online professional learning space to support teacher success and motivation (MacKenzie and Staley 2001). We drew on all of these perspectives as we developed and continue to develop eCALLMS eWorkshops.

Finally, the content of our eWorkshops has been deeply impacted by the literature and frameworks described above (i.e. García and Wei 2014; Lucas and Villegas 2011; Teemant et al. 2014) as well as various conceptual frameworks suggesting what content teachers of multilingual learners should know and be able to do (see Viesca et al. 2016).

Implementation of the eCALLMS eWorkshops

Our first set of eWorkshops across three strands (language in science, language in mathematics and bilingual/second language development), were launched for public use in 2013. Since then we have had hundreds of teachers across Colorado, Finland and Germany participate in our eWorkshops and have continued to launch more. Currently we have 10 eWorkshops available for public use with seven more slotted for release in late spring of 2016 and approximately 15 more will be finalized by the end of the grant period (August 2016). It is simple to use our eWorkshops. With 2-week notice, we can launch any interested group into their own course shell for the eWorkshop where they can collaborate with their selected peers in a password protected learning management system environment (we use Canvas). Canvas offers a free platform that works well for our eWorkshops, so there is no cost to users for access to our content. Teacher educators are welcome to use our eWorkshops as well in their work with pre- and in-service teachers. Anyone interested in our work may get in touch with us through our website (http://ecallms.ucdsehd.net/). However, the remainder of this chapter provides content from our eWorkshops that may be used in classes or professional learning approaches with science teachers who work with multilingual students.

Samples of eCALLMS Content and Materials

Grounded in the format and guiding principles described above, we have designed multiple eWorkshops focused on supporting science teachers to expand their expertise around language and content development for multilingual learners in their classroom. Our program promotes a comprehensive perspective of language development at the word, sentence and discourse level within science classrooms. In this section we offer examples of the work we have designed as well as teacher's work participating in it. For each sample we share two actual responses from teachers who engaged in our eWorkshops and did that particular activity. The teachers are all unique teachers across the samples we share.

Sample 1

eWorkshop Title Inquiry Science for Bilinguals

Guiding Question for the Unit How can open-ended pre-assessments inform me of my learners' assets in language and science?

Context Information During the first unit of the eWorkshop, teachers are introduced to the key concept that pre-assessments should be *biased for the best* (Swain 1984). We look for ways to connect with what students do know, rather than looking for what students do not know. In a pre-assessment with bilingual learners we collect observations in three areas: (1) language use (English and home languages); (2) collaboration, critical thinking, process skills, and; (3) science content. The vocabulary prediction activity is one of several ways to learn about how students use language and what their incoming understandings are regarding the science concepts under investigation.

Make It Work Activity

1. Select a set of key words for pre-assessment. Select words that are essential for understanding, can be used across content areas, are particularly tricky for bilingual learners (homophones, idioms, etc.), and/or lend themselves to interesting conversations about language or content. For example, (a) states of matter: solid,

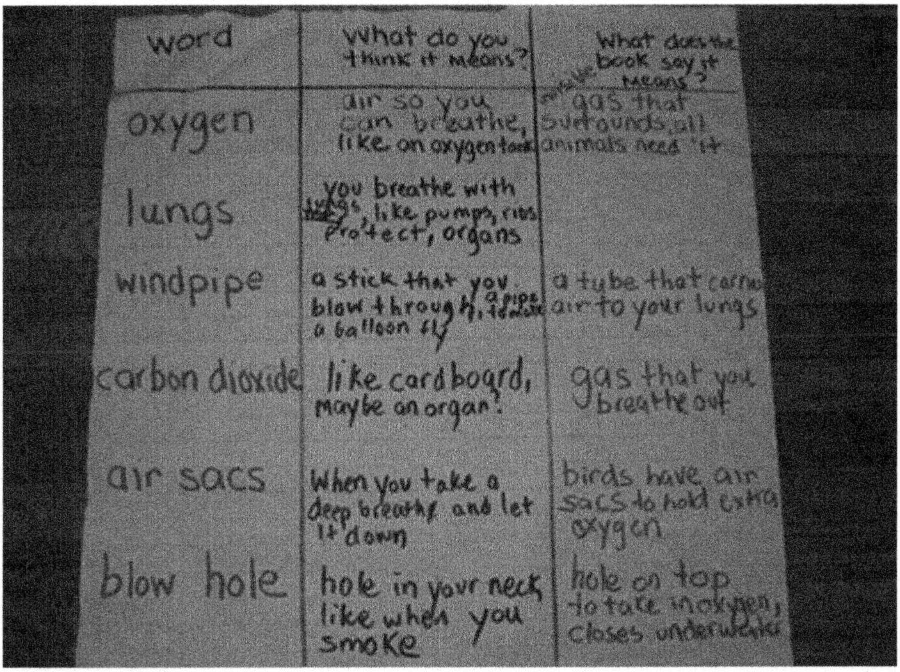

Fig. 7.2 Example vocabulary prediction chart

liquid, gas, vapor, melting, boiling, mass, volume; and (b) Weathering and erosion: weathering, erosion, deposition, glacier, abrasion, sediment, meander.

2. Put students in pairs and give them a vocabulary prediction chart (Fig. 7.2). Do a think-aloud with the first word to show how to make a vocabulary prediction. People usually say the word out loud, connect it to other words they know (these can include words from other languages than English), and look at word parts during this prediction phase. Let students complete the prediction column with their partners. Students may use English or any other language they choose. You, as a teacher, are observing the language use, the critical thinking skills, and the science content knowledge.

Vocabulary Word	Prediction	Changes to my prediction

3. Have each set of students pair-up with another set of students. Each group explains and justifies their prediction. If applicable, students can make changes to their predictions in the third column after the conversation and throughout the unit.

Share Actual Teacher Responses to This Activity

Sample 1 Teacher A I really like this idea of having them predict the vocabulary words first. I had never done this before. I understand the value of not giving students vocabulary words in science until they have first had some exposure to the item or concept and then name it later. Therefore, I have never given the words upfront. However, I thought this was a great way to start to understand some of their preconceptions. I will definitely use it again! …The words that were on the list were Mixture, Property, Solution, Dissolving, Evaporation. What surprised me most is that most of the students thought dissolving meant unable to solve. DIS- Solving. Since they have no background with this word, I thought it was pretty inventive of them to think of that. I was not surprised to see that they predicted property was something they owned (although this made me sad since they have talked about properties in science since Kindergarten) and they all thought solution had to do with solving a problem. This activity that took very little time to plan for, and very little class time, told me a lot about these students' understandings! It is clear that they are not thinking about things in a scientific matter. I am guessing this is coming from the lack of consistent science education K-4. Which means (the classroom teacher) is going to have a larger hill to climb…but at least we are now armed with this information and it is something we can keep in mind while planning future lessons for them.

Sample 1 Teacher B I really liked the conversation that went with the Vocabulary Prediction. The kids got into groups of 2–3 and talked about what they thought each word meant. Kids who were unsure were able to use the support of their group to come up with a prediction. I didn't give feedback to the kids around their predictions, but I did ask them to explain their thinking about their prediction. The kids then joined another group to make groups of 4–6. They shared their thinking again and wrote their prediction in their science journals. I felt the conversation that went along with the predictions was time well spent. This pre-assessment reminded me that I need to make sure I help my students make better connections between what they already know to scientific vocabulary.

Sample 2

eWorkshop Title Inquiry Science for Bilinguals

Guiding Question for the Unit How can we use student observation to launch the inquiry cycle?

Context Information The inquiry eWorkshop is aligned closely with the NGSS Science and Engineering Practice 1: Asking Questions and Defining Problems. In this activity, teachers use students' natural curiosity about the world around them as a launching point for deeper investigations.

Teachers are encouraged to have students observe simple, everyday phenomena, especially where students will be able to manipulate, and experience in a multisensory way. This enables students to use their existing language repertoire to generate observations and questions, which serves three purposes: (1) students can launch their inquiry in any language or register, (2) the teacher is able to get essential assessment information about the language tools available to students, and (3) the teacher is able to assess students' conceptual understanding (especially when the student is encouraged touse pictures to represent observations that are difficult to express).

This *Make It Work* activity was supported by the flexible learning that takes place during the Explore phase: teachers watched a short lecture describing the power of observing in all languages, and chose between several readings detailing possible student-centered hand-on observation experiences or describing teaching methods for improving scientific observations.

At the core of this unit's learning is the idea that expert scientific observation is not dependent on English language skills. All languages are capable of expressing specific, objective details drawing on all senses, thus learning to observe scientifically in any language is a transferrable skill that will lead to stronger bilingual language skills as well as growing scientific understandings.

Make It Work Activity

1. Work with a small group or with your whole class.
2. Pick a simple observation, which could launch an inquiry. For example:
   - A steady trickle of water meandering down a sheet of glass
   - The temperature of ice and water change as it is heated
   - A drop of food coloring mixed in water

   If you are not teaching water, choose an observation that relates to your current topic of investigation.
3. Plan to use a T-chart, I notice/I wonder.

I notice…	I wonder…

4. Model how to observe:
   - Expert – specific details, quantitative if possible, 5 senses, non-judgmental
   - Novice – non-specific, judgmental, inferences
5. Group students to conduct observations. Encourage all languages, dialects, comments, and uses of conventions on the T-chart.
6. Pair learners or group to share their T-charts.

Share Actual Teacher Responses to This Activity

Sample 2 Teacher A We modified [an activity in which students experimented with ways to separate mixtures and solutions] and had them write an "I noticed" and

"I wonder" about that. Most of their answers were fairly similar–"I noticed the water and salt went through the screen," "I noticed I can see a little bit of the salt in the water still, but not all of it." Their "I wonders" often came up with ideas about leaving the water out in the sun or near a heater and wondering if they would be able to see the salt after the water evaporated. I think this activity was another great way for students to get more practice in writing down their thinking. So often we ask them what they think, but don't have them write about it, so when it comes time for them to write a scientific explanation, it is very difficult for them. I think this is a great way for them to practice and for teachers to be able to see some of those pre-conceptions that still exist. (One of the students wondered if the salt would stay melted in the water–they clearly need some more work around the difference of melting and dissolving). I noticed from the student writing that they still need a lot of practice with explaining their thinking. Most of what they wrote was pretty vague (I wonder if we put it in the sun). [eWorkshop Colleague] and I have already discussed their need for precise language so it is a focus of ours. This is just another opportunity to reiterate the importance of it.

Sample 2 Teacher B Something that I noticed was the students' conversations with their groups were strong. They were holding each other accountable for using specific and scientific vocabulary and what they noticed and what they wondered were clear, focused and centered around scientific thinking. However… the specific vocabulary was missing and as [eWorkshop Colleague] said, what they wrote was vague. This will be part of my focus and work [with another teacher] throughout our work this year!

Sample 3

eWorkshop Title Inquiry Science for Bilinguals

Guiding Question for the Unit What strategies facilitate great discourse?

Context Information In this unit of the inquiry eWorkshop, we focus on classroom discourse specific to science. Participation in classroom discourse, either whole-class or small group, is an essential element of enacting the NGSS science and engineering practices of analyzing and interpreting data, constructing explanations, and engaging in arguments from evidence. Discourse is also an essential part of enacting the Common Core State Standards for English Language Arts (CCSS ELA), one of the top concerns for elementary generalist teachers and an area of increasing emphasis for secondary science content teachers.

For our bilingual students, this work takes on added importance. Oral language practice with discourse patterns of science facilitates both the sense-making needed for a deeper understanding of the content, but the language practice that takes place during argumentation from evidence is thought to be a key ingredient in the development of the deeper literacy skills needed in all subjects (Lee et al. 2013).

During the *Explore* phase of this unit's learning, teachers watch a short lecture about drawing conclusions through classroom dialogue, and then chose between several options to further their learning. This work relies heavily on the work of Michaels and O'Connor (2012) in promoting active, engaged discussion. The following *Make it Work* activity, one of three options for the unit, involves practicing the talk moves described in the *Explore* resources.

Make It Work Activity Talk Moves can be helpful in all content areas, including math, literacy, social studies, and art. The purpose of this *Make it Work* activity is to plan for the intentional use of two or more talk moves.

1. Plan the setting for the productive talk
   First, you may want to review the goals of productive talk, as described by Michaels and O'Connor (2012):

   - Individual students share, expand and clarify their own thinking
   - Students listen carefully to one another
   - Students deepen their reasoning
   - Students think with others

   Next consider, how will you communicate these goals to your students and remind them (in student-friendly language) of classroom norms for discussion? Then, decide on a setting for the talk: whole class or small group discussion with the teacher.
2. Review the nine talk moves described by Michaels and O'Connor (2012):

   - Time to Think
     - Use pair/share discussion time or private reasoning time. Ensure that each question or prompt is followed by a small amount of wait time.
   - Say More…
     - "Tell me more about that…"
   - So, are you saying…
     - Paraphrase the student's answer as a question: "So, are you saying…?," allowing the student to respond.
   - Who can rephrase?
     - "Who can use their own words to repeat what _______ said?"
   - Ask for evidence or reasoning
     - Ask questions such as "Why do you think that?" or "Can you offer some evidence to support this claim?"
     - Encourage students to ask others for evidence or reasoning.

- Challenge or counter example
  - "Does it always work this way?"
- Agree Disagree/why?
  - "Do you agree? Why?" "Are you saying the same thing as…?" "Does anyone want to respond to this idea?"
- Add-on
  - "Who can add onto this idea?"
- Explaining what someone else means
  - "Who can explain what ______ means when she says that…"

Because it may be hard to keep all of these talk moves in mind, select one or two that you would like to work on for the discussion. Consider putting these moves on note cards to help you remember the specific language you would like to use during student discussion.

3. Plan to how modify talk moves, as needed, for your emerging bilingual learners:

- Reduce linguistic complexity (not cognitive complexity)
  - "Your example provides some support of your model, but are their other cases demonstrate a need for the model to be refined?" → "Does it always work this way? Think of some other examples."
- Speak slowly and clearly. Give extended wait time after each question prompt.
- Refer to concrete items (realia) or use visual aids.
- Allow student to draw, use visuals to explain thinking.
- Give students time to process: "I'll come back in a few minutes…"
- Allow students to engage in talk moves in the language of their choice

4. Plan to gather evidence of student engagement.
How will you reflect on the effectiveness of your talk moves in building your students' capacity to engage in productive talk? Consider asking a coach to observe you, or recording the discussion to aid in your own reflection and growth.

Share Actual Teacher Responses to This Activity

Sample 3 Teacher A Last week and yesterday I worked on my talk moves with my class. While I'm trying to make talk moves an automatic habit, I was consciously working on it during these two sessions. Last week, I was using talk moves to help my students develop a well planned procedure to answer the question, "Do all solids have the same solubility." Yesterday, I used talk moves to help my students discuss their conclusions about that question. Without using talk moves, I would not have discovered many misunderstandings and misconceptions my students had. To begin with, I was able to clear up a few misunderstandings and misconceptions around what solubility and saturation is [sic] before the students developed a procedure. The procedure and investigation would have just been playtime for the students without a clear understanding around what they were doing and why they were doing it. As a result of the talk moves yesterday, I made adjustments to future lessons to hopefully prevent a few misunderstandings in the future.

Sample 3 Teacher B I had never heard the phrase Talk Moves before, however they are something that I use frequently with my students. This week I focused specifically on using the silent signal, rephrasing student observations (sometimes asking students to rephrase another student), and asking students to cite specific evidence for their observations. Since we are a dual language classroom, I sometimes allow my students to respond in either language, so some of my students were answering or rephrasing in both languages. I have found that rephrasing is often very helpful to bilingual students, particularly when they can't quite think of the word they are looking for in the target language. This week we discussed and observed a distillation lab that we had created in October. Our experiment didn't work out as planned, but the student hypothesis and observations were a great opportunity to use some Talk Moves to help them to broaden their thinking… I will definitely use Talk Moves more intentionally in the future, particularly when expecting students to clarify on their own thinking or process. This will be applicable in all subjects, not just science!

Sample 4

eWorkshop Title The 5E Science Model for Multilingual Students

Guiding Question for the Unit How can I provide comprehensible input for multilingual students in the Explain phase of a 5E Model lesson?

Context Information The 5-E Science Model for Multilingual Students proceeds unit by unit through the 5-Es: Engage, Explore, Explain, Elaborate and Evaluate (Ansberry and Morgan 2007). During this unit, teachers consider how to offer support to multilingual students during the Explain phase of instruction. During this phase, students use their own words as well as content and general academic vocabulary to explain their understanding of the science concepts that they have experienced this far. Interactive Word Walls use visuals, realia, a graphically organized structure, and a student interactive component to provide access to the needed words and phrases for bilingual students. Interactive word walls are connected to the main theme of Unit 3, which is comprehensible input (Krashen 1981). Comprehensible input is the idea that students should be able to understand the concepts and language that are being presented in a lesson. For example, the use of dictionary definitions of scientific vocabulary may offer limited comprehensible input, whereas an interactive word wall is the epitome of comprehensible input.

During the *Explore* section of the eWorkshop, teachers read two short articles (Jackson and Narvaez 2013; Jackson et al. 2011), which explain how to create interactive word walls.

Make It Work Activity Teachers create a word wall on a current science unit following the five steps explained in the articles:

1. Plan the word wall.
    - Determine vocabulary needs.
    - Create a concept map.
2. Create a student work-sheet.
3. Place the word wall (Fig. 7.3).
4. Build the wall in class.
5. Complete student record sheet and word wall together.

Share Actual Teacher Responses to This Activity

Sample 4 Teacher A I have a sub-par word wall going for science. I thought I was doing a great job of at least keeping up with the words. I was getting the words posted under my alphabet, have the kids use a Frayer type organizer to record the words in their notebook glossaries, and I draw "icons" or black line pictures. After reading the article, I understand more that, "the most effective word walls include photographs or the actual item (realia)." The other part that really struck me was the interaction piece. I continually struggle with getting kids to use any resource in the room, vocab walls included. I know (but fail to apply) the idea that "student participation in creating and maintaining word walls is crucial." It can be easier and faster to just do it myself. I DO NOT let my students push the responsibility of other aspects of learning off on to me, so I'm not sure why I've taken over the classroom walls/resources. The other part of the article that struck me was how the word walls are organized. When I scanned the article, I immediately said, "that's not a word wall, that's a concept map." As I actually read the article, I began to see how the concept map is really a higher level word wall and has so many more uses. Perhaps with more purpose, kids will interact with the wall more often! This phrase was one that made [me] a believer in this style of word wall, "because they build schema for individual terms through the use of images and manipulatives while showcasing connections between terms in a unit or lesson."

Sample 4 Teacher B Your post gave me two ideas! 1. What if we let some advanced students [lists students names] design a wall/part of a wall of the classroom. This is just a start, but it could give us some insight to what the kids could envision. We could show them some concept maps of our science topic and let them design one of their own. Then once they do the basics, we could have other kids add in the class as we go. 2. If I'm having trouble with space in the classroom, I could have each kid have a concept map for each unit that they keep and add to in their science notebook. I would have to start it with them and remember to frequently return to it. For some

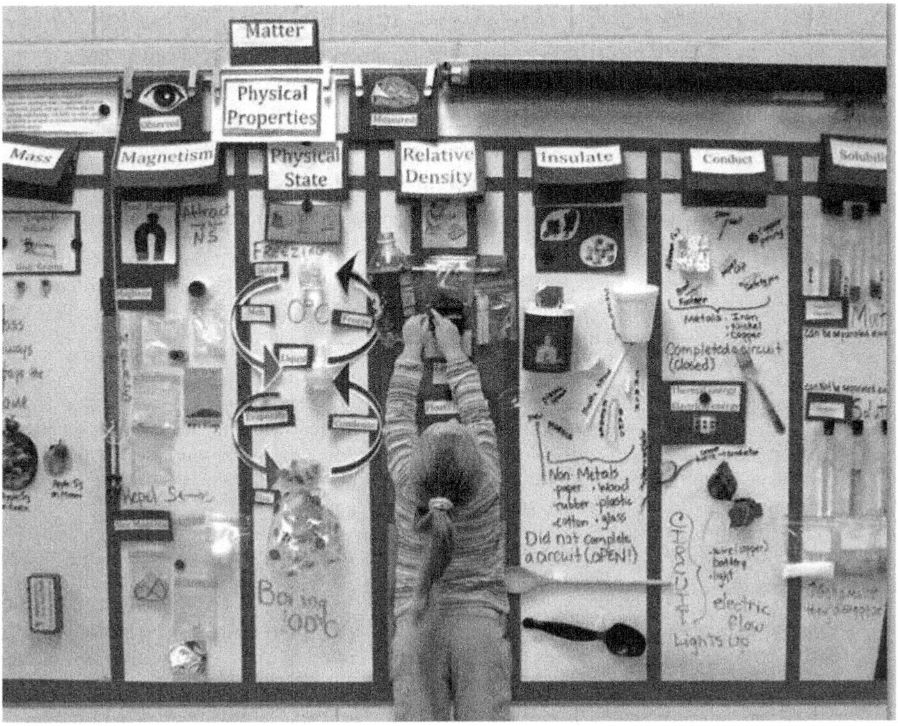

Fig. 7.3 Example of an interactive word wall from Jackson and Narvaez (2013)

kids it will need to be more supported and scaffolded, but for some kids, they could really go in their own direction. We could have little vocab cards and pictures for them to cut out and glue on.

Outcomes

While research on the impacts and outcomes of eWorkshop participation is ongoing, through the annual evaluation of the eCALLMS project, we have valuable evidence of the impact of this approach to professional learning for science teachers (and other content teachers) working with multilingual students. Overall, the eWorkshop participants have been positive about their experiences with the eWorkshops. For instance one eWorkshop tester stated that working in the eWorkshop, "Reminded me of the need to build a gradual release model of linguistic structures into every subject." Another tester stated:

> I am always looking for ways to improve. Most often when we change our thinking it is because we have been presented with new information. I am certainly thinking differently about my science instruction, but I am in a state of disequilibrium.

We think it is positive that our eWorkshop was able to help this teacher think differently about instructing multilingual learners in science classrooms, but we also hope that this teacher will continue on with more of our content to work through that state of disequilibrium.

Initial interviews with school leaders indicate that the eWorkshop model may have a powerful effect when used by professional learning communities in schools. One school leader that led aneWorkshop with four teachers in his/her school stated:

> When I went on walk-throughs, when I observed them…I noticed that they were grouping and they were providing support. So one of them had visuals and the other had different sentence strings for students and the other one was doing Total Physical Response, TPR, with students, so I was glad to see those things in their classroom.

The same group leader stated about teacher participation in an eWorkshop:

> At the beginning there was no differentiation. The supports were not evident. And now when I'm walking in the classrooms, I'm able to see supports, so groupings, visuals… vocabulary development, songs. Various supports that teachers are providing, being aware of students' language development stages.

An administrator at a school where many teachers participated in eWorkshops stated about the teachers:

> They absolutely loved the [eWorkshops]…It was something they said was one of the most valuable professional developments that they had ever done…We had a first-year teacher all the way up to someone who had I think it was 27, 28 years of experience. All of the people in the groups felt the same way…they are looking forward to the next time [they can take aneWorkshop].

Another leader overseeing the use of eWorkshops in his/her school stated:

> In particular, people were interested in the fact of the 'Make it Work' section. I think that took some of the theory we see in the 'Explore' section and it makes it concrete. And I think that was the main thing that attracted both the teachers and the school leadership because that was where there seemed to be a concrete connection to actual classroom practice that came directly out of the articles, the theory…or the PowerPoint that we saw in the 'Explore' section.

This same person also mentioned the value of the brief time commitment and accessibility of the content stating:

> I think also the brevity of it. In other words, that it's not a semester-long graduate course. It's a totally different thing. It's a much more manageable piece that is broken down into weeks so that it looks and feels like less of a commitment, I guess, than signing up for a whole class. I think also the brevity of the 'Explore' section, how there's something that's like easy to…maybe not easy to digest, but at least not so daunting in terms of the content. It's pretty accessible in terms of the content, at least in the length of time it takes to read or watch it.

These perspectives are representative of data we've collected and analyzed over the past 4 years of the project from testers and users including teacher created digital texts in the eWorkshops, surveys, focus-groups and interviews. Based in these data, we feel strongly that the content created through the eCALLMS project for science teachers (and other content teachers) is valuable for teacher professional learning regarding working with multilingual learners. Our research is expanding and growing regarding eWorkshop content and over time we will have data from quasi-experimental studies and other studies looking at teacher motivation and engagement to further define the impacts and outcomes of this work. We will have over 30 eWorkshops publically available for teachers across the globe to use as of August 2016. Our hope is that ongoing work with eCALLMS content can have lasting, positive effects for many teachers, schools and districts.

Conclusion

In this chapter, we have introduced the eCALLMS eWorkshop approach to professional learning for science teachers of multilingual learners. Grounded in research, and designed to impact practice, the eCALLMS eWorkshops have experienced valuable success with science teachers and are worth learning from, using and/or emulating. We have also provided four samples of *Make It Work* activities from various science eWorkshops. We hope these provide valuable tools for various approaches to professional learning for science teachers of multilingual learners. In total, we hope our work offers you either an invitation to join us and use our eWorkshops or at least to benefit from what we have learned and shared with you here.

In summary, we feel that science classrooms have an excellent opportunity to promote strong language development activities, particularly when language is treated like a verb and mapped meaningfully onto the hands-on and engaging activities that can so easily take place in strong science content instruction. We strive to help teachers to create "languaging" experiences for students in sciencethrough an active inquiry approach in their own practice. We also strive to support teacher professional learning by creating flexible learning opportunities where teachers make choices and engage in work that is most relevant to them and their students. Overall, we hope that the ideas and resources provided here will help to continue to improve the quality of instructions for science teachers and their multilingual learners through the ideas and resources we have provided.

References

Ansberry, K., & Morgan, E. (2007). *More picture-perfect science lessons: Using children's books to guide inquiry: K-4*. Arlington: National Science Teachers Association.

Borko, H. (2004). Professional development and teacher learning: Mapping the terrain. *Educational Researcher, 33*(8), 3–15.

Brisk, M. E. (2006). *Bilingual education: From compensatory to quality schooling*. Mahwah: Lawrence Erlbaum Associates.

Cady, J., & Reardon, K. (2009). Delivering online professional development in mathematics to rural educators. *Journal of Technology and Teacher Education, 17*(3), 281–298.

Carr, V. B. (2010). The viability of online education for professional development. *AASA Journal of Scholarship & Practice, 7*(3), 6–16.
Carr, N., & Chambers, D. P. (2006). Cultural and organizational issues facing online learning communities of teachers. *Education Information Technology, 11*, 269–282. doi:10.1007/s10639-006-9024-2.
Carter, K. (2004). Online training: What's really working? *Technology & Learning, 24*(10), 32–36.
Cavanaugh, C., & Dawson, K. (2010). Design of online professional development in science content and pedagogy: A pilot study in Florida. *Journal of Science Education Technology, 19*, 438–446. doi:10.1007/s10956-010-9210-2.
Cho, S., & McDonnough, J. T. (2009). Meeting the needs of high school science teachers in English language learner instruction. *Journal of Science Teacher Education, 20*(4), 385–402.
Cochran-Smith, M., & Lytle, S. L. (1999). Relationships of knowledge and practice: Teacher learning in communities. *Review of Research in Education, 24*, 249–305.
Cochran-Smith, M., & Lytle, S. L. (2009). *Inquiry as stance: Practitioner research in the next generation*. New York: Teachers College Press.
Desimone, L. M., Porter, A. C., Garet, M. S., Yoon, K. S., & Birman, B. F. (2002). Effects of professional development on teachers' instruction: Results from a three-year longitudinal study. *Educational Evaluation and Policy Analysis, 24*(2), 81–112.
Fishman, B., Konstantopoulos, S., Kubitskey, B., Vath, R., Johnson, H., & Edelson, D. (2013). Comparing the impact of online and face-to-face professional development in the context of curriculum implementation. *Journal of Teacher Education, 64*(5), 426–438.
García, O. (2009). *Bilingual education in the 21st century: A global perspective*. Malden: Wiley-Blackwell.
García, O., & Wei, L. (2014). *Translanguaging: Language, bilingualism and education*. New York: Palgrave Macmillan.
García, O., Kleifgen, J. A., & Flachi, L. (2008). *From English language learners to emergent bilinguals*. New York: The Campaign for Educational Equity.
Jackson, J., & Narvaez, R. (2013). Interactive word walls: Create a tool to increase science vocabulary in five easy steps. *Science and Children, 51*, 42–49.
Jackson, J., Tripp, S., & Cox, K. (2011). Interactive word walls: Transforming content vocabulary instruction. *Science Scope, 35*, 45–49.
Krashen, S. D. (1981). *Second language acquisition and second language learning*. Oxford: Pergamon Press Inc.
Lead States, N. G. S. S. (2013). *Next generation science standards: For states, by states*. Washington, DC: The National Academies Press.
Lee, O., Quinn, H., & Valdés, G. (2013). Science and language for English language learners in relation to next generation science standards and with implications for common core state standards for English language arts and mathematics. *Educational Researcher, 41*(4), 223–233.
Lucas, T., & Villegas, A. M. (2010). The missing piece in teacher education: The preparation of linguistically responsive teachers. *National Society for the Study of Education, 109*(2), 297–318.
Lucas, T., & Villegas, A. M. (2011). A framework for preparing linguistically responsive teachers. In T. Lucas (Ed.), *Teacher preparation for linguistically diverse classrooms: A resource for teacher educators* (pp. 55–72). New York: Routledge.
Lucas, T., Villegas, A. M., & Freedson-Gonzalez, M. (2008). Linguistically responsive teacher education: Preparing classroom teachers to teach English language learners. *Journal of Teacher Education, 29*(4), 361–373. doi:10.1177/0022487108322110.
MacKenzie, N., & Staley, A. (2001). Online professional development for academic staff: Putting the curriculum first. *Innovations in Education and Teaching International, 38*(1), 42–53. doi:10.1080/147032901300002846.
Michaels, S., & O'Connor, C. (2012). *Talk science primer*. Cambridge, MA: TERC.

Mitchell, K. (2013). Race, difference, meritocracy, and English: Majoritarian stories in the education of secondary multilingual learners. *Race Ethnicity and Education, 16*(3), 339–364. doi:10.1080/13613324.2011.645569.

National Science Teachers Association. (2009). *NSTA position statement: Science for English language learners.* Retrieved from http://www.nsta.org/docs/PositionStatement_ScienceForEnglishLanguageLearners.pdf

O'Dwyer, L. M., Masters, J., Dash, S., De Kramer, R. M., Humez, A., & Russell, M. (2010). E Learning for educators: Effects of on-line professional development on teachers and their students. *inTASC REPORT: e-Learning for educators*, http://www.intasc.org

Reeves, T. D., & Pedulla, J. J. (2011). Predictors of teacher satisfaction with online professional development: Evidence from the USA's e-Learning for educators initiative. *Professional Development in Education, 37*(4), 591–611. doi:10.1080/19415257.2011.553824.

Russell, M., Kleiman, G., Carey, R., & Douglas, J. (2009). Comparing self-paced and cohort based online courses for teachers. *Journal of Research on Technology in Education, 41*(4), 443–446.

Smith, S. U. (2014). Frameworks shaping an online professional development program for K-12 teachers of ELLs: Toward supporting the sharing of ideas for empowering classroom teachers online. *TESOL Journal, 5*(3), 444–464.

Sung, L. C. (2009). Online professional development for K-12 inservice teachers from views of Online Learning Forum (OLF) designers. In *Information technology: New generations, 2009. Sixth international conference on information technology*, Las Vegas, USA (pp. 1120–1127). IEEE, http://ieeexplore.ieee.org/xpl/login.jsp?tp=&arnumber=5070775&url=http%3A%2F%2Fieeexplore.ieee.org%2Fxpls%2Fabs_all.jsp%3Farnumber%3D5070775.

Swain, M. (1984). Teaching and testing communicatively. *TESL Talk, 15*(1/2), 7–18.

Teemant, A., & Hausman, C. S. (2013). The relationship of teacher use of critical sociocultural practices with student achievement. *Critical Education, 4*(4). Retrieved from http://ojs.library.ubc.ca/index.php/criticaled/article/view/182434

Teemant, A., Leland, C., & Berghoff, B. (2014). Development and validation of a measure of critical stance for instructional coaching. *Teaching and Teacher Education, 39*, 136–147. doi:10.1016/j.tate.2013.11.008.

Valdés, G., Poza, L., & Brooks, M. D. (2015). Language acquisition in bilingual education. In W. Wright, S. Boun, & O. García (Eds.), *The handbook of bilingual and multilingual education*. Malden: Wiley.

Verplaetse, L. S. (1998). How content teachers interact with english language learners. *TESOL Journal, 7*(5), 24–28.

Viesca, K. M., Hamilton, B., Davidson, A., & The eCALLMS Team (2016). Supporting linguistically responsive teaching: e-Learning communities for academic language learning in mathematics and science (eCALLMS). In C. P. Proctor, A. Boardman, & E. H. Hiebert (Eds.), *Teaching emergent bilingual students: Flexible approaches in an era of new standards.* New York: Guilford Press.

Chapter 8
Preparing Science Teachers for English Learners: A Targeted and Integrated Approach to Pre-service Teacher Education

Lara K. Smetana and Amy J. Heineke

Recent educational reform efforts, including the Next Generation Science Standards (NGSS) (NGSS Lead States 2013), promote a vision of learning and doing science that requires high quality teachers for *all* students. Today's educators must be able to bring science content to life in meaningful ways for students. This involves bridging school science experiences with students' individual cultures, experiences, and everyday lives (Birmingham et al. 2015; Birmingham and Calabrese Barton 2014; Rascoe 2013). Moreover, to meet the needs of all students in all classrooms, science teachers must be prepared to provide rigorous classroom instruction with students from increasingly diverse cultural and linguistic backgrounds (Gándara and Hopkins 2010). Nevertheless, unprepared and underprepared teachers continue to enter classrooms across the United States (U.S.), lacking the knowledge and skills to meet the large and growing number of students who are in the process of acquiring English in content-area classrooms (Cohen and Clewell 2007; Giambo and Szecsi 2005).

Specifically considering the new era of science education guided by the NGSS, where communication and cognition take center stage in curriculum and instruction, teachers need knowledge and skills to support students' academic language development across scientific disciplines (Lee et al. 2013; Walqui and Hertiage 2012). As language and cognition develop together over time, an explicit focus on language development is integral to provide excellent and equitable science instruction and support students' academic achievement (NSTA 2009; VanLier and Walqui 2012). Despite the centrality of language, the NGSS were not written with culturally and linguistically diverse students in mind, particularly students who are labeled

L.K. Smetana (✉) • A.J. Heineke
Loyola University Chicago, Chicago, IL 60611, USA
e-mail: lsmetana@luc.edu; aheineke@luc.edu

A.W. Oliveira, M.H. Weinburgh (eds.), *Science Teacher Preparation in Content-Based Second Language Acquisition*, ASTE Series in Science Education,
DOI 10.1007/978-3-319-43516-9_8

as English learners (ELs) based on standardized tests of English-medium listening, speaking, reading, and writing (Walqui and Hertiage 2012). In this way, teacher education programs hold the responsibility to prepare science teachers for today's changing classrooms across the U.S.

In this chapter, we describe Loyola University Chicago's *Teaching, Learning and Leading with Schools and Communities* (TLLSC), a pre-service, field-based teacher preparation program that has enthusiastically accepted this responsibility to prepare educators to work with culturally and linguistically diverse students.[1] We work closely with the Chicago Public Schools, where over 70,000 students (more than 16 %) are classified as ELs with over 140 languages represented. From the first day of their preparation program, teacher candidates confront challenges and opportunities of culturally and linguistically diverse classrooms, including ways in which ELs have historically been marginalized within public schools in terms of inequitable access to curriculum, instructional materials and facilities, and appropriate assessments (Gándara et al. 2003). With a university mission focused on social justice, TLLSC takes an asset-based approach to having ELs in the classroom, recognizing language as a resource that contributes to learning and to the classroom community (Ruiz 1984). Goals around providing rigorous instruction that supports all students' language development are integrated across candidates' entire preparation program to ensure that they are confident in pedagogical skills, informed about policies effecting ELs, and able to apply current research on second language acquisition as they begin their careers. We also ensure that all candidates are eligible for the state's ESL endorsement upon completion of their specific programs of study.

To support the work of fellow teacher educators, this chapter provides programmatic design considerations and examples from current course experiences and assignments aimed to prepare all science teachers to promote discipline-specific language and literacy simultaneous to conceptual understanding for ELs in pre-Kindergarten-through-grade-12 (PK-12) classrooms. We begin with a brief overview of the TLLSC program and then introduce the program's *field-based apprenticeship model* (Rogoff 1994, 1995), intended to foster candidates' professional development and growth across 4 years. To illustrate the targeted and integrated nature of preparing science teachers for ELs, we utilize descriptive vignettes to detail learning experiences and assessments, as well as provide evidence of outcomes from candidates enrolled in the program. Explicit reference is made to ways in which the program aligns the principles and declarations in NSTA's (2009) position statement on science for ELs, as well as most current research-based recommendations. Finally, we discuss implications based on our practical experiences and research findings and offer recommendations for science teacher educators and researchers.

[1] See http://www.luc.edu/education/for more information about the TLLSC program.

Preparing Effective Teachers and Leaders: Program Overview

Launched in 2013, Loyola's TLLSC program is an innovative, comprehensive teacher preparation program at the forefront of redefining how to prepare science teaching professionals for diverse, urban educational settings. Breaking from traditional models of teacher education in which instruction occurred primarily in the university classroom until the student teaching experience (See Fig. 8.1), TLLSC integrates academic knowledge with authentic teaching and learning experiences within a variety of Chicago school and community contexts (Ball and Forzani 2009; Rogoff 1994, 1995; Ryan et al. 2014; Zeichner 2009). Moving teacher preparation out of the confines of the university, teaching and learning take place primarily in partner schools, community spaces, and cultural institutions where university, school, and community partners share the responsibility of preparing PK-12 teachers to serve students from diverse backgrounds. This chapter will focus specifically on the TLLSC undergraduate programs in which approximately 80 % of instructional time is spent away from the university campus and in culturally and linguistically diverse educational settings.

While TLLSC provides one overarching framework for all candidates regardless of area of specialization, professional learning and experiences become increasingly tailored to elementary or secondary science education as candidates progress through three phases of the 4-year program.

- *Exploration phase – Semesters 1, 2, and 3*: Candidates develop knowledge and skills applicable across teaching contexts and specializations, such as language development, culturally relevant teaching, inclusive classroom environments, and instructional planning. Simultaneously, candidates engage in professional learning across varied school and non-school educational sites (e.g., museums) and with a wide array of grade levels and content areas to explore the rich opportunities across the broad field of education.
- *Concentration phase – Semesters 4, 5, and 6*: After exploring education across contexts, candidates declare their program of study before moving into the concentration phase. Here, candidates build pertinent knowledge and skills within a focal area of expertise, such as elementary or secondary science. Working alongside faculty and other education professionals, candidates are purposefully apprenticed into science teaching and learning.
- *Specialization phase – Semesters 7 and 8*: To culminate their programs of study and deeply prepare for professional teaching, candidates engage in 1-year internships teaching science in culturally and linguistically diverse classroom contexts. The 1-year internship provides candidates with in-depth, rigorous professional experience in science education.

In each of the three phases of teacher development, candidates actively participate in integrated, field-based courses and experiences, referred to as *modules*, organized sequentially across academic semesters, referred to as *sequences*. Ranging in temporal length from 3 to 8 weeks, modules are designed and implemented based

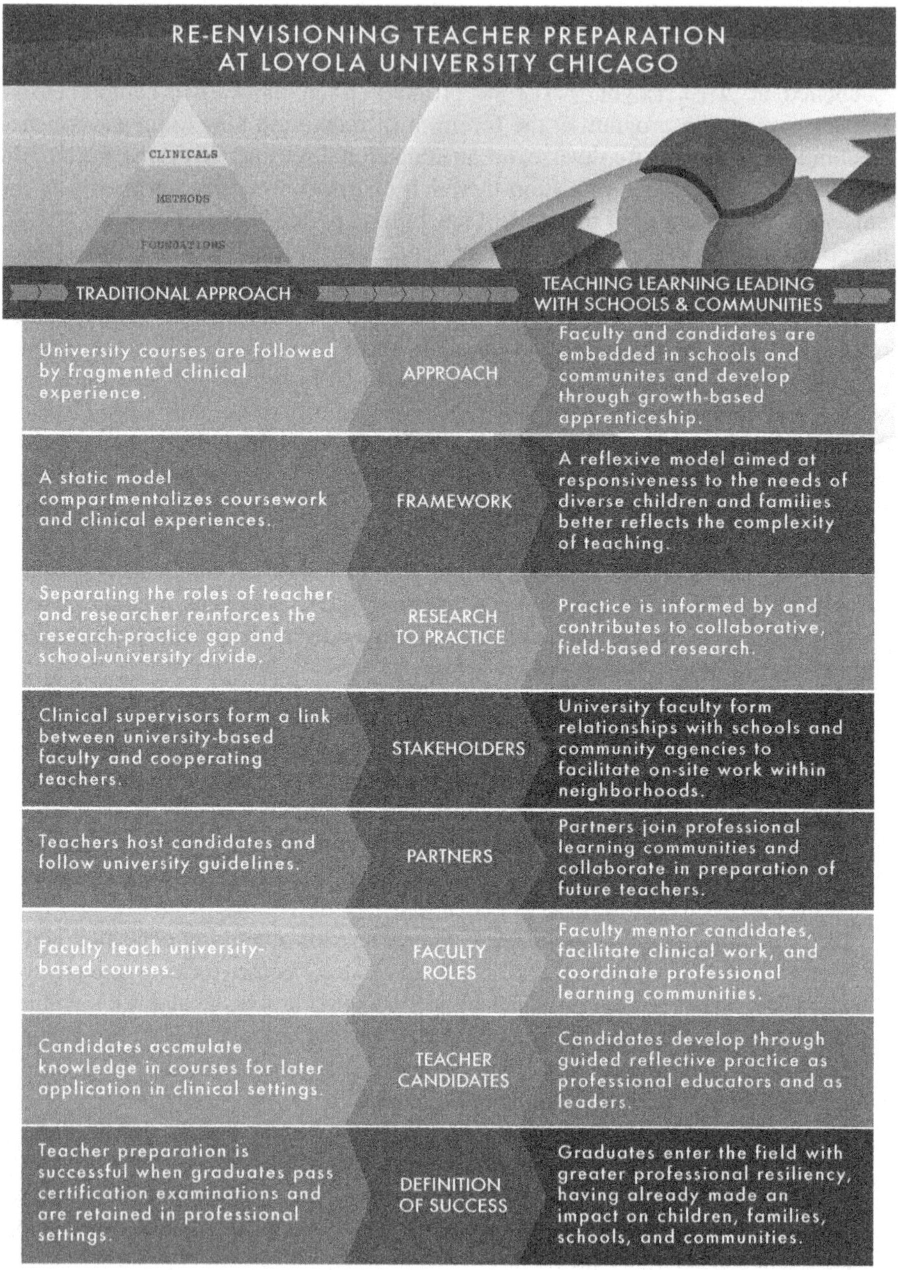

TRADITIONAL APPROACH		TEACHING LEARNING LEADING WITH SCHOOLS & COMMUNITIES
University courses are followed by fragmented clinical experience.	APPROACH	Faculty and candidates are embedded in schools and communites and develop through growth-based apprenticeship.
A static model compartmentalizes coursework and clinical experiences.	FRAMEWORK	A reflexive model aimed at responsiveness to the needs of diverse children and families better reflects the complexity of teaching.
Separating the roles of teacher and researcher reinforces the research-practice gap and school-university divide.	RESEARCH TO PRACTICE	Practice is informed by and contributes to collaborative, field-based research.
Clinical supervisors form a link between university-based faculty and cooperating teachers.	STAKEHOLDERS	University faculty form relationships with schools and community agencies to facilitate on-site work within neighborhoods.
Teachers host candidates and follow university guidelines.	PARTNERS	Partners join professional learning communities and collaborate in preparation of future teachers.
Faculty teach university-based courses.	FACULTY ROLES	Faculty mentor candidates, facilitate clinical work, and coordinate professional learning communities.
Candidates accmulate knowledge in courses for later application in clinical settings.	TEACHER CANDIDATES	Candidates develop through guided reflective practice as professional educators and as leaders.
Teacher preparation is successful when graduates pass certification examinations and are retained in professional settings.	DEFINITION OF SUCCESS	Graduates enter the field with greater professional resiliency, having already made an impact on children, families, schools, and communities.

Fig. 8.1 Program comparison

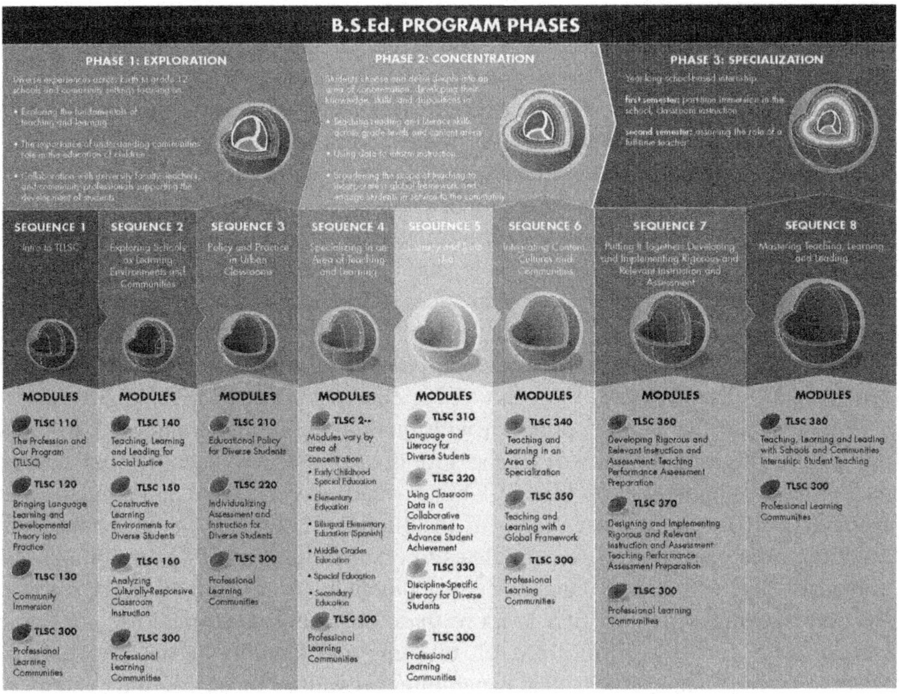

Fig. 8.2 Program phases and sequence of courses

on the needed learning time for content and field experiences rather than the university's semester schedule, as had traditionally been the case. Following the principles of Understanding by Design (Wiggins and McTighe 2005), the program spirals enduring understandings, knowledge, skills, and dispositions across modules and sequences. In this way, candidates progressively develop mastery and expertise as science educators working in diverse and complex school and community settings. Finally, each sequence ends with professional learning communities that bring together science education faculty and candidates to collectively make sense of semester experiences and learning. See Fig. 8.2.

Preparing Science Teachers of English Learners: An Apprenticeship Model

Collaborating with faculty across program areas, we designed the TLLSC program drawing from our shared theoretical and conceptual frameworks and extant research on teaching and teacher education. Situated within the sociocultural paradigm, we conceptualized teacher learning as occurring through candidates' participation in the social and cultural activities of daily educational practice (Lave and Wenger 1991;

Vygotsky 1978). Thus, we designed a *field-based apprenticeship model*, where university- and school-based teacher educators collaborate with candidates while engaging in authentic educational practices (Heineke et al. 2013; Nasir and Heineke 2014; Rogoff 1994, 1995; Ryan et al. 2014). In this way, we aimed to produce expert and effective teachers capable of positively influencing diverse students' learning, development, and achievement. To this end, we conceptualized teacher expertise as developing through apprenticeship across multiple planes of practice: (a) *institutional*, candidates' preparation situated in the institutional contexts of education, including communities, schools, and classrooms; (b) *interpersonal*, candidates' guided participation in shared practices of teaching and learning, mediated by expert teachers and teacher educators, and (c) *individual*, candidates' application of learning and appropriation into classroom practice with diverse students (Ball and Forzani 2009; Rogoff 1995; Vygotsky 1978; Zeichner 2009). See Fig. 8.3.

We designed the TLLSC program with the apprenticeship framework to yield our goals of high-quality novice teachers for diverse students, specifically the large and growing population of ELs in Chicago and across the U.S. (Gándara and Hopkins 2010). Across these planes of institutional, interpersonal, and individual practice, we aimed to prepare candidates to effectively support culturally and linguistically diverse students' language development in science and other classrooms. Recognizing that teaching ELs is more than "just good teaching" (de Jong and Harper 2005, p. 1), we prioritized EL teaching and learning across programs of study to build general professional capabilities, as well as those specific to elementary and secondary science classrooms (de Jong et al. 2013; Manzo et al. 2011). In this way, TLLSC aligns with the principles that NSTA (2009) has put forth to ensure all students have the opportunities to excel in science. Specifically, these guidelines call for curriculum and instruction that is standards-based and anchored to investigations that promote inquiry, builds upon students' prior knowledge and experiences, allows for authentic engagement in the practices of science, incorporates literacy skills and the development of academic language literacy, and recognizes and respects all learners' linguistic and cultural experiences.

Tapping into recent research and literature, we defined teacher expertise for ELs, including the enduring understandings, knowledge, skills, and mindsets needed by all teachers to positively influence students' social, emotional, cultural, linguistic, and academic development and learning (Herrera 2010). Running beneath all professional practice, *enduring understandings* of language development and learning allows candidates to take a linguistic lens in all educational decision-making within their classrooms and schools (Lucas et al. 2008). Candidates then need deep *knowledge* of their content area and how language develops and operates within that content area, as well as the unique backgrounds, abilities, and needs of ELs (e.g., Herrera 2010; Moll and González 1997; Valdés et al. 2005; VanLier and Walqui 2012; Walqui and Hertiage 2012). Applying knowledge in classroom practice, candidates require expert *skills* to support students' language development, including the design and implementation of instruction, learning environment, and assessments with an expert linguistic lens (e.g., Heritage et al. 2015; Herrera 2010; Lee et al. 2013). Finally, EL student learning, development, and achievement requires

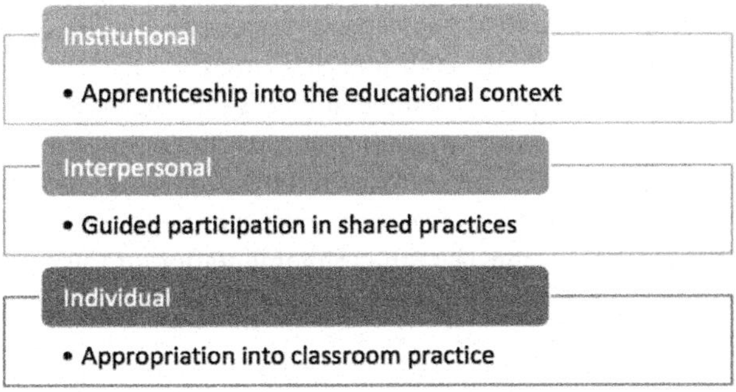

Fig. 8.3 Apprenticeship framework for teacher learning (Rogoff 1995)

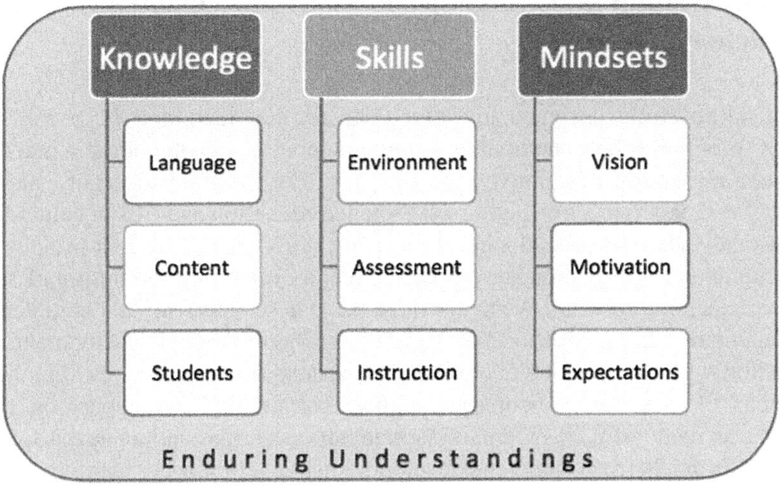

Fig. 8.4 Teacher expertise for English learners

teachers have asset-based *mindsets* with clear vision, purposeful motivation, and high expectations that ELs can engage with rigorous academic content while simultaneously developing language (e.g., deJong 2011, Ruiz 1984) (Fig. 8.4).

Historically, teacher education programs have not offered experiences specifically for ELs, resulting in the unpreparedness of the majority of U.S. teachers for this student sub-group (Cohen and Clewell 2007; Gándara and Maxwell-Jolly 2006). When universities do recognize the importance of teacher preparation for ELs, they often go about implementation in ways that maintain exclusionary silos of EL teaching in PK-12 schools, such as maintaining candidates' learning about ELs in one stand-alone course apart from other courses (Heineke et al. 2013). Noticing the limitations these traditional models, we employed a *targeted* and *integrated* approach to develop teacher expertise for ELs: *targeting* the needed knowledge and skills in specific modules and field experiences and *integrating* those understandings across the 4-year, field-based program (García et al. 2010; Heineke et al. 2013; Lucas et al. 2008; Valdés et al. 2005). In the TLLSC program, teacher educators designed learning experiences to: (a) *target* enduring understandings central to teaching culturally and linguistically diverse students in the exploration phase, (b) *extend* those understandings, knowledge, and skills to grade- and content-specific instructional practice in the concentration phase, and (c) *integrate* candidates into authentic practice with teaching and learning in the specialization phase. In the next section, we share programmatic vignettes of the TLLSC program in practice to demonstrate the implementation of the targeted and integrated approach to preparing all elementary and secondary science teachers for ELs.

Implementation and Outcomes in Practice: Vignettes of Teacher Learning

While purposefully designed to foster deep teacher learning, the power of the TLLSC program is best captured in its implementation and outcomes in practice. In this section, we use descriptive vignettes from our program to detail candidates' learning and experiences engaging with science education and ELs in culturally and linguistically diverse educational settings. In each vignette, we first provide pertinent contextual information for teacher educators regarding the design of science teacher education for ELs. We then utilize these descriptive snapshots to detail the implementation in practice to support candidates' professional development in collaboration with our school and community partners. Finally, we weave in the integral lens of candidate outcomes, summarizing findings on impact on teacher learning as demonstrated by data systematically collected and analyzed from Fall 2013 to Spring 2016. To show the developmental flow of teacher learning, we organize the vignettes sequentially across the TLLSC program, as candidates progress through phases and develop expertise as science educators for ELs.

Targeted Learning in Semester 1, Language Development Across Educational Settings

Candidates begin their program with a semester sequence, *Introduction to TLLSC and the Teaching Profession*, which immerses them in a range of instructional sites including schools, classrooms, informal learning environments, and local communities. Within these varied sites, candidates are introduced the profession of teaching with a focus on collaborative relationships within and among schools, families, and communities, developing an early understanding and appreciation of the role communities play in educating students. Candidates gain exposure to the specific knowledge and skills necessary for educators to embody the dispositions of the profession, including supporting the language development of all youth, and begin to connect those roles and responsibilities directly to the learning and development of PK-12 students.

Of specific interest to this chapter is the second of three instructional modules of this sequence, entitled *Bringing Language Learning and Developmental Theory into Practice*, which is designed to deepen candidates' understanding of the roles and responsibilities of educators in light of the learning and development of youth from birth through adolescence. Candidates learn about the cognitive and linguistic attainments of students across these developmental levels, as well as the relationship between cognitive, linguistic, and academic development. Unique to the TLLSC program is how this site-based study of child and adolescent development links theory with practical applications. Local museums and cultural institutions allow for observation of youth outside of the formal classroom setting as well as for consideration of the ways in which social, cultural, and environmental contexts shape children and adolescents' development. Theoretical perspectives of linguistic development, including second language acquisition and pedagogy, are brought to life as candidates consider language demands within these various learning contexts. For instance, a session at a partner museum begins with a discussion of how we all use language in different ways and for different purposes within different contexts. After an introduction to World-class Instructional Design and Assessment (WIDA)'s Principles of Language Development (WIDA Consortium 2010) and a review of terms such as academic language, language functions, language demands, language scaffolds and supports, candidates head off in small groups into the museum exhibitions where they take observational notes about themselves as learners as well as about the visitors they encounter in these spaces. Each group reports back on the prompts provided to guide their observations, such as describing how museum-goers utilize language to engage in social and academic dialogue and analyzing for particular language demands in museum exhibits.

Through participation in this sequence, candidates are able to identify specific examples from site visits to explain how their knowledge of language development, second language acquisition, and developmental and learning theories applies in both in-school and out-of-school settings. In this first sequence, candidates also develop strategies for supporting PK-12 youth to use language and other modes of communication to express ideas, information, and concepts while engaged in a variety of personally meaningful and purposeful activities. For example, in his final report, Bob shares that "At the Planetarium I noticed a few of the Montessori students talking to themselves while they figured out one exhibit. This is a great example of Vygotsky's views on self-talk. It shows how they were talking themselves through a problem that they were having trouble with in a way that an adult might talk through a problem with a child." Another group notes the language demands from an exhibition they explore at The Field Museum. "In the Egypt exhibit students were reading complex signs with figurative language, metaphorical language, and song lyrics. There were also signs explaining differences in language between cultures...a teacher could provide prompts that facilitated connecting with the language." Having visited exhibits within the Chicago Children's Museum, another group draws attention to how spaces require young student to use language to

WaterWays
at the Chicago Children's Museum

Learning:

- simple machines - **WaterWays exposes children to hands-on examples of simple machines including pulleys, wheels, levers and pumps; creating opportunities for discussion about the mechanics of the machines and their importance to our everyday life. (supports** cognitive, physical **and** language **development and aligns with Piaget's preoperational stage)**
- play with dams, boats and channel systems **teaches children the principle ideas of cause and effect and forces them to communicate on which dams to open or close to allow boats to float through the channel. (supports** cognitive **and** language **development)**
- painting with water **provides an artist medium to engage in the activity even if they are not scientifically inclined.**

Fig. 8.5 Candidate's report on opportunities for language development at a museum exhibit

acquire, assess and communicate information. The figure above is taken from Mathew's report, in which he reviews how language and play created authentic opportunities for young children to develop language abilities and scientific understandings. Overall, participation in this sequence engages candidates in foundational learning around language, providing candidates a theoretical framework with which to understand and conceptualize language development (Fig. 8.5).

Targeted Learning in Semester 3, English Learners in Urban Classrooms

In the third semester of the TLLSC program, typically occurring during the fall semester of sophomore year, candidates round out the exploration phase by continuing to hone knowledge and skills around policy and practice for culturally and linguistically diverse students. All candidates across program areas enroll in an instructional sequence entitled *Policy and Practice in Urban Schools*, which consists of two field-based modules that explore the macro- and micro-level influences on daily classroom practice in urban schools. Field sites for teacher learning are urban elementary and high schools, including both public and Catholic schools, with rich culturally and linguistic diversity and large numbers of students labeled as ELs. Candidates are placed at one of six school sites for the duration of the semester, with the goal to build relationships with cooperating teachers and students simultaneous to context-specific knowledge and skills. Across these six schools on the north side of Chicago, students come to school speaking over 110 different languages from across the globe. Situated in this diverse locale, instruction for all candidates focuses on educational policies and individualized assessment and instruction. Candidates investigate the language diversity of partner schools, assess the language needs of identified ELs, and present and discuss these findings to peers and school faculty.

In the first 4-week module, entitled *Educational Policy for Culturally and Linguistically Diverse Students*, candidates confront questions about how policies manifest in daily practice in urban schools. Central to their future roles as educators in U.S. classrooms, the module focuses on four key policies influencing daily practice in schools: (a) the current standards, including Common Core State Standards (CCSS) (National Governors Association Center for Best Practices and Council of Chief State School Officers 2010) and NGSS, (b) language policy guiding practice with EL and bilingual students, (c) special education policy, and (d) the International Baccalaureate (IB), a curricular movement in Chicago and across the globe. Tapping into field-based settings, module goals are not simply to learn about various court cases, laws, and requirements, but to see how educators enact policies in practice to support student learning and achievement (Heineke et al. 2015). In each module session, candidates collaboratively learn about target policies with the instructor, move into classrooms with policy-in-practice checklists to guide observations and interactions with students and teachers, and return to deconstruct and discuss policy in practice in the various and unique classroom contexts ranging across grade levels and content areas. For example, Jennifer investigates policy in practice at a Catholic high school, where some students struggled to engage with the content lesson due to language proficiency. Tapping into her knowledge of Illinois policy requiring to bilingual supports for ELs, she notices the lack of supports for academic language development, specifically noting that "students' learning would be enhanced if the lesson were taught partially in Spanish." Unique to the Catholic school setting without the same policy structures as public schools, she goes on to consider the complex task of the teacher in discerning language needs when students lack formal labels of language proficiency. With strategic situation in linguistically diverse schools, candidates like Jennifer consistently return to the challenges and opportunities to support students' language development simultaneous to the implementation of various policies (Walqui and Hertiage 2012).

In the second 8-week module, entitled *Individualized Assessment and Instruction for Culturally and Linguistically Diverse Students*, candidates work one-on-one with students to appreciate, discern, and utilize the individual backgrounds and needs of students to plan instruction and support student achievement. With the primary purpose of the module focused on ELs, candidates collect multiple forms of sociocultural, cognitive, linguistic, and academic data while engaged in authentic, language-rich tasks with students (Herrera 2010; O'Malley and Pierce 1996). Through this individualized work with ELs, formalized through case study research with students, candidates draw from their experiences to deepen their understandings of foundational theories, principles, and applications of language and language development introduced in Sequence 1. Candidates first learn and explore students' funds of knowledge, the unique cultural and familial resources emergent from rich experiences in homes and communities (Moll and González 1997). After introducing and solidifying the asset-based mindset for ELs and other culturally and linguistically diverse students, candidates then learn firsthand from students' first and second language development as candidates and youth collaboratively engage in language-rich academic tasks targeting students' listening, speaking, reading, and writing (O'Malley and Pierce 1996). Connecting to knowledge of first language development and second language acquisition theories from Sequence 1, described above, faculty introduce candidates to the WIDA tools, which include English language development (ELD) standards (WIDA Consortium 2012), assessments, and instructional supports for classroom practice. Candidates use their case study findings to incorporate data-driven, individualized instructional supports for language development.

Strategically situated after the policy-focused module, this 8-week experience pushes candidates to recognize the need to value, recognize, and respond to students' cultures and languages in content instruction. Situated at an elementary school with 65 % of students labeled as ELs from a variety of linguistic backgrounds, Fatima works one-on-one with Agu for the case study project. She purposively selected Agu to work with and learn from, after realizing that she had falsely assumed he was an English-proficient, African American student due to his skin color. Her funds of knowledge interview gleaned information about his native language, experiences in coming to the U.S., and his likes and interests outside of school. After collecting data on his abilities and needs in listening, speaking, reading, and writing via authentic, language-rich tasks, Fatima puts forth a series of instructional recommendations to tap into Agu's funds of knowledge to increase his classroom engagement and support academic language development. For example, she proposes the idea of using his love of skateboarding to explore related content concepts. Following her fieldwork with Agu, she spans out and applies her learning more broadly, recognizing the diversity within the label of EL, as well as the important realization that: "The culture and language of these students are assets in the classroom."

Through participation in this sequence, candidates explore how macro-level policies manifested in teachers' and students' practice in urban classrooms. Evidence throughout this exploratory semester include: (a) policy case study, where candidates investigate one focal policy from multiple lenses and layers in practice, (b) student case studies, where candidates work one-on-one with individual students labeled as ELs and as having special needs, and (c) teacher study, which merges the macro- and micro-level foci for candidates to critically consider the central role of the teacher in decision-making and advocacy for diverse students. Overall, as a part of participating in this school-based sequence of learning, candidates uncover the nuances of contemporary policies influencing classroom practice, specifically CCSS and NGSS, as well as observing and engaging with the challenges and opportunities of implementing policies in practice with diverse students (Heineke et al. 2015). Through their one-on-one work with labeled ELs, candidates also grasp the diversity within the label and the centrality of individual students' backgrounds and needs guiding instruction, including rich and diverse funds of knowledge, language proficiency levels, linguistic abilities across domains (i.e., listening, speaking, reading, writing), and corresponding supports and scaffolds in classroom practice (Heineke et al. 2013; Nasir and Heineke 2014). Teacher studies specifically demonstrate candidates' recognition of their active role in implementing policies in such a way to positively influence students' language development and academic achievement.

After rounding out this school-based sequence and the exploration phase of the TLLSC program, candidates now shift to the concentration phase where they begin to hone knowledge, skills, and dispositions within their selected program of study. Drawing from these foundational learning experiences within the policy-in-practice sequence, elementary and secondary candidates now begin to sharpen professional knowledge and skills for ELs and language development specifically in the context of science education.

Integrated Learning in Semester 4, Language Development in Science Instruction

In their fourth semester, candidates begin to specialize in a specific area and are enrolled in their first content-specific courses. Elementary education candidates enroll in an instructional sequence entitled *Specializing in an Area of Teaching and Learning: Integrated Instruction in Elementary Classrooms*, which consists of three field-based modules that emphasize the common practices shared across science and social studies. Instruction takes place within a mix of school and non-school settings, including elementary classrooms, the university's biodiesel and urban agriculture laboratories, and local science- and history-focused museums. School-based class sessions typically begin with the class meeting together for an introduction to content related to the day's topics and discussion of assigned readings. Then, candidates move into classrooms to interact with teachers and students, bringing to life the topics previously introduced. Finally, candidates re-group to debrief their classroom visits and make sense of their experiences.

During the science-focused module, teacher candidates gain familiarity with the goals of the Framework for K-12 Science Education and the dimensions of the NGSS. They also dig further into WIDA's ELD standards by considering the explicit connections between the content and language standards. As candidates work alongside practicing teachers and university faculty to help elementary students in small- and whole-group classroom instruction, they consider the language demands and opportunities embedded in the NGSS science and engineering practices, and how youth use language to construct and communicate meaning in their science classes. They also begin to incorporate principles and strategies for teaching culturally and linguistically diverse learners that they were introduced to in the exploration phase. Key goals related to language development within this sequence include being able to: (a) identify the language functions and demands within the NGSS and in class activities and assignments, i.e. the specific vocabulary and discourse of science; and (b) gather information about the ways in which students make sense of the world and create opportunities for them to use the experiences they bring from their home and community environments to form scientific explanations.

First, science, like all content areas, has a specific set of vocabulary that requires students to code-switch from everyday uses of language to the language of science (Brown and Ryoo 2008; Moje et al. 2001). Science also has a particular discourse and the practices of science require multiple language functions (Lee et al. 2013). For instance, the practice of planning and carrying out investigations requires students to use language functions such as designing, describing, planning, and organizing. The practice of developing and communicating evidence-based claims, at the heart of scientific argumentation, is developed as students make sense of core ideas in science. As the class explores the NGSS, candidates use the partner school's curriculum plan to practice with developing language-based content objectives, or those lesson objectives that blend language-intensive practices with science core ideas and cross-cutting concepts. Then, they consult WIDA's Can-Do descriptors (WIDA Consortium 2009) to determine what supports they might provide for students at varying levels of language proficiency. A reflection assignment following this lesson asks candidates to deconstruct one of the supports or scaffolds they have learned about and explain how and why they would implement it in their classroom. Julianne, for example, writes:

> Aside from individual vocabulary words, I've noticed that my EL students have also struggled with writing sentences afterwards to explain what they did during their experiment/exploration process, so in order to help support students in this regard, I've also made a poster, which stays up on the wall year round, with sentence starters for retelling. These sentence starts help students restate what their claim was, what evidence they found to support that claim, as well as how they believe that evidence supports their claim. Ultimately, I think it's most important to realize that yes, all of these supports I've just described will undoubtedly help an English learner, but it will help every other student in the classroom as well.

The second goal recognizes that all learners bring linguistic and cultural experiences from their home and community environments to the classroom and the most effective science educators and recognize how students' knowledge, interests and experiences serve as the foundation for developing science understandings and science identities (NSTA 2009; Rosebery and Warren 2008). To support the uncovering and incorporating of students' linguistic and cultural experiences, candidates design and conduct science talks with students around current topics that integrate NGSS core ideas and cross-cutting concepts. Science talks honor students' diverse experiences and language practices while also reveal information about students' ideas and sense-making in regard to scientific phenomena (Rosebery and Ballenger 2008). Candidates pair up and work with small groups of students to conduct the science talk, with one candidate leading the talk while the other audio records and takes notes. Then, they switch and the second candidate leads a new science talk with a different group. Afterward, the full class regroups to debrief the activity with the instructor. Through these discussions and a subsequent essay, elementary candidates reflect on what they learned about students' ideas about the selected topics and their sense-making. While most candidates are nervous at first about conducting a lesson consisting of open-ended questions and conversations with young students, they emerge with a newfound appreciation for this type of formative assessment. After finishing a talk with her students about the ingredients in household products, Korinne excitedly reports "Students were very interested and constantly talking but on topic! Each student came around to answers by taking a little bit from each other." Reflecting on her experience, Janice also notes the opportunities for collaboration that her talk with students about Mars opened up. "The talk illustrated how mutually beneficial thinking out loud together can be for all students involved. Not only are they helping to synthesize their own knowledge, but they are also learning from one another". Similarly, reflecting on her talk with students about the Flint, MI water crisis, Aida writes, "Conducting this science talk was a very valuable experience for me. I learned how to facilitate a discussion using talk moves such as 'What else can you tell me?' and 'What do you all think?'...I learned how to come up with questions that are clear and open-ended...I actually saw how students have so many funds of knowledge from prior experiences." Together, these examples illustrate how the talks aid candidates to develop deeper understandings of some of the ways language is used within science, and the knowledge and skills to implement science instruction that incorporates literacy skills such as speaking, listening, and persuasive argument, thereby promoting both students' science and language proficiency (NSTA, 2009).

Overall, as a part of participating in this science-focused module, elementary candidates demonstrated competence in developing and defending science instruction based upon the NGSS and that includes research-based instructional strategies to meet the needs of culturally and linguistically diverse students. Demonstrated by reflections and module assignments, candidates broadened and deepened their conceptions of the work of scientists and what it means to engage in the core actions of science. Figure 8.6 depicts a page from Sarah's journal in which she uses graphics to help communicate her developing understandings about the role of language in science.

After engaging with science education with a specific lens on language development, elementary candidates move into their junior year with the clear recognition of the interplay of content and language learning to make meaning of future field experiences with culturally and linguistically diverse students.

Integrated Learning in Semester 6, Sheltered Science in IB Settings

The sixth semester sequence, *Integrating Content, Cultures and Communities*, typically occurs during the junior year and, for secondary science candidates, is based within culturally and linguistically diverse local middle and high school science classrooms. Note that a similar sequence takes place for elementary candidates

Fig. 8.6 Excerpt from candidate's response to the role language plays in the science classroom

within elementary classrooms, although will not be described in this chapter. The sequence begins with candidates honing skills designing and implementing instruction within their area of specialty (e.g., secondary biology) and then broadens to a focus in the second instructional module on interdisciplinary teaching through the IB framework, increasingly used throughout Chicago area schools. Throughout the semester, candidates engage in deep learning of how to integrate science content and pedagogy and how to make curriculum more responsive to students' immediate and future needs, including language needs. Key goals related to language development within this sequence include being able to: (a) design instruction that provides students with opportunities to practice using language functions essential to developing the practices and understandings about core ideas of science; and (b) plan, implement and reflect upon rigorous science instruction that promotes students' language proficiency and conceptual understanding. Major course assignments for secondary science candidates include working collaboratively with middle or high school teachers to develop and implement a multi-week interdisciplinary curriculum unit that effectively integrates NGSS, CCSS, and WIDA standards and promotes both science and language learning for all students.

An early semester experience involves candidates using knowledge and skills developed in the exploration phase to take inventories of students' language assets and needs within biology, chemistry, and physics classrooms to then plan instructional units to meet these needs. Class sessions also focus on identifying the various language demands and opportunities for linguistic supports and scaffolds within the partner science classrooms. Through participating in these high school science classrooms, candidates build understanding of science instruction that incorporates language and literacy (i.e. reading, writing, speaking, listening, viewing, and representing), and how such instruction supports students in promoting both science and language learning. During one class session, candidates observe, discuss, and reflect upon questions: *What language demands were associated with the class activities and assignments? What language functions were central to the class activities and assignments? What specific vocabulary of science did the class encounter? What were the different ways in which students were able to communicate their learning? What other ways might have been offered to them? What language scaffolds were provided, or could have been provided? How effectively were learning activities structured and differentiated to allow for all students to participate?* As the semester progresses, candidates become increasingly attentive to the need for differentiation. For instance, Nathan writes in his journal about how an "effective strategy for teaching science simply involves removing language barriers," which can be done through strategies such as vocabulary supports, word walls, and graphic organizers. Specifically seeking to maintain scientific learning and academic rigor for all learners, he noted, "In a science classroom, it would be helpful to break down lab procedures or scientific texts for students by providing scaffolding, such as vocabulary note cards or modifying handouts to include less complex language."

Later in the semester, candidates plan and implement instruction collaboratively with cooperating teachers to: (a) record and discuss language functions that facilitate the enactment of scientific practices, and (b) consider which class activities allow students the opportunity to practice and further develop these functions. For instance, class discussion and journal prompts pose questions such as: *How are students provided opportunities to practice constructing and communicating explanations, or distinguishing opinion from evidence as they evaluate conclusions?* Reflecting on a physics lab in one journal entry, Benjamin writes about the importance of moving from gathering evidence to providing explanations.

> Analysis is key in making sure students are correctly applying scientific principles to their lab work. Without analysis, lab work becomes just glorified note taking and bookkeeping. When students have to analyze their data and lab work, it means that students are actually looking at their lab work, and requires that they be able to use language to communicate effectively... It is important that students gain the necessary language skills to be able to adequately explain their surroundings.

Candidates also attend to which language scaffolds they feel promote both students' science and language learning, and how these can be altered to match language proficiency. *What ways do you and your cooperating teacher use visualizations, digital images and graphic organizers to help make abstract biology, chemistry and physics concepts more concrete for students?* For example, Benjamin explains that data collected by each lab group "is presented in tables that may make it easier for ELs to use symbolic knowledge to complete the assessment." Building on experiences conducting case studies in Sequence 3, candidates take opportunities to work one-on-one with students to gain their perspectives about the effectiveness of the scaffolds and supports provided for classroom learning activities. For example, Nathan writes about the importance of manipulatives during a mitosis lesson. "A lot of scientific concepts occur in a three-dimensional space, so being able to see a visual alongside learning the content helps students."

Other semester experiences focus on appropriate assessment within the secondary science classroom with emphasis on differentiated performance assessments that tap into students' background knowledge. When designing unit plans, candidates must use understandings of learning theories and research to defend instruction and assessment decisions. For instance, they provide rationales for the different options they offer students to demonstrate and communicate their learning. Finally, as part of the course's interdisciplinary unit plan assignment, candidates develop, implement, and share the outcomes of formative and summative assessments that allow students to demonstrate their learning in multiple and authentic ways. Here, they make use of the WIDA Can-Do descriptors to differentiate accordingly for varying proficiency levels. Finally, at the end of the course, candidates explain and reflect upon how effectively their instruction and assessment plans provided students with opportunities to engage in scientific practices as well as use and develop language functions. As an example, Nathan planned for students to verbally explain the phases of mitosis by making small group oral presentations of their pipe cleaner models. In his rational statement, he explains, "by having students take in language and then work to produce language, they foster recognition networks and delivery networks in their neural systems."

Overall, through class discussions, reflective journals, interactions with partner teachers and students, and module assessments, candidates recognize the roles and responsibilities that science teachers have in promoting academic language development, particularly for ELs. Benjamin and Nathan, for instance, show evidence of this in their regular inclusion of both content and language objectives and the identification of academic language demands as well as appropriate instructional supports. Reflecting on the language needs of students in his class, Nathan writes "Students who are learning language will see success if they are interacting with that language frequently and authentically." Overall, by the end of this sequence, candidates demonstrate increased ability to differentiate instruction and assessment to meet the learning needs of students and are ready to move on to a full year of student teaching

Conclusions and Recommendations

In this chapter, we described the design and implementation of one university's teacher education program that prioritized preparation for ELs and situated professional learning in schools and communities with culturally and linguistically diverse students. Now entering its fourth year of collaborative implementation with faculty, teachers, and leaders in the urban hub of Chicago, data have demonstrated the efficacy of the field-based model and targeted and integrated approach to EL teacher preparation (e.g., Birmingham et al. 2015; Heineke et al. 2013, 2015; Nasir and Heineke 2014; Ryan et al. 2014). Across the past 6 years of programmatic design and implementation, we have come to recognize the challenges and opportunities of utilizing a field-based program for preparing teachers for work with ELs in science classrooms. With the growing cultural and linguistic diversity across the U.S., as well as the large number of students labeled as ELs (Gándara and Hopkins 2010), opportunities center around the professional development of elementary and secondary classroom teachers who design rich science and language learning environments and instruction for all students. Nevertheless, challenges arise in this plight, including the preparedness of teacher educators in both classrooms and universities to facilitate this learning with candidates. We close the chapter by using our challenges and opportunities to draw conclusions and offer recommendations for science teacher educators to develop and implement language-focused professional learning and development for beginning and experienced teachers. Here, we emphasize ways for science and language educators to leverage collaborations across their local schools and communities.

Challenges of this collaborative approach to science teacher education center on the traditional structures guiding daily practice in universities and schools of education, as well as on the traditional boundaries between teacher education faculty. The same silos that often characterize PK-12 school organization, such as the separation between ESL teachers responsible for supporting students' language development and science teachers focused on content-specific teaching and learning, also manifest at the university (Heineke 2014). EL-related faculty teach EL-specific courses to candidates who have typically opted into an ESL endorsement, whereas science-related faculty teach science-related courses to candidates who have chosen science education as their desired field of study. To successfully design and implement the TLLSC program, we first had to shift and redefine our collaborative faculty mindset, prioritizing the needs of PK-12 youth and schools, which included the demand for all teachers to have deep and extensive knowledge and skills for ELs (de Jong and Harper 2005; Gándara and Maxwell-Jolly 2006). This shared commitment supported the initial dissolving of the traditional content boundaries and led to partnerships across teacher education faculty. Whereas EL faculty maintain responsibility for the targeted EL learning experiences, they collaborate with science education and other faculty to ensure purposeful and effective integration of EL knowledge and skills across programs of study. This requires science education and other content-area faculty to acknowledge that we must be proficient in the same breadth of knowledge and skill sets that we ask of our beginning teachers. Just as candidates commit to accepting the responsibility of promoting all students' academic literacy, so must all teacher educators accept the responsibility of ensuring that their courses prepare candidates for this charge. Likely, this commitment necessitates science education faculty to seek out professional development of our own. In sum, readers should recognize that what we describe in this chapter was in no way straightforward, quick, or easy: the purposeful re-design of our teacher preparation programs, as well as the strategic targeting and integrating of EL-specific knowledge, skills, and mindsets, took and continues to require ample time, effort, collaboration, and negotiation.

Despite the challenges that we faced and maneuvered throughout the design and implementation of our field-based teacher preparation program, we want to emphasize the opportunities that made, and continue to make, the work worthwhile. Primarily, we have found that candidates and teacher educators learning alongside classroom teachers, school leaders, and community experts yield valuable experiences for all involved, while simultaneously improving both teacher and PK-12 education (Ryan et al. 2014). Explicit in our approach is a commitment to mutual benefit, where we ensure that our school and community partners gain as much from the TLLSC program and university partnership as do our teacher candidates and educators (Kruger et al. 2009). We have committed partners who enthusiastically welcome our faculty and candidates every semester, recognizing the benefits for youth and classroom teachers, including having extra sets of hands in the classrooms and teachers learning from candidates' contemporary and comprehensive expertise. Additionally, we offer opportunities for schools to improve and build capacity, specifically recognizing the needs for teachers and school leaders to develop knowledge, skills, and asset-based mindsets around ELs and supporting students' language development (Lucas et al. 2008; Valdés et al. 2005). Tapping into our human and material resources at the university, we provide an array of professional development opportunities for our partners, including bi-annual conferences broadly focused on EL teaching and learning, workshops on language development targeting content area teachers, and clinical vouchers for hosting candidates that can teachers can exchange for ESL endorsement coursework. Participating in these professional development opportunities has also proven valuable for science education and other content-area faculty whose preparation for teaching PK-12 and adult learners did not consist of the sort of instruction described in this chapter. In this way, we seek to build broader educational capacity for ELs through both our current partners and future teachers.

Drawing from these challenges and opportunities, our recommendations for science teacher educators center on designing opportunities for collaborative professional learning for all stakeholders, including teacher candidates, university faculty, cooperating teachers, and school leaders. First, begin conversations with teacher education faculty to collaboratively evaluate your program's effectiveness in preparing elementary and secondary science teachers for today's culturally and linguistically diverse classrooms, schools, and communities. Next, once faculty have collaboratively committed to preparing all science teachers for ELs, define the necessary knowledge, skills, and mindsets and strategically design instruction that targets teacher learning specifically for ELs and also integrates and applies EL expertise into science education. Then, build partnerships with schools, community organizations, and cultural institutions to leverage the external expertise in science and EL education in order to provide candidates with authentic and rigorous teaching and learning experiences. Finally, recognize the opportunities to not only support the learning and development of teacher candidates, but to simultaneously build capacity for science education for ELs in local classrooms, schools, and communities. In sum, through field-based, science teacher education for ELs, teacher educators hold the great opportunity to improve science education in PK-12 schools.

References

Ball, D. L., & Forzani, F. M. (2009). The work of teaching and the challenge for teacher education. *Journal of Teacher Education, 60*(5), 497–511.

Birmingham, D., & Calabrese Barton, A. (2014). Putting on a green carnival: Youth taking educated action on socioscientific issues. *Journal of Research in Science Teaching, 51*(3), 286–314.

Birmingham, D., Smetana, L. K., & Coleman, E. R. (2015, April). *Developing science identities: What role does a teacher preparation program play?* Paper presented at the annual meeting of the National Association for Research in Science Teaching, Chicago, IL.

Brown, B., & Ryoo, K. (2008). Teaching science as a language: A "content-first" approach to science teaching. *Journal of Research in Science Teaching, 45*(5), 529–553.

Cohen, B. C., & Clewell, B. C. (2007). Putting English language learners on the educational map. In *Education in focus: Urban institute policy brief.* Retrieved from http://www.urban.org/UploadedPDF/311468_ell.pdf

de Jong, E. J. (2011). *Foundations for multilingualism in education: From principles to practice.* Philadelphia: Caslon Inc.

de Jong, E. J., & Harper, C. A. (2005). Preparing mainstream teachers for English language learners: Is being a good teacher good enough? *Teacher Education Quarterly, 32*, 101–124.

de Jong, E. J., Harper, C. A., & Coady, M. (2013). Enhanced knowledge and skills for elementary mainstream teachers of English language learners. *Theory Into Practice, 52*(2), 89–97.

Gándara, P., & Hopkins, M. (2010). The changing linguistic landscape of the United States. In P. Gándara & M. Hopkins (Eds.), *Forbidden language: English learners and restrictive language policies* (pp. 7–19). New York: Teachers College.

Gándara, P., & Maxwell-Jolly, J. (2006). Critical issues in developing the teacher corps for English learners. In K. Téllez & H. C. Waxman (Eds.), *Preparing quality educators for English language learners: Research, policies, and practices* (pp. 99–120). Mahwah: Lawrence Erlbaum.

Gándara, P., Rumberger, R., Maxwell-Jolly, J., & Callahan, R. (2003). *English learners in California schools: Unequal resources, unequal outcomes.* Educational Policy Analysis Archives.

García, E., Arias, B. M., Harris-Murri, N. J., & Serna, C. (2010). Developing responsive teachers: A challenge for a demographic reality. *Journal of Teacher Education, 61*, 132–142.

Giambo, D., & Szecsi, T. (2005/06). Opening up to the issues: Preparing preservice teachers to work effectively with English language learners. *Childhood Education, 82*(2), 107–110.

Heineke, A. J. (2014). Dialoging about English learners: Preparing teachers through culturally relevant literature circles. *Action in Teacher Education, 36*(2), 117–140.

Heineke, A. J., Kennedy, A., & Lees, A. (2013). Preparing early childhood professionals for the culturally and linguistically diverse classrooms and communities of Illinois. *Early Childhood Research and Practice, 15*(2). Accessible online at: http://ecrp.uiuc.edu/v15n2/heineke.html

Heineke, A. J., Ryan, A. M., & Tocci, T. (2015). Teaching, learning, and leading: Preparing teachers as educational policy actors. *Journal of Teacher Education, 66*, 382–394.

Heritage, M., Walqui, A., & Linquanti, R. (2015). *English language learners and the new standards: Developing language, content knowledge, and analytical practices in the classroom.* Cambridge, MA: Harvard Education Press.

Herrera, S. (2010). *Biography-driven culturally responsive teaching*. New York: Teachers College Press.

Kruger, T., Davies, A., Eckersley, B., Newell, F., & Cherednichenko, B. (2009). *Effective and sustainable university-school partnerships: Beyond determined efforts by inspired individuals.* Canberra: Teaching Australia.

Lave, J., & Wenger, E. (1991). *Situated learning: Legitimate peripheral participation*. Cambridge: Cambridge University Press.

Lee, O., Quinn, H., & Valdes, G. (2013). Science and language for English language learners in relation to next generation science standards and with implications for common core state standards for English language arts and mathematics. *Educational Researcher, 42*(4), 223–233.

Lucas, T., Villegas, A. M., & Freedson-González, M. (2008). Linguistically responsive teacher education: Preparing classroom teachers to teach English language learners. *Journal of Teacher Education, 59*, 361–373.

Manzo, R. D., Cruz, L., Faltis, C., & de la Torre, A. (2011). Professional development of secondary science teachers of English learners in immigrant communities. *AMAE Journal, 5*, 41–48.

Moje, E. B., Collazo, T., Carrillo, R., & Marx, R. W. (2001). "Maestro, what is 'quality'?": Language, literacy, and discourse in project-based Science. *Journal of Research in Science Teaching, 38*(4), 469–498.

Moll, L. C., & González, N. (1997). Teachers as social scientists: Learning about culture from household research. In P. Hall (Ed.), *Race, ethnicity, and multiculturalism: Policy and practice* (pp. 89–114). New York: Routledge.

Nasir, A., & Heineke, A. J. (2014). Teacher candidates and Latina/o English learners at Fenton elementary school: The role of early clinical experiences in urban teacher education. *Journal of the Association of Mexican American Educators, 8*, 60–71.

National Governors Association Center for Best Practices & Council of Chief State School Officers. (2010). *Common core state standards*. Washington, DC: National Governors Association Center for Best Practices & Council of Chief State School Officers.

National Science Teachers Association (NSTA). (2009). *Position statement: Science for English language learners.*

NGSS Lead States. (2013). *Next generation science standards: For states, by states*. Washington, DC: The National Academies Press.

O'Malley, M., & Pierce, L. V. (1996). *Authentic assessment for English language learners: Practical approaches for teachers*. Boston: Addison-Wesley.

Rascoe, B. (2013). Negotiating science content: A structural barrier in science academic performance. In M. Atwater, M. Russell, & M. Butler (Eds.), *Multicultural science education: Preparing teachers for equity and social justice* (Vol. 119, p. 136). Dordrecht: Springer Science & Business Media.

Rogoff, B. (1995). Observing sociocultural activity on three planes: Participatory appropriation, guided participation, and apprenticeship. In P. M. J. Goodnow & F. Kessel (Eds.), *Sociocultural studies of mind* (pp.139–164). Cambridge: Cambridge University Press.

Rogoff, B. (1994). Developing understanding of the idea of communities of learners. *Mind, Culture, and Activity, 1*, 209–229.

Rosebery, A. S., & Ballenger, C. (2008). Essay: Creating a foundation through student conversation. In A. Rosebery & B. Warren (Eds.), *Teaching science to English language learners* (pp. 1–12). Arlington: NSTA Press.

Rosebery, A. S., & Warren, B. (2008). *Teaching science to English language learners: Building on students' strengths*. Arlington: NSTA Press.

Ruíz, R. (1984). Orientations in language planning. *NABE: The Journal for the National Association for Bilingual Education, 8*(2), 15–34.

Ryan, A. M., Ensminger, D., Heineke, A. J., Kennedy, A. S., Prasse, D. P., & Smetana, L. K. (2014). Teaching, learning, and leading with schools and communities: One urban university re-envisions teacher preparation for the next generation. *Issues in Teacher Education, 22*(2), 139–153.

Valdés, G., Bunch, G., Snow, C., Lee, C., & Matos, L. (2005). Enhancing the development of students' language. In L. Darling-Hammond & J. Bransford (Eds.), *Preparing teachers for a changing world: What teachers should learn and be able to do* (pp. 126–168). San Francisco: Jossey-Bass.

VanLier, L., & Walqui, A. (2012). *Language and the common core state standards*. Understanding Language Initiative, Commissioned Paper, Stanford University, CA. Available at http://ell.stanford.edu/

Vygotsky, L. S. (1978). *Mind in society: The development of higher psychological processes*. Cambridge, MA: Harvard University Press.

Walqui, A., & Hertiage, M. (2012). *Instruction for diverse groups of English language learners*. Understanding Language Initiative, Commissioned Paper, Stanford University, CA. Available at http://ell.stanford.edu/

WIDA Consortium. (2009). *WIDA can do descriptors*. Madison: State of Wisconsin.

WIDA Consortium. (2010). *WIDA principles of language development*. Madison: State of Wisconsin.

WIDA Consortium. (2012). *WIDA English language development standards*. Madison: State of Wisconsin.

Wiggins, G., & McTighe, J. (2005). *Understanding by design* (2nd ed.). Washington, DC: Association of Supervision and Curriculum Development.

Zeichner, K. M. (2009). Changing the epistemology of teacher education. *Teacher Education & Practice, 22*(4), 488–491.

Part II
In-Service Teacher Preparation

The Project

The Inheritances Books are a collaborative project between Ichabod Crane High School's Illustration and ELL students. It is made possible by a grant from the Berkshire Taconic Community Foundation.

The Student-Artist

Isabelle Tennier, grade 11, 16 years old.

Chapter 9
The Integration of English Language Development and Inquiry Science into a Blended Professional Development Design

Susan Gomez Zwiep and William J. Straits

Introduction

The directive from science education reform documents is clear; all students should have opportunities to participate in scientific inquiry throughout their K-12 education (American Association for the Advancement of Society 2009; National Research Council 2012). However at the elementary level, science is often overlooked or underemphasized due to the pressure to perform well on math and language arts assessments (Dorph et al. 2011). Access to science is further diminished in schools with large populations of English Language Learners (ELLs) where the urgency to develop English proficiency is an additional pressure on teachers and students (Brown and DiRanna 2012). Instructional policies often exclude ELLs from equal access to quality science instruction in an effort to hasten their English language development. This restricted access to science affects a significant number of students; in California, where this study took place, more than 22 % (1,413,549) of K-12 students and nearly 35 % of K-4 students are English Language Learners (California Department of Education 2015).

Contrary to this approach, a substantial and growing body of research suggests that English Language Development (ELD) and science instruction are complementary (Gomez Zwiep and Straits 2013; Gomez Zwiep et al. 2011; Lee et al. 2013; Stoddart et al. 2002; Yore et al. 2006). Inquiry science can provide a learning environment where collaboration and peer-to-peer talk is a natural part of how students make meaning. Given the hands-on nature of inquiry science, it also can lower the linguistic burden for students while they engage in this learning (Lee et al. 2006). Furthermore, the integration of scientific inquiry and second language acquisition can promote higher-order thinking (Stoddart et al. 2002) that is often absent when

S.G. Zwiep (✉) • W.J. Straits
California State University, Long Beach, CA, USA
e-mail: Susan.GomezZwiep@csulb.edu; W.Straits@csulb.edu

A.W. Oliveira, M.H. Weinburgh (eds.), *Science Teacher Preparation in Content-Based Second Language Acquisition*, ASTE Series in Science Education,
DOI 10.1007/978-3-319-43516-9_9

lessons are dependent upon an ELL's English-based literacy skills for complexity. The potential for language learning during science instruction is great, but what should an ELD-focused, inquiry science lesson look like and could it actually achieve this potential?

Setting

We began our work with a large, urban, California school district in 2008. At the onset of our project the district was serving a culturally and linguistically diverse population with large numbers of English Language Learners (56 % of K-4 students) and students living in poverty (81 % of K-4 students) (California Department of Education 2015). For the majority of ELL's in the district Spanish was the primary language (98 %, California Department of Education 2015). The district served neighborhoods that were more likely to function in Spanish, both socially and in commerce, limiting student access to English outside of the school day. Subsequently, the district had a significant number ELLs entering Kindergarten with little to no English proficiency. These students typically mastered Basic Interpersonal Communication Skills (Cummins 2008) by 2nd grade and gained intermediate fluency by 4th grade. However, ELLs often failed to develop the necessary English to engage in academic tasks (Cognitive Academic Language Proficiency) with many stalling at the intermediate-advance levels of proficiency at the end of middle school. This trend was common across all schools in the district.

The district was in danger of federal sanctions due to its failure to make Adequate Yearly Progress (AYP) towards statewide proficiency goals. An analysis of state testing data from years prior to our work indicated that the majority of all students were failing to make adequate academic progress in Language Arts, Mathematics, and English acquisition. English Language Learners were a particular concern as this sub-group consistently fell below the AYP minimum across the district at all grade levels. In response, the district mandated increased instructional time for subjects weighted heavily on state exams (i.e., English Language Arts and Mathematics). As a consequence, students received very little, if any, instruction in science. This was particularly true of ELLs who, in addition to increased Language Arts and Mathematics, received additional instruction in English Language Development.

As we began our work, the program and its goal of improving science and language learning for English Language Learners was explained to each elementary site within the district, in an effort to recruit schools. Schools were then invited to participate based on evidence of a complete, site-based commitment to the program. This commitment included the principal's and all K-4 teachers' participation in professional development and a willingness to replace the current English Language Development curriculum and to provide daily instructional time for science in grades K-4. The level of commitment ranged among participating schools but overall there was a significant level of buy-in by both teachers and site administrators from the beginning.

The teachers at our three participating schools all entered the program at similar places in their proficiencies with science and ELD instruction. For several years, the school district had invested extensive resources towards ELD professional development and virtually no resources to science in the elementary grades. Therefore, the teachers had a reasonable depth of knowledge related to ELD when the project began, but quite the opposite for their science teaching knowledge. Encouragingly, one of the main reasons teachers and principals were willing to participate in the professional development was an awareness that their science teaching needed improvement in order to more fully serve their students.

Prior to the implementation of our science/ELD blended program, the district used a popular ELD program correlated with the state's English Language Arts and English Language Development standards. This program focused on developing academic vocabulary and language skills through the use of multi-leveled reading selections and also relied heavily on teacher modeling correct forms of English and academic language use. The reading selections included fiction and non-fiction text, but topics were not aligned with content standards. In 2008, after years of using this ELD program, a total of 60 elementary teachers (grades K-4), three elementary school principals, six district second language acquisition coaches, and more than 2000 students participated in a bold change – abandoning their adopted ELD program and embedding ELD instruction into inquiry science lessons.

Teacher Professional Development Model

A professional development team comprised of district personnel, higher education faculty, and a state-wide professional development organization collaborated to assist the teachers and administrators who participated in this professional development project. The overall structure of this 3-year, professional development effort included intensive 2-week long, summer institutes that focused on a language socialization approach to second language acquisition theory and practice (Duffy 2002) and on science content and inquiry-based science pedagogy, along with site-based lesson study teams, called Teacher Learning Collaboratives (DiRanna et al. 2009) held throughout the school year. The major components of the project include professional development related to science content, science pedagogy, and second language acquisition theory and strategies.

At the beginning of the project, the professional development team approached the development of a science/ELD blended lesson design with different foci for lesson planning. Science educators advocated for the use of Bybee's (1997) 5E lesson design (i.e., engage, explore, explain, elaborate, and evaluate) as the lesson planning template. Specifically, we proposed using a version of the 5E design that included an additional section for teachers to explicitly state the science concept developed at each phase, from students' prior knowledge to the final learning goal of the lesson (DiRanna et al. 2009). This science lesson template would emphasize conceptual understanding, hands-on activities, and student interaction and support the creation of lessons that began with the elicitation of students' prior knowledge of a concept and then provide a series of experiences that allow students to build on that initial understanding. Although this focus included specific points in each lesson where students would discuss their thinking with peers and their teacher, vocabulary and specific language functions and forms were not emphasized during lesson planning.

On the other hand, the ELD professionals on our team focused on traditional ELD lessons that made the language of instruction (oral and written) accessible to learners through the use of specific language forms (e.g., grammatical features or word usage) and language functions (i.e., the task or purpose, such as compare). Within a lesson, language was made accessible through the use of explicit instruction, modeling, and scaffolding by the teacher (Duffy 2002). Language forms and functions were scaffolded with predetermined sentence frames that students could use to build language (for example, "I think ______ because ______."). Sentence frames provided necessary support for students to generate sentences and express their thinking as students often possess vocabulary specific to the content, but lack the words or phrases necessary to construct sentences. In addition, ELD lessons often front-loaded language, pre-teaching specific grammatical structures and vocabulary prior to their use in a cognitive task. In ELD lessons, language instruction was often embedded in content-based lessons, but conceptual understanding of that content was not always emphasized; the goal was the development of English language skills (Echevarria et al. 2008).

Our science education philosophy was grounded in inquiry instruction where concepts and language unfold out of student-centered learning experiences, while our ELD philosophy relied more on highly-facilitated instruction where the teacher frames, directs, and monitors student language use, accommodating for varied English language proficiency levels. However, in reconciling these two philosophies learning opportunities were created that provided access to rigorous science content for English learners while simultaneously developing their proficiency in English language. The richness of the blend was due to several factors. First, science practices and thinking skills mirrored functional language purposes (e.g., describing, comparing, citing information). Second, science content provided a highly-contextualized setting for language development. Finally, science provided important opportunities for students to engage in and demonstrate complex thinking, even if students were not yet proficient in English. However, success in this approach was dependent on several considerations. Vocabulary, along with specific language functions and forms, needed to be carefully examined for what, when, and how they would be used. Determinations of which new words should be embedded in the lesson and which new words should be front-loaded (pre-taught) were based on the instructional goals of the lesson. And throughout the lesson student thinking needed to be prioritized; as such, the science should not be simplified in an attempt to simplify language.

Year 1

During our first summer institute, teacher professional development occurred in three sessions: science content, science pedagogy, and second language acquisition. While the science pedagogy and second language acquisition sessions were designed to improve the implementation of science and language development learning experiences for students, the science content sessions were designed to increase teacher understanding of both scientific ideas and science practices. Approximately half of each summer institute was dedicated to deepening teachers' science content knowledge through these content sessions, which focused on teacher learning and were delivered at an adult learner level to provide rich, and challenging, inquiry learning experiences. These inquiry experiences provided context for authentic language use while participants struggled to make meaning of challenging content.

The placement and relationship among the three different sessions changed each year to increase their connection and explicit use by facilitators. In the initial year of the project, the three components were presented as separate elements to teachers. However, even at this early stage, the facilitators purposefully merged specific elements to model the integration of science learning and language elements during each session. For example, science content sessions utilized models and strategies presented in pedagogy sessions, including the using the 5E lesson design and integrating facilitated questioning strategies and linguistic supports, such as the use of realia, partner talk, sentence frames, and other linguistic supports.

During the school year, teachers participated in three rounds of grade-level specific, lesson study. The lesson study rounds were each 2 days long: 1 day for collaborative planning and 1 day for collaborative teaching and reflection. The first day of the lesson study supported teachers in planning a 5E science lesson and then blending into the lesson specific second language acquisition elements to build students' proficiency in English. On the second day, teachers collaboratively taught the lesson twice, with time to modify the lesson between teaching rounds based on their analysis of student work produced during the lesson.

Year 2

Once participating teachers had a foundation of science and language pedagogy, the second year of the institute was more explicit in the merging of science and language. Although, many teachers began the program with established expertise in using appropriate strategies for English Language Learners, they did not always know how to use these techniques within a science context. Summer institute sessions were designed to demonstrate how typical language development strategies could support students' language and learning as students discuss and debate ideas about scientific phenomena. In the second summer institute, the second language acquisition sessions used material from content sessions as context for discussion

Science Objective:					
Science Standard:					
Language Objective:					
ELD Standard:					
	Teacher Does	Student Does Differentiated by Language Level			Science Concept and Language Function
		Low	Medium	High	
Engage					
Explore					
Explain					
Elaborate					
Evaluation					

Fig. 9.1 Science/ELD blended lesson design template. This modified 5E lesson plan template identifies science concept development, language function, and teacher and student actions. Student actions are differentiated based on language proficiency level

and exploration. To accomplish this second language acquisition experts participated in the science content sessions, noting the types and forms of language used by participants as they explored topics such as properties of matter, force, and motion. A sequence of planning was made explicit: science first, language second. This sequence was followed in both the presentation of session materials and the protocols for teacher-developed lessons. This sequence ensured that the science content was accurate and appropriate for the grade-level, avoided artificial and awkward language that may have obstructed student thinking, and utilized teachers' natural discussion during their collaboratively-planned science lessons to identify student language needs.

In our second year, the science/ELD blended lesson design template (Fig. 9.1) was formalized. This 5E-based template included columns for teacher actions and student actions, as well as places for teachers to identify the science concept and primary language function developed during each phase of the lesson (Gomez Zwiep et al. 2011). Language functions were added to the lesson template to encourage teachers to pre-think which language functions would naturally emerge during the inquiry and would require support. Identifying the function of language that students would be using during the science lesson allowed teachers to select and use the appropriate linguistic scaffolds such as sentence frames and graphic organizers. For example, if students were creating descriptions teachers would employ strategies to support describing; if students were comparing and contrasting, teachers would employ a different set of linguistic scaffolds. The student action column was divided into sections to focus teachers on the varied English proficiency levels in their classrooms. These sections provided a place for teachers to plan specific strategies based on language function and specific to each proficiency level. For participating teachers, the use of language functions were a familiar part of ELD lessons; teachers had great experience with their use. Here, teachers applied this expertise as they designed science lessons that provided an authentic and natural use of a language function within a context-rich environment.

During the lesson study sessions, facilitators pushed teachers to be more specific about student responses, such as specifying language frames at each stage in the lesson. For example, if students were asked to recall information about the properties of rocks in the engage phase of a lesson, teachers might discuss the prior knowledge students might have about rocks and how students might communicate that knowledge; the facilitator would then encourage additional discussion about means for eliciting and representing these student ideas, such as using drawings, graphic organizers, or sentence frames, and what these might look like at different levels of proficiency. This allowed for more specific language support, and made it easier for teachers to engage students with limited English skills in more scientifically-rich conversations and activities.

Year 3

The blended design developed further during the final year of the project. As teachers grew in their sophistication, our discussions of science and language became increasingly seamless. To a great extent the professional development and the work of teachers was "science/ELD" and not, "science" and "ELD." We continued to provide much-needed science content, but our science pedagogy and language development pedagogy sessions were largely devoted to facilitated planning time for teachers to develop, through a multi-step process (see Appendix), their own blended lessons and units. The content sessions included an explicit focus on the nature of science, emphasizing scientific forms of discourse, such as the use of evidence and reasoned arguments. These sessions emphasized how language is used within the scientific community to validate or discredit new ideas through public debate (written and oral) (Osborne 2014). These "scientific" forms and uses of language were introduced to teachers, helping to further solidify the link between science and language.

During this final year, it was decided that within our blended lessons students needed more room to express their ideas and that more room was needed for the use of primary language and "imperfect" language (Lee et al. 2013). Initially, we thought it was necessary for language functions and frames to be identified in each phase of the 5E lesson design. However, as teachers developed their expertise and students were exposed to quality science instruction, the role of imperfect language became evident. Opportunities for less structured language were created within the summer professional development sessions and included in the lesson study protocol used during the school year. Allowing students more freedom in how they communicated their thinking created deeper science understanding and promoted language development opportunities. Primary language and "imperfect" language was given more room in the first phases of the 5E lesson to allow natural language and space for student thinking while science understanding is developing. Sentence frames and other linguistic supports that focused on correct grammatical structure were removed from the first two phases (engage and explore) of the lesson. However, formal second language supports are still included in the later phases (explain and elaborate) of the lesson to help students articulate their thinking.

Materials

Although this professional development project aimed to support K-4 teachers in teaching science/ELD blended lessons, at the outset the professional development team did not have a vision for what science/ELD blended lessons should look like. The planning tools we presented to teachers were living documents with elements added, amended, and completely thrown out during the course of the project. We made herky-jerky progress. And, in the end, had a blended science/ELD lesson planning tool that teachers found highly effective (Appendix).

Outcomes

In an effort to better understand how this project impacted teachers' practice, we analyzed teacher-generated lesson plans, observed classrooms as teachers implemented these lessons, and conducted semi-structured interviews with participating teachers and principals throughout the project. Selective coding (Charmaz 2002) was used to sort, synthesize, and conceptualize the emergent qualitative data by adopting frequently appearing initial codes relevant to the focus of the study. Coded data that posed coherent sets of ideas, were organized into categories. These categories were revisited as new data provided alternative vantage points for re-interpretation. Ultimately, those categories that sustained coherent and plausible interpretations were organized as key insights. These insights provide perspective on the impact the blended program had on teachers, students, and the overall school culture.

Enhanced Status for Science

It is an understatement to say that prior to the implementation of the blended program, science was not a priority at our participating elementary schools. In fact, teachers reported that, when new science textbooks were adopted in 2008, at the end of the 7-year curriculum cycle, they turned in brand new science textbooks, never opened. "We all joked when we were turning them that some of us let the kids take them home for a week before we turned them in so that they would look more used." However, the status of science changed with the implementation of the new program. This was in part because the program required science be part of ELD which long had dedicated instructional minutes during the school day. As one teacher explained,

> English language development has always been a key focal point. It is so engrained in us that we need 45 minutes a day, no matter what. Putting both of them [science and ELD] together makes science one of the top priorities. Before it was we had half an hour a week to teach science, social studies and P.E. Now science is taught everyday.

Science became a "top priority" primarily because of its link to student English language development and the importance of English proficiency for success during state testing. Because high-stakes tests are often given in English, the performance of English learners is often indicative of their English proficiency rather than the skill or knowledge being assessed. Given this and the long history of emphasizing ELD at our participating schools, it was not surprising that connecting science to ELD heightened the importance of science in the eyes of teachers and administrators. What was surprising to participating teachers was students' responses to science.

Students at the participating schools were excited about science and looked forward to their science lessons. While the hands-on, process of discovery is often intrinsically motivating for students, there was more to the additional appeal of science for students than engaging lessons. From the student perspective, the program was seen as a switch to science rather than a different approach to ELD, lifting away their perceived negative stereotypes related to the label "English Learner." Prior to the project, students who were considered proficient in English were given full access to the curriculum; students designated as English learners received additional instruction in English at the expense of participating in other subjects, such as science. As one teacher reported, "One of my students told me, 'I don't go to ELD anymore, now I get to go to science instead.'" Additionally, perhaps due to the fact that science is now seen as a privilege, students have fewer behavioral problems during science lessons compared to other instructional times during the day. "Now I don't have any real behavior issues. Now I just say, 'Is that how scientists act?' and they get back into it. They're really intense." This came as a surprise to teachers who had previously expressed fear of keeping students on task and behaving appropriately during science, as a major factor discouraging them from hands-on science.

Increased Use of Oral Language

A critically important impact of students' excitement for science was that students became excited about talking about their ideas and new learning in science. Across the board, teachers and school administrators were overwhelmed with the students' increased use of English. Teachers reported this increase in both oral and written English, but seem most impressed by students increased use of oral language. Teachers were noticeably delighted when they described the change in their students' willingness and ability to communicate in English. "It is much more exciting so kids are willing to talk more, in English." "You should see the vocabulary they [students] use now, 'we predicted today, we did some observations.'" This increase in English use extended beyond science and beyond the classroom. Principals and teachers across the participating school sites described an increase in English use in other content areas and in non-classroom settings such as recess or in the office when speaking to support staff. School principals reported that students often want to tell them about the science activity, book, or lesson they are learning about in class, "When I am walking around the cafeteria or see the children walking out of the library they can't wait to tell me about the different planets, rocks and minerals, or erosion." Principals also noted an increased English language use beyond academics. "We had a group of students in the office trying to settle a dispute that occurred on the playground and they were using English even though the office staff are fluent in Spanish. That was a first around here." This increased use of oral language, both within and outside the classroom, was perhaps the most apparent and wide-ranging impact of blending science and ELD instruction.

Changes in Teacher Perceptions

Our close work with teachers provided important insights to teachers' creation and implementation of science/ELD blended lesson plans. Many of these are not earth-shattering for teacher educators, but were enormously enlightening for individual teachers as they grew in their understanding of effective teaching and their ability to critique their own practice. In particular, teachers grew to be more effective in and critical of their planning for instruction and structuring of lessons. As stated by a teacher,

> It is how I teach it that is going to give me the desired outcome. If I expect the child to know this then I need to guide them to that place and not expect it to come out of the blue somewhere in my lesson. It makes sense, but I never thought about it that way before.

Participating teachers shared additional ways they had grown as professionals, most prominently in terms of their expectations of students and the affect these new expectations have on their pedagogy. Many teachers interviewed described a shift in thinking about what a child with limited English is capable of learning, becoming more focused on how they structure learning in their classrooms and less focused on the label of a student. "Even my low EL learners can verbalize these things [science understandings]. You have to expect them to because sometimes it is just the language and not that they aren't thinking these things in their minds." Teachers often commented on the belief that their students can have a good understanding of the science, but be limited in their ability to express that thinking by their language ability. In other words, a limited student response might represent limited English skills rather than limited conceptual understanding. For example, although our teachers believed sentence frames to be essential scaffolds for students with limited language skills; they grew to understand that the sentence frames they provided limited student responses and resulted in student work that failed to display the range of content understanding. This critical insight led teachers to explore additional measures of student understanding (especially for students with beginning language skills) that were not as language dependent – developing assessments that included graphic organizers, pictures, and asking students to physically manipulate materials.

Student Achievement

Of course, teachers and administrators didn't just tell us about the benefits of the program, they also told their peers at other schools and the science/ELD blended lesson design spread across the district. But this spread, not to mention the sustained enthusiasm of our participants, may not have occurred if it not for the program's impact on student achievement data. Student achievement was measured using existing state-mandated tests in English Language Arts and English Language Development. The number of students in the sample varied depending on the number of students who were present and completed the assessment each year. However, all students at the three treatment and two comparison schools with a valid score on these assessments were included in the sample. Comparison schools were chosen based on similar student demographics (socio-economic status, ethnicities, percentage of ELLs) and previous performance on state assessments. In the analysis of student achievement data, a response variable of mean improvement from a baseline year was used. Baseline was determined by the year a student started at the school site. There is not one baseline, but rather multiple ones corresponding with each student's arrival at the site (i.e., when they began the program). This provided a richer sample for analysis than using a single baseline for analysis allowing all students who had a score on any measure to be included in the sample. For example, for a 1st grader who began in the school year 2006–2007 the analysis followed the improvement from 2007 to 2008 (Year 1 improvement) to 2010–2011 (Year 4 improvement). Since the analysis used student proficiency levels, an ordinal variable, non-parametric statistics were used. Group statistics and Mann–Whitney U tests were performed on state assessment data to compare differences between the comparison and treatment schools. A Bonferroni correction was used to help reduce the overall type 1 error rate to 5 %.

This project began when, in desperation, schools were willing to remove their district's established and widely used curriculum in favor of a novel approach to both elementary science instruction and English language development. In so doing, we essentially "stole" instructional minutes from second language acquisition to make room for science, a subject that, prior to our project, was rarely taught. Honestly, we would have considered our project a success if participating students simply continued to develop their English language skills at a rate similar to those of students who used the state-adopted English Language Development program. Instead, results from student assessments indicate that the English language proficiency of students in the blended program, when compared to students participating in the traditional ELD program, actually improved. Although gains were modest, improvement was seen across multiple indicators and through different means of data analysis (overall proficiency, sub-skills, multiple years of treatment).

There were statistically significant gains in student performance on the California English Language Development Test for students with 1, 2, 3 and 4 years of participation ($U=4.226$, $U=5.205$, $U=5.134$ and $U=5.321$, respectively and $p=0.000$ for all years). Significant improvements were seen in student performance on the California State Test, English Language Arts for 1, 2 and 4 years of participation ($p=0.000$, 0.004, and 0.040 respectively). In particular, participation in the blended program appears to have had a positive affect on students' oral language development (i.e., listening and speaking). For many in our district, this was the proof that mattered and it has helped to sustain science instruction at our participating schools and others in the district.

Summation

This project developed a successful method for improving K-4 students' English language skills. Success was a direct result of the blended lesson design's focus on creating opportunities for students to work collaboratively, discussing and debating their ideas with evidence from scientific investigations. Student-to-student dialogue is a major component of the blended lesson design, as scientist-to-scientist dialogue is a central component of the scientific enterprise. These opportunities need to be carefully crafted to allow students space to explore new science concepts and using manipulatives and other realia (Lee et al. 2013; Snow 2010). Teachers should provide language scaffolds, but these carefully crafted language supports should not interfere with the scientific inquiry central to the construction of new scientific knowledge. Vocabulary essential for participation during science investigations (hard, blue, smooth) is front-loaded and language frames are provided, but these are designed to support authentic scientific inquiry and maintain the central role of student thinking within instruction. Learning occurs best when students feel safe to share their developing science ideas, with whatever communication skills they possess, including possibly imperfect language. The lesson design should provide linguistic supports that allow space for student ideas to develop and promote communication while still acknowledging student contributions for their value within scientific discourse. This provides a more authentic and rich environment for both science and language development. In this project, we did more than simply replace science topics for the existing topics in the ELD instructional materials; we attempted to integrate the best of both science instruction and English language development.

The successful blending of inquiry science and language development requires a significant level of skill and knowledge. Science specific pedagogical content knowledge is needed to identify the optimal moments to support language within science while keeping the inquiry and rigor of the science intact. Which language forms or functions are necessary for students to fully engage in the science learning and which would stifle their explorations are decisions best made by teachers who possess great knowledge of second language development and command a deep understanding of science content. Therefore, although some measure of success could be achieved without it, extensive teacher professional development is needed to optimize use of the blended science/ELD lesson model. With this support, science classrooms can be rich language learning environments where ELLs use and develop language to make sense of scientific phenomena around them.

Appendix: Planning Sequence of 5E/ELD Lesson Design

In teachers' use of the science/ELD blended design template a specific sequence was followed. This sequence begins with identifying the development of a science concept through the 5Es (Step 1). Teachers then develop details of an inquiry science lesson designed to achieve each conceptual goal independent of language objectives (Step 2). Finally, teachers modify the lesson by adding appropriate ELD support (Step 3). This sequence is illustrated in the tables below. For Steps 2 and 3, only the Engage phase of the lesson is shown. For further details regarding this sequence see, Gomez Zwiep et al. 2015).

Step 1: Plan conceptual storyline of each E

	Teacher	Student	Science concept
Engage			SC: Sounds can be heard all around us
			Sounds have different qualities
Explore			Sounds are made by vibrations. Changing the vibrations can change the sound
Explain			Vibrations cause the sounds to be created. Different kinds of sounds can be made from the vibrations
Extend			Sounds can be high or low (precursor to pitch)

Step 2: Develop science lesson sequence and predict student responses

Teacher	Student responses	Concept
I want everyone to close your eyes and listen to all of the different sounds that you hear		SC: Sounds can be heard all around us
		Sounds have different qualities
Give students 30 s to a minute to listen for sounds. (If the school area is particularly quiet, make some sounds like crumpling up a piece of paper or banging a trash can.)		
What were the sounds like?	Bird, Boys, Talking, Bugs, Cars	

Step 3: ELD supports: Identify appropriate language function match; insert appropriate language scaffolds; adjust Expected student responses for proficiency levels of students in the class

ENGAGE: Teacher	Low	Med	High	Science concept/ Language function
I want everyone to close your eyes and listen to all of the different sounds that you hear Give students thirty seconds to a minute to listen for sounds. (If the school area is particularly quiet, make some sounds like crumpling up a piece of paper or banging a trash can.)				SC: Sounds can be heard all around us Sounds have different qualities LF: Describing
Turn to your partner and tell him or her what sounds you heard *Partner A will tell partner B one thing they heard* *Then, partner B will tell partner A one thing they heard* *Keep going until you have shared all the things you heard*		The birds outside, Students next door, Flies buzzing	I heard whispering, It was loud yelling, The buzzing was tiny.	
(Students take turns sharing with their partners).	Bird, Boys, Talking, Bugs, Car	It was soft		
What were the sounds like? Turn to your partner and describe the sounds *Who can share with the class something their partner shared?* Record the types of sounds on the board/ graphic organizer as students share			I heard students in the class next door.	
Bird chirping, talking, cars, buzzing, loud, quiet, soft 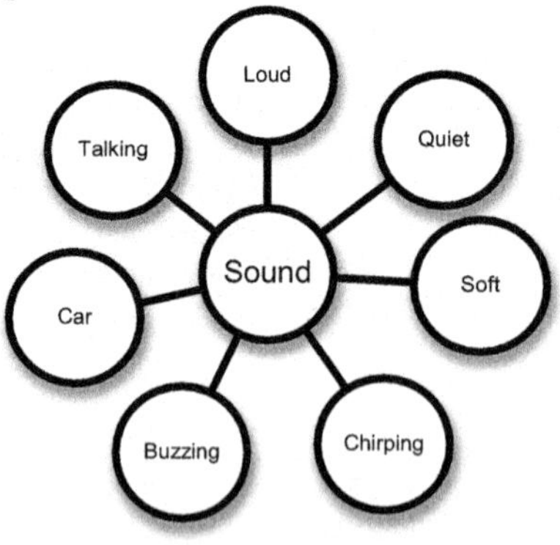	Student share ideas from their partner talk.			
Great! So there are different types of sounds around us. Let's find out more about sounds.				

References

American Association for the Advancement of Science (AAAS). (2009). *Benchmarks for science literacy: Project 2062*. New York: Oxford University Press.

Brown, Z. A., & DiRanna, K. (2012). *Equal access to content instruction for English learners: An example from science*. San Francisco: Region IX Equity Assistance Center at WestEd.

Bybee, R. W. (1997). *Achieving scientific literacy from purposes to practices*. Portsmouth: Heinemann.

California Department of Education. (2015). *Data and statistics*. Retrieved November 18, 2015, from http://www.cde.ca.gov/ds/

Charmaz, K. (2002). Qualitative interviewing and grounded theory analysis. In J. F. Gubrium & J. A. Holstein (Eds.), *Handbook of interview research: Context and method* (pp. 675–694). Thousand Oaks: Sage.

Cummins, J. (2008). BICS and CALP: Empirical and theoretical status of the distinction. In *Encyclopedia of language and education* (pp. 487–499). New York: Springer.

DiRanna, K., Topps, J., Cerwin, K., & Gomez Zwiep, S. (2009). Teaching learning collaborative: A process for supporting professional learning communities. In S. Mundry & K. E. Stiles (Eds.), *Professional learning communities for science teaching: Lessons from research and practice* (pp. 34–54). Arlington: NSTA Press.

Dorph, R., Shields, P., Tiffany-Morales, J., Hartry, A., & McCaffrey, T. (2011). *High hopes–few opportunities: The status of elementary science education in California*. Sacramento: The Center for the Future of Teaching and Learning at WestEd.

Duffy, G. G. (2002). The case for direct explanation of strategies. In C. C. Block & M. Pressley (Eds.), *Comprehension instruction: Research-based best practices*. New York: Guilford Press.

Echevarria, J., Vogt, M., & Short, D. J. (2008). *Making content comprehensible for English learners: The SIOP model*. Boston: Pearson.

Gomez Zwiep, S., & Straits, W. J. (2013). Inquiry science: The gateway to English language proficiency. *Journal of Science Teacher Education, 24*(8), 1315–1331.

Gomez Zwiep, S., Straits, W. J., Stone, K., Beltran, D., & Furtado, L. (2011). The integration of English language development and science instruction in elementary classrooms. *Journal of Science Teacher Education, 22*(8), 769–785.

Gomez Zwiep, S., Straits, W., & Topps, J. (2015). Building inquiry –based science lessons: An authentic context for English language development. *Science and Children, 53*(2), 67–73.

Lee, O., Buxton, C., Lewis, S., & LeRoy, K. (2006). Science inquiry and student diversity: Enhanced abilities and continuing difficulties after instructional intervention. *Journal of Research in Science Teaching, 43*(7), 607–636.

Lee, O., Quinn, H., & Valdes, G. (2013). Science and language for English language learners in relation to next generation science standards and with implications for common core state standards for English language arts and mathematics. *Educational Researcher, 42*(4), 223–233.

National Research Council. (2012). *A framework for K-12 science education: Practices, crosscutting concepts, and core ideas*. Washington, DC: National Academy Press.

Osborne, J. (2014). Teaching scientific practices: Meeting the challenge of change. *Journal of Science Teacher Education, 25*(2), 177–196.

Snow, C. (2010). Academic language and the challenge of reading for learning about science. *Science, 328*, 450–452.

Stoddart, T., Pinal, A., Latzke, M., & Canaday, D. (2002). Integrating inquiry science and language development for English language learners. *Journal of Research in Science Teaching, 39*, 664–687.

Yore, L. D., Florence, M. K., Pearson, T. W., & Weaver, A. J. (2006). Written discourse in scientific communities: A conversation with two scientists about their views of science, the use of language, role of writing in doing science, and compatibility between the epistemic views and language. *International Journal of Science Education, 28*(2–3), 109–141.

Chapter 10
Doing and Talking Science: Engaging ELs in the Discourse of the Science and Engineering Practices

Rita MacDonald, Emily Miller, and Sarah Lord

Introduction

The Next Generation Science Standards (NGSS; NGSS Lead States 2013), and the Common Core State Standards (CCSS; National Governors Association Center for Best Practices and Council of Chief State School Officers 2010) shift teaching and learning across the US to focus on disciplinary, language-rich practices, with broad implications for teaching English learners (ELs). The NGSS calls students to engage deeply and actively in the exploration and discussion of ideas by enacting three interacting dimensions: practices, core ideas, and cross-cutting concepts. Three-dimensional science learning engages students in scientific and engineering practices as they explore phenomena to develop interdisciplinary science ideas in relation to cross-cutting concepts. Similarly, the CCSS (which include standards for literacy in science, as well as in other technical subjects) increased emphasis on critical thinking, problem solving, and analytic tasks in core academic subjects. Together, these standards "implicitly demand students acquire ever-increasing command of language in order to acquire and perform the knowledge and skills articulated" (Council of Chief State School Officers 2012, p. ii). Yet, at a time when the EL population continues to be the most rapidly growing segment of the K-12 student population, instruction of ELs is too often characterized by three persistent problems of practice, each of which we observed in our pre-intervention visits to classrooms:

1. In whole group work, teachers used primarily IRE (teacher *Inquires*, student *Responds*, teacher *Evaluates* by indicating whether that response is correct or

R. MacDonald (✉) • E. Miller • S. Lord
University of Wisconsin-Madison, Madison, WI, USA
e-mail: rkmacdonald@wisc.edu; emilycatherine329@gmail.com; salord@wisc.edu

A.W. Oliveira, M.H. Weinburgh (eds.), *Science Teacher Preparation in Content-Based Second Language Acquisition*, ASTE Series in Science Education,
DOI 10.1007/978-3-319-43516-9_10

not) interaction patterns (Schegloff 2007) that focused attention on teacher ideas rather than on student ideas.
2. In collaborative groups, student discourse tended to be focused on procedures and task accomplishment, rather than on meaning-making, and either excluded ELs altogether or placed ELs in the role of listener.
3. Language development was viewed primarily as vocabulary instruction.

Classroom practices such as these are not likely to foster the rich academic discourse through which students learn to reason deeply and critically, express their reasoning, and challenge and critique that of others, nor are they likely to include ELs in that critical discourse. The need for resources to support effective engagement of ELs in these essential academic discourse practices is critical.

This chapter shares findings and materials from the pilot of a professional development (PD) approach that offered science teachers a set of resources to support their facilitation of students' collaborative and discourse-rich reasoning in science, along with the development of the language needed for these critical functions—all of this in ways fully inclusive of ELs as sense-makers along with their classmates.

Participants were four teachers in two schools in a Midwestern school district. Two taught science as part of their Grade 4 curriculum, and two taught science in Grade 7. Although the state had not adopted the NGSS, district administrators had expressed a desire to improve the science outcomes of ELs in the district.

Teacher Preparation Model

Stages of PD and Related Inquiry

All of the teachers were new in at least one significant dimension related to their teaching. Three of the four were teaching science for the first time and had not minored in a science-related field in their preservice training; one of these had just begun her first year of teaching. The fourth teacher had taught science before but was new to the fourth grade. All four were unfamiliar with their science curricula. None of the science curricula in use was inquiry-based. No teacher had more than 6 ELs in classes that averaged 24 students (a common distribution in many non-urban school districts), and the ELs ranged in English proficiency levels from Beginner to Advanced, based on teachers' reports of the annual ESL assessment results.

All teachers participated in a half-day PD on the NGSS and three-dimensional science learning, and on the integrated nature and enactment of the science and engineering practices. The teachers were observed teaching one science lesson, and then interviewed about their learning objectives and foci in the observed lesson and their reflections on student engagement and sense-making. Following this initial observation, teachers participated in a 2-day PD focused on (a) an assets-based approach to EL inclusion in science, (b) the development of opportunities for collaborative sense-making, (c) enactment of the language-intensive science and

Table 10.1 PD and inquiry stages and activities

Participants
2 Grade 4 science teachers
2 Grade 7 science teachers
Half-day PD: NGSS three-dimensional science and enactment of science and engineering practices
Classroom observation 1
Post-observation interview 1
Two-day PD
Assets-based approach to EL inclusion in science
Developing opportunities for collaborative sense-making
Science and engineering practices 2 and 6 (developing models and constructing explanations)
Discourse engagement strategies & resources
Structured lesson reflection 1
20-min phone call
Structured lesson reflection 2
20-min phone call
Structured lesson reflection 3
20-min phone call
Classroom observation 2
Post-observation interview 2
Focus group to explore aspects of resources provided

engineering practices of modeling and explanation (Practices 2 and 6), and (d) resources for the facilitation of students' collaborative sense-making discourse. Subsequently, teachers spoke monthly over 3 months with PD providers via telephone to discuss a Structured Lesson Reflection document the teachers had sent the PD providers the day before. The 20-min phone calls were used to probe various aspects of teachers' ideas and reflections more deeply, and to provide information or suggestions as requested by teachers. After approximately 3.5 months, all teachers were observed and interviewed once again, using the same protocol and similar questions. As a final stage of the information gathering, teachers were interviewed in more detail about their reactions to the resources and PD provided (Table 10.1).

Theoretical Foundation of the PD

Figure 10.1 depicts the assets-based approach to EL inclusion in which this project was grounded. This approach recognizes that ELs come to their science classrooms with multiple ideas about how the world works (green strand), as well as with knowledge about one or more languages in addition to varying degrees of effectiveness with English (blue strand). Given these strengths, they are well able to engage in scientific reasoning and discussion of their reasoning. If educators are successful

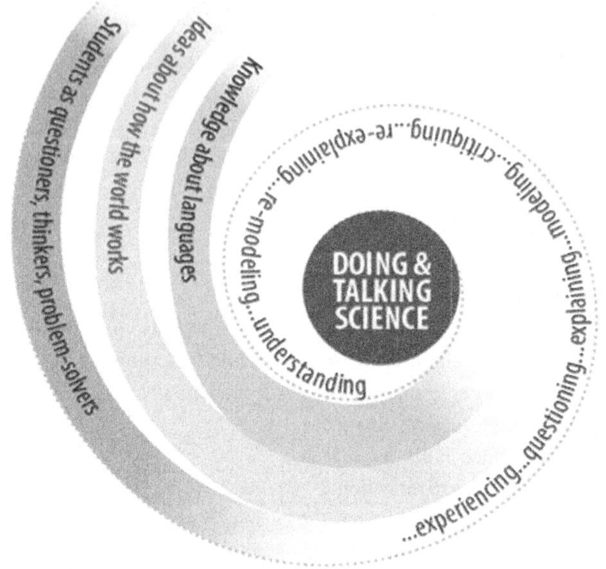

Fig. 10.1 Assets-based inclusion of ELs in explanation and modeling in science

in tapping into and leveraging those assets and capacities by positioning students as questioners and thinkers and themselves as facilitators of student reasoning (purple strand) and by engaging ELs with their classmates in the collaborative sense-making practices of science (the words spiraling around the center), both ELs' knowledge of science and their linguistic effectiveness in science will be strengthened. (See also Lee et al. 2013).

The project's focus on ELs as sense-makers in science, along with their English-fluent classmates, is grounded in a *language as action* perspective (van Lier and Walqui 2013). This contrasts with an accumulation model that considers the development of academic English as the building up of progressively more complex syntax and vocabulary to (eventually) accomplish a broader range of functions. This accumulation of necessary linguistic resources is seen as an inner, cognitive event that progresses slowly and sequentially—a perspective often aligned with a deficit model. In an accumulation model, students first come to know (language) and then they do (science). The *language as action* approach views the process quite differently: By doing (science) together, students come to know (language). In other words, language is seen as action and developed through action, and more specifically, through action that occurs among individuals in a shared context. In this sociocultural approach, meaning does not reside solely in language, but is a larger construct developed through negotiated and shared experiences during which participants construct and represent meaning together, only in part, through language (Gee 2005; Rogoff 2008; MacDonald & Molle 2015). Put simply, meaning is not stored language; meaning is stored experience.

In this project, as in the *language as action* approach, shared activity is seen as the engine that drives language development. All students, including those still developing English, are given the opportunity to engage in collaborative reasoning, and are expected and supported to be active sense-makers. In this approach, ELs have the opportunity and support to be initiators of ideas, along with their classmates, rather than simply passive responders. Language development for all students is thus deeply contextualized within interactive sense-making, and instructional attention is focused on students' effectiveness at marshaling the diverse sense-making resources (linguistic and other) they command, rather than on the correctness of their language. For the rapidly growing number of ELs in US classrooms who may require years of English language development before their language is fully proficient, this is an important and supportive shift. ELs can, and do, engage in important reasoning and learning with imperfect language and it is this "doing" that supports their progress toward more effective and, eventually, more correct or more appropriate English.

These affordances of the *language as action* approach align well with the language expectations of the NGSS and three-dimensional science, as illustrated by the following comments:

- "For all students, the emphasis should be on making meaning, on hearing and understanding the contribution of others and on communicating their own ideas in a common effort to build understanding …." (Lee et al. 2013, p. 3).
- "Essentially all of the science and engineering practices require student discourse to be a central element of classroom activity, and, properly managed by the teacher, such discourse includes all students and pushes every student to refine and extend language abilities." (Quinn 2015, p. 14).
- "Only an emphasis on language as action … engages students in the meaningful learning of new disciplinary practices while simultaneously strengthening their language uses in those practices." (Heritage et al. 2015, p. 32)

Efforts to strengthen students' reasoning in science are not easily supported using an atomistic view of academic English as the accumulation of complex syntax and vocabulary. Indeed, as stated by Heritage et al. (2015), "teaching form and function in isolation from real, meaningful, discourse-based communication has not produced generative, transformative learning for ELLs" (p. 31). The *language as action* perspective does, however, focus attention on students' meaning-making and their linguistic effectiveness during interaction with one another around important ideas in science. These examples (Miller and MacDonald 2015) illustrate the important differences in the approaches.

Language goals based on the *form and function* or *accumulation* model:

- Students will compare landforms using descriptive language.
- Students will describe the molecular changes that occurred using the past tense '-ed' form.

Language goals based on the *language as action* approach:

- Students will collaboratively develop a model that explains and predicts patterns in the changes to the land caused by wind and rain.
- Students will collaboratively construct an explanation of the effect of thermal energy on molecular movement.

PD Components

To support teachers in working with the *language as action* perspective and our assets-based approach to EL inclusion in collaborative reasoning in science, the 2-day PD was spent considering elements of the approach and practicing the use of a small set of resources.

Enacting the Science and Engineering Practices Although the disciplinary core ideas of the NGSS are familiar to many, and their relationship to cross-cutting concepts fairly straightforward, the science and engineering practices are less familiar to teachers and require significant changes in science instruction (Windschitl et al. 2011; Lee et al. 2013). The PD focused specifically on two high-leverage practices for ELs that were to be implemented jointly: explanation (because of its language demand) and modeling (to demonstrate the use of models as scaffolds during meaning-making). Facilitators modeled classroom enactment of meaning-making by placing teachers in small groups to consider phenomena shown on video, collaboratively develop models depicting their reasoning about causal forces, and then explain their reasoning, using the models as references. During teachers' explanations, facilitators modeled the Teacher Moves as examples of probing and deepening reasoning.

Creating Opportunities for Collaborative Reasoning Following this demonstration of a collaborative meaning-making activity, PD focused on the role shifts required for both teachers and students when working to strengthen student reasoning in science, summarized in Table 10.2. Given teachers' lack of relevant curricular support materials, considerable PD time was devoted to discussing the benefits of using locally relevant, easily observable phenomena (accessible to ELs) around which to center student reasoning opportunities. A list of such phenomena and their relationship to NGSS disciplinary core ideas and cross-cutting concepts was generated. Teachers were given time and support in selecting a phenomenon with which to initiate an upcoming science unit.

Changing Classroom Interaction Patterns Strengthening students' collaborative reasoning and the language through which much of it is expressed and deepened calls for changes to typical classroom interaction patterns. Much more student talk is needed than typically occurs in many classrooms. The commonly used IRE pat-

Table 10.2 Consideration of changes to teacher and student roles in science

Teacher role: Shape the discussion to promote collaborative meaning-making	**Student role:** Work with classmates to understand unseen forces behind phenomena
Create the need to interact meaningfully	Be responsible for following ideas; listen carefully and track the idea's development
Facilitate students' collaborative meaning-making	Check for accurate understanding of others' statements; persist until clear mutual understanding is achieved
Model effective language as needed and discuss reasons for linguistic choices	Consider the ideas of others as sensible first, and then take up the idea or discard the idea based on evidence
Design for ELs to be initiators as well as responders in meaning-making interactions	Compare evolving explanations to other information; does it make sense? Is something missing?
Promote student-to-student interactions	Respond to ideas; support or challenge or build on ideas
Support perseverance in understanding and meaning-making	

tern may move a class quickly through a review of known information (known, that is, to the teacher) or may move a class toward the teacher's predetermined goal, but it provides few opportunities for students to talk. Student input is usually constrained to very truncated responses, and those for only the few students able to formulate responses very quickly—a group from which ELs are frequently excluded. IRE exchanges offer few opportunities for students to use language to express and engage in extended reasoning. This project focused on three ways to increase student opportunities for meaningful language use: (1) the use of small group work to focus on challenges in reasoning, rather than on task accomplishment, and in ways that ensure full participation by all group members, including ELs; (2) the use of Teacher Moves to promote more extensive discussion and include additional students in reasoning-focused whole-class interactions; and (3) the use of Teacher Moves and Student Moves to promote increased student-to-student reasoning-focused interchanges during whole class time and small group work. Teachers' enactment of this approach was further supported during brief monthly contacts.

Materials

Teacher Moves: Discourse Facilitation Moves for Teachers

Although teacher education literature has focused attention on supporting teachers in learning more student-focused interaction patterns (Chapin et al. 2003; Michaels and O'Connor 2012; Windschitl et al. 2011), these resources are not yet well known by teachers. Given their critical role in our approach, a small set of discourse

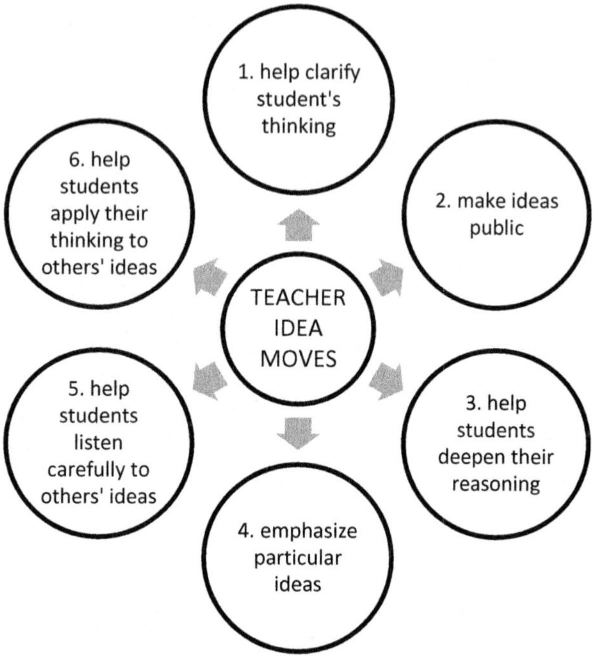

Fig. 10.2 Meta-cognitive framework for teacher moves

facilitation moves was created and organized to form a compact, six-category set of Teacher Moves. A graphic illustrating the six different purposes by which the Teacher Moves were organized was developed (shown below in Fig. 10.2), to serve as both a meta-cognitive framework and a quick visual reminder to teachers of their options when student ideas were on the table. In each category, examples were provided, some of which are shown in Table 10.3, below.

Teacher Moves serve three purposes, which can be considered sequential:

1. Clarify individual student ideas and surface them for consideration by the group (Moves 1 and 2)
2. Probe and deepen expressed reasoning (Moves 2, 3, and 4)
3. Promote student-to-student interchanges (Moves 5 and 6)

The Teacher Moves all support teacher efforts to extend additional invitations for student talk and reasoning. By not closing down interactions with the typical IRE third move of Evaluation, but instead asking another question or bouncing the idea to another student, the teacher provides additional opportunities for students to reason and to express their reasoning (Greeno 2015). During the PD, it was suggested that teachers take up one or two moves at a time, focusing on the sequential nature and allowing themselves and their students time to adjust to new expectations for classroom interaction.

Table 10.3 Examples of teacher moves

Teacher moves	Examples
1. Help clarify students' thinking	Provide individual thinking time and pair activities to help students express the "first draft" of their idea
	Charge student pairs with questioning and supporting one another until ideas expressed are understood
	Provide 10–20 s of wait time both before and after student responses
	"Can you show us what you mean?" "Can you draw that?" "Can you say more about that?"
2. Make idea public and available for discussion	"Tell us more about what you're thinking."
	Revoice an idea to repair or model clearer language, but ensure that the ownership of the idea remains in student's hands. "Did I say your idea correctly? Is that what you were thinking, or was it different?"
3. Help students deepen their reasoning	"Can someone give me an example of that?"
	"How could we test that?"
	"What do we need to know more about now?"
4. Emphasize particular ideas	Attend to all ideas, and be explicit about putting some on hold.
	Re-broadcast generative ideas by revoicing, or by asking a student to paraphrase. This allows additional processing time for all.
	"That's interesting. Can you say that again for us?" "Will someone re-tell that idea for us?" "So, are you saying that…?"
5. Help students listen carefully to others' ideas	"Who can restate that for us?"
	"Who wants to explain the reasoning Group A used?"
	"How is that idea different from Mary's?"
6. Help students apply their thinking to others' ideas	"You look uncertain. What can you ask X to find out more?"
	"How does that idea connect to what Group A talked about?"
	"Which explanation is most like your group's? Talk to them and find out how they are different."

Student Moves: Discourse Engagement Moves for Students

Recent research and writing on academic discourse has focused attention on strengthening students' linguistic expression of complex thinking. In particular, the work of Zwiers et al. (2014) has provided examples of discipline-specific Constructive Conversation Skills posters to provide students with reminders of linguistic structures they could use to enact important academic tasks, such as the generation of multiple approaches and the negotiation of ideas. The Student Moves tool developed for this project was focused more broadly on seven general responses students could make to an idea. To support all students in exercising their agency as speakers in collaborative reasoning interactions, the Student Moves tool, like the Teacher Moves tool, included a graphic representation of the meta-cognitive framework (Fig. 10.3) and linguistic examples to accomplish these seven types of responses to ideas (Table 10.4). To support ELs' inclusion, the language examples were written for three broadly conceptualized levels of English language

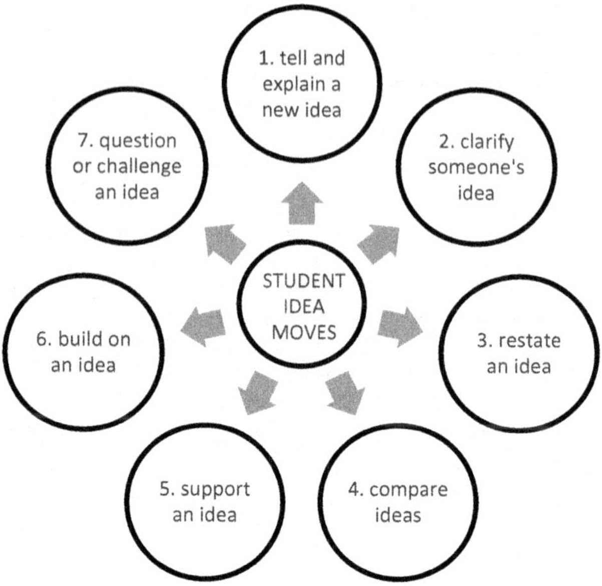

Fig. 10.3 Meta-cognitive framework for student moves

Table 10.4 Linguistic examples of student moves

Student Moves	Examples
1. Tell and explain a new idea	"I think…"
	"The evidence for that is…"
	"Since both situations are similar, we could…"
2. Clarify an idea	"Say again, please."
	"What did you mean when you said…"
	"I wonder if what you're saying is…"
3. Restate an idea	"He said…"
	"In other words, …"
	"The suggestion was made that we…"
4. Compare ideas	"Same thing."
	"Our idea is better because…"
	"The other method would be a better test of …"
5. Support an idea	"Good idea because…"
	"Remember, in our book it said…"
	"The advantage of that method would be …"
6. Build on an idea	"Let's try it."
	"That's what we should do next."
	"That idea would help us figure out whether …"
7. Question or challenge an idea	"I don't think so."
	"But what about…"
	"Isn't there a more efficient way to…"

proficiency, based on the Reference Performance Level Descriptors designed to include or translate English language proficiency definitions across most US states (Cook and MacDonald 2014).

Since academic discourse requires interaction among speakers, one of whom is often the teacher, the Student Moves were designed to work in tandem with Teacher Moves by providing linguistic resources for students to respond to the teacher's discourse facilitation and to collaborate with one another during small group work. During the PD, it was suggested that teachers first introduce the meta-cognitive framework to students, then a few sentence frames, and that they generate additional sentence frames with students, as well as capturing examples of student-generated moves when they occurred. It was also suggested that students have some personal representation of the Student Moves available, rather than being dependent on classroom posters, to increase their ownership of the Student Moves and support their independent action in small group work.

Implementation

Participating teachers devoted considerable time and effort to creating or adapting classroom activities to provide meaningful opportunities for collaborative, extended student reasoning.

- Grade 4 teachers adapted a scripted ball and ramp activity originally intended to demonstrate ideas of force and motion by adding an additional variable (changes in ramp height) and asking students to model and explain the forces at work. After students compared models and explanations, they were asked to collaboratively develop ways to test their ideas. One teacher noted how pleased she was to hear her ELs debating alternate ideas with their peers.
- Grade 4 teachers adapted an activity that involved shooting materials into the air with levers of different length (focused originally on providing data with which to practice graphing skills, with little focus on reasoning about the relationships) to enable students to reason further about relationships between potential and kinetic energy.
- Grade 7 teachers focused attention on a local phenomenon (the daily, early morning observation of clouds of water vapor over a heavily forested bluff) to introduce a unit on transpiration in plants. Over consecutive days, student groups discussed and developed models that they shared and then revised. One teacher noted how actively his ELs (who had formerly been in an EL-only group) participated with others, and how patient and helpful their peers were when ELs were introducing and explaining their ideas.
- Grade 7 teachers introduced a unit on ecosystem carrying capacity by creating a predator-prey game, in which wolves were the predators, linking to local concerns about wolf predation. All students played various parts (vigorously and noisily!) and noted the differential outcomes when ratios of predators, decomposers, etc. were changed. During the activity, students were heard explaining

excitedly to one another their understandings that if they partnered up in various ways, they could prolong their survival. Although both teachers were new to science, they expressed a strong belief that this activity resulted in greater engagement with the science than would the reading assignment suggested as an opening activity in their textbook. One teacher also noted how pleased she was that all of her students learned the associated vocabulary "simply through using it," and that she no longer believed she had to pre-teach vocabulary to her ELs.

As teachers gained confidence in developing phenomenon-based activities, they began to use a greater variety of Teacher Moves. Reflecting on her own learning, one teacher noted that when she failed to allow sufficient wait time for students to think and respond, her attempts to support student reasoning were always unsuccessful. She remarked that she used to believe that classroom interactions needed to happen at a rapid pace, so she would not lose students if they got bored, and also noted that her nervousness as a new teacher made it difficult for her to endure silent moments. However, at the end of this project, she observed that her prior belief and practice were interfering with her students' opportunity to think deeply and critically, and she resolved to work toward adjusting her practice. This teacher's remarks also serve as an illustration of a changed perspective on engagement: from engagement as behavior to engagement as reasoning.

Outcomes

Initially, like their district and school colleagues, teachers used primarily teacher-fronted lessons based on textbook chapters or on scripted activities that demonstrated rather than explored science constructs, and used classroom interaction patterns characterized by whole group lecture and classic IRE/F interactions. Following the 2-day PD, teachers in the participant group began to make significant changes.

Changes in Classroom Structures and Activity

Grade 7 teachers began to place students into small groups focused on the collaborative development of questions, models and possible explanations. Grade 4 teachers, who had already placed students in functional, task-focused groups, changed the focus of small group work from completing worksheets to students' collaborative development of explanations. Additionally, teacher comments indicated a change in what they considered engagement. Initially, engagement was seen as students being on-task and not disruptive, but later comments suggested that teachers considered engagement to be students' cognitive engagement with the ideas being discussed. All teachers reported an increase in student engagement in science.

Changes Specific to ELs

In one classroom in which all ELs had previously been grouped together with an ESL paraprofessional, ELs were integrated into small working groups with their English-fluent peers. All teachers reported being pleased at how well ELs were able to participate in the activities with minimal additional scaffolding by the teacher, and two of the four remarked how pleased they were to find that ELs were able to learn new vocabulary through using it in the midst of activities, and that they no longer felt the need to use class time to pre-teach vocabulary. In describing the changes observed, teachers remarked:

> One of our struggling ELs took the risk to share an idea he was not certain about, and then kept talking to work through his thinking again – all in front of the whole class.
>
> Our ELs view themselves differently because they're able to talk about ideas now. That's made a huge impact on their perceptions of themselves as learners. They've always been smart, but now I think they feel smart.

Discourse Facilitation Tools and Opportunities for Student Reasoning

Teacher Appropriation of the Discourse Facilitation Tools Review and coding of interview transcripts and field notes revealed an interactive relationship among the use of the separate discourse tools and teachers' success at creating meaning-making opportunities, as depicted in Fig. 10.4.

When able to create effective opportunities for student reasoning (experiences and driving questions that stimulated rich, extended discussion of ideas), teachers were more likely to use a variety of Teacher Moves to probe and deepen students' reasoning. When the attempted meaning-making opportunities were less rich

Fig. 10.4 Interaction of student reasoning opportunity with discourse moves

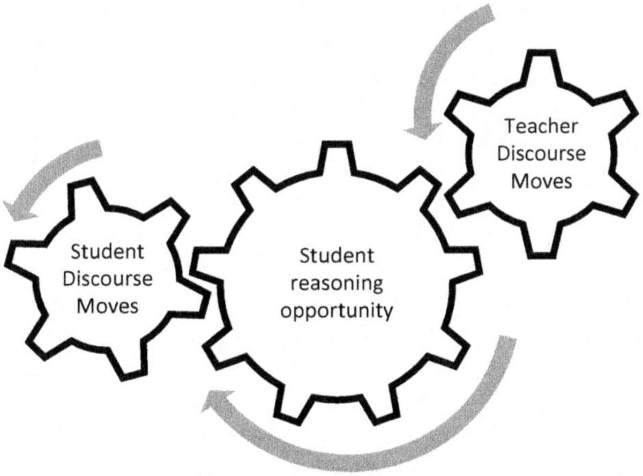

(e.g., fewer or less complex ideas about which to reason collaboratively), teachers used fewer and less varied Teacher Moves, sometimes simply repeating "Why?" in response to students' proffered explanations. The interactive nature between opportunity to reason and Teacher Moves is apparent. With nothing meaningful to explore or about which to reason collaboratively, the student collaborative reasoning process is too short-lived to require Teacher Moves, and teacher attempts to gain experience in using these discourse facilitation moves fall flat.

Conversely, if rich opportunities for sustained collaborative reasoning are provided but are met only with surfacing, introductory discourse moves (e.g., the repetition of "Why?"), the resulting classroom discussion resembles the "popcorn" pattern in which individual ideas are neither examined nor set in relation to one another for further exploration by students. Although even this introductory Teacher Move does result in increased exposure of student ideas (a desired outcome), if not followed by Teacher Moves that lead students to consider and react to others' ideas, classroom interaction does not move in the desired direction of strengthening students' collaborative reasoning. Thus, teachers' opportunities for successful experience with the Teacher Moves requires the creation, first, of a student experience that has the potential to stimulate sustained reasoning. With such a component in place, teachers have the opportunity to practice and develop effectiveness in their discourse facilitation skills.

Challenges in Providing Meaningful Opportunities for Student Reasoning The teachers in our project (none of whom were able to draw on robust experiences in teaching science) found it challenging to develop effective meaning-making opportunities for students. Lacking relevant curricular support materials, teachers visited multiple websites and resources to find and vet activities to fit their curriculum. Grade 4 teachers modified the scripted activities in the school's commercial science activity kits to stimulate the deep exploration of a phenomenon and the modeling of possible causal forces. However, the challenge of leveraging meaningful phenomena was especially difficult for the Grade 7 teachers, who taught three or four other subjects in addition to science, had only a traditional textbook series focused on the delivery and subsequent testing of information, and had few materials suited for hands-on student activity. In the third and final month of the project, these teachers were able to streamline the lesson revision process somewhat by collaborating to identify the big ideas and cross-cutting concepts to which their textbook units might be linked, and to search out activities based on the disciplinary core ideas and questions from The Framework for K-12 Science education (National Research Council, 2012).

The teachers in the project occasionally encountered difficulty in reasoning about science ideas. At times, the teachers appeared at a loss when attempting to negotiate the multiple student ideas expressed to develop deeper conceptual understanding. When they were not confident about the science concepts, they expressed uncertainty about which ideas to revoice or probe more deeply or set in relation to one another, and the facilitation around ideas reverted to IRE or declarative knowledge or definitions. Our hypothesis is that this breakdown of facilitation correlated with a lack of deep understanding of the phenomenon they were exploring with students. Because of the paucity of teacher materials that provided the conceptual frameworks involved in explaining phenomena, teachers would have benefited from additional resources and guidance to help fill in those gaps.

The amount of difficulty teachers encountered before they experienced some effectiveness in creating opportunities for student reasoning resulted in a sequential nature in their enactment of the discourse tools. It was only after teachers had achieved some degree of effectiveness in creating opportunities for student reasoning that they began to experiment more frequently with the variety of Teacher Moves. Thus, the PD pilot did not provide adequate opportunity for teachers to experience the interactive relationship among the three components: (1) reasoning opportunity, (2) Teacher Moves, and (3) Student Moves. The Grade 4 teachers did introduce the Student Moves to their students early in the process, using only the linguistic element of the Student Moves (the sentence frames) to establish norms for respectful classroom conversation. At the end of the 3 months, they had just begun to incorporate these moves into small group activities. Grade 7 teachers did not use the Student Moves at all, and at the end of the project, one teacher noted these as the next step and regretted that the PD resources to support their introduction of Student Moves in their classrooms would no longer be available.

This slower-paced and more sequential aspect to teachers' experimentation with the tools is in sharp contrast to what occurred with a small group of mathematics teachers engaged in a parallel pilot of these resources in a different district. This group of highly experienced mathematics teachers, familiar with their curricula and grade-levels, experienced the same need to develop meaning-making opportunities, but progressed more quickly to the point of effectiveness with this component and began almost immediately to practice their use of the Teacher Moves and to introduce Student Moves as tools for collaborative small group reasoning. These teachers were able to experience the benefits of all three components (opportunity, Teacher Moves, and Student Moves) working interactively, and noted the power of the positive classroom experience in heightening their commitment to the work of developing their discourse facilitation skills. One mathematics teacher discussed both her initial struggle in using the approach, and her increased confidence in their students' understanding:

> I work a lot harder now. Sometimes, it's just easier to go by the textbook and say, "OK, this is why it works—let me show you." But there's no connection, there's no meaning behind it. And that's the hardest thing, I think: to change that teacher behavior of having to control the conversation, and just give it up to the group to talk until they figure it out. There was one day students spent at least 20 minutes in a discussion of one idea, and it about killed me to spend that much time talking about it, but now, you could ask any kid on my team and they could explain it and tell you exactly why it's that way. I have never felt so confident that my students understand things, ever.

This contrast between the experiences of the science and mathematics teachers suggests that revision of the PD approach to include additional resources that would enable science teachers to more quickly experience the interaction of the three components should be considered.

Integration of Science and Engineering Practices

Although the science and engineering practices were unknown to teachers at the project onset, teachers quickly began to integrate Explanation and Modeling (Practices 2 and 6) into their lessons. All four teachers began to create opportunities for students to collaboratively consider phenomena and driving questions that elicited multiple possible explanations to be further examined, and began to incorporate the Teacher Moves to surface, probe, and deepen student explanations. All four teachers began to integrate the practice of modeling into their meaning-making focused instruction. Although not yet fully leveraging models as explanatory devices (Mayer and Krajcik 2015), teachers did note the value of drawn models as supports to which ELs could refer when not yet able to convey intended meanings solely with words. Thus, teachers' emergent integration of modeling into their lessons enabled ELs to more frequently and successfully join their peers in collaborative meaning-making.

Summation

Science teachers using this collaborative meaning-making approach with minimal support made significant changes in shifting their practice to focus on active engagement of students in the exploration and discussion of ideas, in ways that engaged ELs as sense-makers along with their classmates. The brief pilot of this approach, with its three interacting components (opportunity for student reasoning, teacher discourse facilitation moves, and student discourse engagement moves), offers resources and insights to help science teachers meet the critical need for materials and methods to enact the three-dimensional vision of science in ways that include the rapidly growing number of ELs in US classrooms.

Positive Effects of the Pilot

In relation the three persistent problems of practice noted in the introduction, the positive effects of this pilot suggest that:

- When given resources like the Teacher Moves and some support for the development of meaning-making opportunities for students, teachers can begin to change their interaction patterns to more actively engage student in interactions, and deepen student reasoning about ideas;
- When given a tool such as the Student Moves, with its meta-cognitive framework and language examples, ELs can and do join in collaborative meaning-making, acting as initiators of ideas rather than simply as passive responders. Similarly, when it is clear that ELs' ideas are being solicited and valued, non-EL classmates

persevere in efforts to comprehend ELs and to assist them in their explanations, thus enacting the negotiation of meaning-making that drives language development for ELs.
- Teachers enacting this approach can come to understand through their lived experience that ELs' language development is not dependent on decontextualized pre-teaching of language components, but can be supported in the midst of science instruction by engaging all students in the rich, collaborative discussion of ideas.

Possible Shortcomings of the Approach

Observing teachers' gradual appropriation of these resources has also pointed out shortcomings in the PD approach, which should be addressed by those wishing to follow up on this. The approach underestimated the degree of difficulty teachers would experience in creating opportunities for student reasoning. Although examples were provided, these were not sufficient to enable teachers to move quickly enough into trying out the Teacher Move and Student Moves. Thus, the integration of the three components did not occur quickly enough to enable these teachers to experience the benefits of all three components working interactively, which had heightened and seemed to hasten the development of confidence in the discourse facilitation efforts of a separate group of mathematics teachers in a related PD pilot. Future efforts might include the provision of sample activities related to grade-level units, to jump-start teachers' experience with the interaction of the three components. Additionally, it would be helpful to have at hand resources that provide accessible explanations of the science constructs related to a number of phenomena. Lack of teachers' confidence in their own science understanding may affect both their confidence in adopting interaction moves that open up the floor to student ideas as well as their ability to marshal those ideas toward a deeper understanding.

Further Considerations

Unaddressed in this pilot, but important to consider in more extended versions, is the need to provide teachers constructs by which to monitor and support students' English language development. The need to develop teachers' language awareness is present in any approach to content instruction for ELs. Those working from a "language as accumulation" approach are likely to focus on increased correctness. For those working from a *language as action* perspective, a different lens is needed. It should not be focused on correctness, but on effectiveness; it should support students in using English to more effectively explain and argue in support of their ideas. The components and dimensions of increasing effectiveness are worthy of continued consideration and exploration.

Pedagogical approaches that focus on increasing ELs' effectiveness in English while developing their science knowledge are critical to ELs' achievement in the new standards. The NGSS focus attention on a process-based goal, but they do not provide the pathway toward that goal. For that goal to be achieved, new approaches to teaching and learning are needed to inform curricula that are fully inclusive of ELs. Teachers need support and resources to enact the changes described in the NGSS, and to consider the additional and critical aspect of students' language development. The approach shared in this chapter can help teachers mediate the new standards into practice, for all students. This confidence was first expressed by science teachers at the end of a presentation of this approach at an NSTA conference. Several teachers remarked, "We know this is how we're supposed to teach, but nobody has shown us how to do it. This shows us how!"

Acknowledgments We would like to thank the National Science Foundation for its support of this work, funded by grant number DRL-1346491; H. Gary Cook (Wisconsin Center for Education Research) for serving as Principal Investigator and constant inspiration; and our colleagues Melissa Braaten (University of Wisconsin-Madison), Okhee Lee (New York University), and Judit Moschkovich (University of California-Santa Cruz) for their valuable contributions during the early development of these resources. We extend our gratitude to the unnamed school district leaders, teachers, and students with whom we learned during this research.

References

Chapin, S., O'Connor, C., & Anderson, N. (2003). *Classroom discussions: Using math talk to help students learn*. Sausalito: Math Solutions Publications.

Cook, H. G., & MacDonald, R. (2014). *Reference performance level descriptors: Outcome of a national working session on defining an "English proficient" performance standard*. Washington, DC: Council of Chief State School Officers. Retrieved from http://www.ccsso.org/Resources/Publications/Reference_Performance_Level_Descriptors.html

Council of Chief State School Officers. (2012). *Framework for English language proficiency development standards corresponding to the common core state standards and the next generation science standards*. Washington, DC: CCSSO.

Gee, J. P. (2005). Language in the science classroom: Academic social languages as the heart of school-based literacy. In R. K. Yerrick & W. M. Roth (Eds.), *Establishing scientific classroom discourse communities: Multiple voices of teaching and learning research* (pp. 19–37). Mahwah: Lawrence Erlbaum.

Greeno, J. (2015). Classroom talk sequences and learning. In L. Resnick, C. Asterhan, & S. Clarke (Eds.), *Socializing intelligence through academic talk and dialogue* (pp. 255–262). Washington, DC: AERA.

Heritage, M., Walqui, A., & Linquanti, R. (2015). *English language learners and the new standards: Developing language, content knowledge, and analytical practices in the classroom*. Cambridge, MA: Harvard University Press.

Lee, O., Quinn, H., & Valdés, G. (2013). Science and language for English language learners in relation to Next Generation Science Standards and with implications for Common Core State Standards for English language arts and mathematics. *Educational Researcher, 42*(4), 223–244.

MacDonald, R., Molle, D., & TESOL International. (2015). Creating meaning through key practices in English language arts: Integrating practice, content, and language. In L. C. de Oliveira,

M. Klassen, & M. Maune (Eds.), *The common core standards in English language arts for English language learners: Grades 6–12* (pp. 39–52). Alexandria: TESOL International.

Mayer, K., & Krajcik, J. (2015). Designing and assessing scientific modeling tasks. In R. Gunstone (Ed.), *Encyclopedia of science education* (pp. 291–297). Heidelberg/New York/London: Springer Dordrecht. p. 294.

Michaels, S., & O'Connor, C. (2012). *Talk science primer*. Cambridge, MA: TERC. Retrieved from http://inquiryproject.terc.edu/shared/pd/TalkScience_Primer.pdf

Miller, E., & MacDonald, R. (2015). *Rethinking language goals in science: Can three-dimensional learning allow us to shift our thinking around science learning and language goals?* Colorín Colorado TE.L.L.-EGRAM, Feb. 2015. Retrieved from www.colorincolorado.org/article/63672/

National Governors Association Center for Best Practices & Council of Chief State School Officers. (2010). *Common core state standards*. Washington, DC: National Governors Association Center for Best Practices/Council of Chief State School Officers.

National Research Council. (2012). *A framework for K-12 science education: Practices, crosscutting concepts, and core ideas*. Committee on a Conceptual Framework for New K-12 Science Education Standards. Board on Science Education, Division of Behavioral and Social Science and Education. Washington, DC: The National Academies Press.

NGSS Lead States. (2013). *Next generation science standards: For states, by states*. Washington, DC: National Academies Press.

Quinn, H. (2015). Science and engineering practices for equity: Creating opportunities for diverse students to learn science and develop foundational capacities. In O. Lee, E. Miller, & R. Januszyk (Eds.), *NGSS for all students* (pp. 7–20). Arlington: NSTA Press.

Rogoff, B. (2008). Observing sociocultural activity on three planes: Participatory appropriation, guided participation, and apprenticeship. In K. Hall, P. Murphy, & J. Soler (Eds.), *Pedagogy and practice: Culture and identities* (pp. 58–74). Thousand Oaks: Sage.

Schegloff, E. (2007). *Sequence organization in interaction: A primer in conversation analysis* (Vol. 1). Cambridge: Cambridge University Press.

Van Lier, L., & Walqui, A. (2013). *Language and the common core standards*. Stanford: Stanford University, Understanding Language Initiative at Stanford University. Retrieved from http://ell.stanford.edu/sites/default/files/pdf/academic-papers/04-Van%20Lier%20Walqui%20Language%20and%20CCSS%20FINAL.pdf

Windschitl, M., Thompson, J., & Braaten, M. (2011). Ambitious pedagogy by novice teachers? Who benefits from tool-supported collaborative inquiry into practice and why. *Teachers College Record, 113*(7), 1311–1360.

Zwiers, J., O'Hara, S., & Pritchard, R. (2014). *Common Core Standards in diverse classrooms: Essential practices for developing academic language and disciplinary literacy*. Portland: Stenhouse Publishers.

Chapter 11
Academic Language and Literacy in Every Setting (ALLIES+): Strengthening the STEM Learning Ecosystem

Susan O'Hara, Robert Pritchard, Deborah Pitta, Renee N. Newton, Uyen H. Do, and Lisa Sullivan

In order to succeed in school, all students need opportunities to develop the specialized academic language that is associated with content learning. For English learners (ELs) in particular, the development of academic language is one of the most important factors in academic success; where academic language is weak or missing, it is increasingly cited as a major contributor to gaps in achievement between ELs and native speakers of English (Anstrom et al. 2010; Francis et al. 2006). Academic-language development is also associated with student achievement as demonstrated by the correlation between measures of English-language proficiency and content-assessment scores (Cook et al. 2011; Echevarria et al. 2012).

Academic-language development is particularly problematic for ELs who enter the educational system in grades 4–8. With comparatively fewer years to master the English language than those who enter in the primary grades, these students have the dual task of learning complex course content and developing English-language proficiency (O'Hara et al. in press). In all classes and grade levels then, as ELs simultaneously learn, comprehend, and apply content-area concepts through their second (or third) language, they need skillful teachers armed with the knowledge and expertise necessary to facilitate language and literacy development in English (Achinstein et al. 2012; Genesee et al. 2006).

S. O'Hara (✉)
Resourcing Excellence in Education, University of California Davis, Davis, CA, USA
e-mail: spohara@ucdavis.edu

R. Pritchard • D. Pitta • R.N. Newton • U.H. Do • L. Sullivan
University of California Davis, Davis, CA, USA
e-mail: pritchard@csus.edu; debipitta@gmail.com; rnnewton@ucdavis.edu; uhdo@ucdavis.edu; lhsullivan@ucdavis.edu

A.W. Oliveira, M.H. Weinburgh (eds.), *Science Teacher Preparation in Content-Based Second Language Acquisition*, ASTE Series in Science Education,
DOI 10.1007/978-3-319-43516-9_11

The task of teaching science content to ELs is especially complex and challenging in light of the science and engineering practices of the Next Generation Science Standards.

> These science and engineering practices are language intensive and require students to engage in classroom science discourse. For example, students must read and write, as well as view and represent visually, as they develop their models and explanations. They must speak and listen as they present their ideas or make reasoned arguments based on evidence. Science and engineering practices offer rich opportunities and demands for language learning at the same time as they promote science learning. Hence, these practices merit special attention in science classrooms that include ELs (Lee and Llosa 2015, p. 162).

For schools and districts, a related and equally important challenge is to develop a system of support for teachers within schools and across districts that will promote ongoing professional learning as part of an integrated professional development program that can have long-term impact on student learning. To address this challenge, we implemented the Academic Language and Literacy in Every Setting (ALLIES+) project. The overarching goal of ALLIES+ was to develop, implement, and test a user-centered, capacity-building approach for facilitating such a system. Toward that end, we sought to engage educators from both classroom and expanded learning[1] settings to work together in a professional learning community designed to develop a common language across these contexts, improve instructional coordination, reinforce key concepts, and provide more seamless learning environments for students. The specific goals and objectives of the ALLIES+ were:

1. Develop a high-quality, collaborative professional development model for teachers, administrators and expanded learning staff targeting high-leverage practices for promoting academic language and science learning;
2. Build capacity of principals and expanded learning coordinators to support teachers and expanded learning staff in the enactment of these practices;
3. Build capacity of instructional leadership teams within partner schools to support and sustain this work.

Setting

The ALLIES+ project was implemented in Youngstown(pseudonym), a school district that covers 150 square miles of rural, agricultural, and suburban areas in Northern California. The student population is 21.2 % EL, 61.5 % qualified participants in the federal School Lunch Program, 38.8 % Hispanic, and 16.6 % Asian. The predominant languages, other than English, are Spanish and Punjabi. The district serves nearly 14,000 students. The district was a participant in a grant that targeted teachers, administrators, and expanded learning educators of fourth through

[1] Expanded learning settings in this instance included after school and summer learning program staff.

eighth grade students at two district schools. Financial support from the grant provided stipends and/or release time for participants to attend professional learning sessions that focused on the enactment of a set of core teaching practices to develop the academic language and literacy of ELs in science classrooms and expanded learning programs.

Theory of Change and Design Principles

The research literature contains numerous examples of professional development efforts that have failed to impact student learning or that could not sustain their impact over time due to a failure to articulate a theory of change on which to base professional development (Casteel and Ballantyne 2010). Determined to avoid that pitfall, we adopted a multi-tier strategy in developing the ALLIES+ intervention that is aligned with our theory of change and attends to three key design principles for building instructional capacity for academic-language and literacy development.

Targeting High-Leverage Practices

Our first design principle addresses the need to focus any instructional improvement process on a set of targeted, high leverage instructional practices (Windschitl et al. 2013; Fogo 2011; O'Hara et al. 2014). This design principle is predicated on the importance of providing instructional leaders and teachers with a common language around the instructional shifts needed to help ELs meet the challenges of the Common Core State Standards (2010).

Learning In and from Practice

The second design principle focuses on video examples of practice as a key resource for learning, because video can illustrate high-leverage practices in action, provide opportunities to distinguish stronger and weaker versions of them, and afford opportunities to examine the elements of these practices as they unfold in classrooms. Our professional learning model was predicated on the importance of providing video examples of teaching, and time for both teachers and expanded learning staff to practice new instructional shifts aligned with the ALLIES+ practices.

Building Capacity to Develop Sustainable Learning

The third design principle focuses on the importance of building the organizational infrastructure and conditions (e.g., knowledgeable leaders, instructional tools, facilitative organizational structures, and collegial professional relationships) to grow, sustain and spread the use of high-leverage practices that support the academic-language and literacy development of ELs (Jaquith 2013). (See Fig. 11.1) This design principle is premised on four central ideas: (1) instructional leadership is most effective when leadership is shared among a team of people who have different roles and expertise; (2) a shared understanding of the purpose for and value of academic language and literacy in content area teaching is essential for the implementation of new practices; (3) capacity can be built within a school to stimulate, support, and sustain learning about the use of core academic-language and literacy practices and (4) generating site-based capacity to use core academic-language and literacy practices and reflect upon their use creates the conditions for ongoing learning and sustained use of these core practices. In our project we focused attention on building school-based instructional leadership teams to drive the development of the conditions that were needed to support participants in enacting the ALLIES+ instructional practices in their teaching.

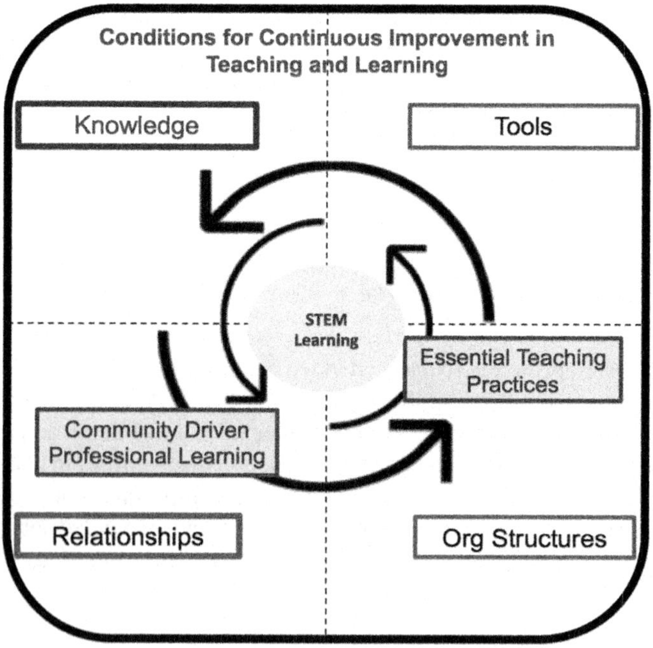

Fig. 11.1 Capacity building approach to proferssional growth

The Professional Development Model

ALLIES+ was designed so that participants became active learners in their own professional development and were provided with the resources – including the time, materials, and intellectual support – they needed to develop and implement more effective and innovative lessons. From January to May, 2015 a team of five university and public school educators provided five workshop sessions totaling 30 h to 24 design team members.[2] The first session took place in a Friday afternoon/all day Saturday format. The subsequent sessions, scheduled at 4–6 week intervals, took place on 4 week-day afternoons. Two representatives from the local County Office of Education also provided technical assistance support to the teams through presentations at some of the professional development workshops and participation in Professional Learning Communities (PLCs) at the school sites.

The workshops focused on how science content and learning activities could be modified to improve academic conversations in classroom and expanded learning settings. The workshops provided time for team members to share ideas, collaborate across classroom and expanded learning settings, and co-design learning activities and inquiry cycles that focused on areas of student need that teams identified. Over the 30 h, the facilitators balanced instruction with support as approximately one third of the face-to-face time was spent engaging participants in explicit instruction and guidance in the use of the ALLIES+ practices, one third afforded participants individual and collaborative experimentation time, and the final third – called studio time – was allocated to the participants designing lessons and sharing them with the group. Balancing explicit instruction, and both individual and collaborative experimentation, was achieved by the facilitators' regular attention to the interests and needs of participants. Additionally, as the skill sets of participants varied, peer support was critical to the group's knowledge development.

In addition to the five professional development sessions, the teams also established PLC's at their school sites with the goal of coming together to plan, design, implement and modify lessons. During the 1-h PLC meetings held between each workshop, school teams discussed successes and challenges as well as any modifications that needed to be made to their lessons. Each team also developed an inquiry question and identified specific teaching strategies and student evidence to examine. The inquiry question drove the process of continual reflection and quality improvement. The design teams' questions were:

- How can we improve students' ability to communicate verbally what they have learned? (school A)
- How can we improve students' ability to communicate using scientific, academic language? (school B)

---

[2] Design teams comprised science teachers, expanded learning coordinators and line staff, and site administrators.

Our professional development model had four unique elements to foster participants' ability to create and implement innovative lessons and help them to develop a repertoire of instructional strategies to meet the needs of ELs in science classes: (1) Sessions were designed in collaboration with the district and aligned with the district's strategic goals; (2) Teachers and district instructional leaders worked side-by-side learning how to implement ALLIES+ practices in support of ELs' academic-language development; (3) Teachers and expanded learning educators were provided with "studio time" to rehearse new instructional practices in a low-risk environment; and, (4) Ongoing inquiry was sustained over time focusing teachers' attention on experimenting with new practices, engaging in cycles of inquiry utilizing artifacts of practice, discussing and adapting lessons plans, and analyzing student work – all supported by the district's instructional leaders and the professional development team.

Materials

We launched the program by developing a set of tools, videos, and instructional resources that serve to illustrate the ALLIES+ practices and facilitate enactment of these practices in science classes in grades 4–8. These resources included a networked website for ALLIES+ participants and school partners with access to all workshop materials on Trello boards for site level use (see appendices for examples of resources described below).

The materials emerged from research on effective instruction to foster the academic-language and literacy development of ELs (Anstrom et al. 2010; August et al. 2010; Baker et al. 2014; Brisk and Proctor 2012; Echevarria et al. 2011; Jiménez et al. 2015; Moschkovich 2012; Kibler et al. 2015; Van Lier and Walqui 2012; Wong Filmore and Filmore 2012; Zwiers et al. 2014). From the research review we generated a list of effective instructional practices. Next, we analyzed a set of videos of exemplary teaching from classrooms in which practices that specifically addressed the academic-language development of ELs were being enacted. The teachers in these classrooms were randomly selected from schools with which we had partnered and volunteered to videotape lessons in which they were engaged in academic-language instruction. We used these to develop a description of the instructional practices that best reflect their enactment in teaching. Then, we repeated this process with an additional set of videos of classrooms to further refine the language of what we began to call the ALLIES essential practices.

This process revealed three essential practices identified as high impact for academic-language and literacy development: Foster Academic Interactions (structuring and strengthening student-to-student interactions that use academic language and literacy); Fortify Academic Output (structuring, strengthening, and supporting the quantity and quality of students' production of original, extended oral and written academic messages which require complex language); and Interact with Complex Text (developing students' overall abilities to practice with and process the language of complex texts).

These essential, high impact practices, although central to effective academic language instruction, alone do not get to the core of academic language teaching. Effective academic language teachers enact another set of instructional practices in support of these essential, high impact practices. We labeled these Cross-Cutting Practices: Facilitate Acquisition of Academic Language, Foster Metacognition, and Monitor and Guide Language Learning. Finally, in preparation for enactment of high-impact and cross-cutting practices, teachers employ the foundational practice: Design Instruction of Academic Language and Literacy Development. This practice focuses on how clearly and directly a teacher aligns academic-language objectives with content objectives, which in turn should align with the lesson's texts and tasks.

Our research also revealed, not just a list of practices, but ways in which the essential instructional practices support one another. For professional learning purposes we organized the practices into three "frames", each consisting of a high impact essential practice supported by three cross-cutting practices and a foundational practice that are common across the three frames. (See Fig. 11.2 for an example.)

High-Impact Practice	**Foster Academic Interactions** • Build conversation skills • Provide extended and supported opportunities for student-to-student interactions		
Cross-Cutting Practices	**Facilitate Acquisition of Academic Language** • Provide multiple, rigorous and supported opportunities for students to acquire and use all three dimensions of academic language • Use a variety of communication strategies to make target academic language understandable	**Foster Metacognition** • Visibly enact metacognitive processes students are expected to use in support of academic language learning • Deconstruct metacognitive strategies that support academic language learning	**Monitor and Guide Language Learning** • Monitor academic language learning and adjust instruction, supports and tasks to meet the needs of students • Provide written and/or oral feedback to promote academic language use
Foundational Practice	**Design Instruction of Academic Language and Literacy Development** • Set academic language and literacy learning targets that are aligned with ELA/Literacy CCSS and the target high-impact practice • Structure and connect tasks that support the academic language and literacy learning targets • Design supports to help students meet the academic language demands of texts and tasks		

Fig. 11.2 Foster academic interactions teaching frame

Because the science and engineering practices of the Next Generation Science Standards (2013) are language intensive and require students to engage in classroom science discourse, the professional development team, in consultation with district representatives, decided to focus on Foster Academic Interactions. The emphasis in the first session was helping design teams develop an understanding of this practice, so we introduced them to videos depicting classroom teachers' use of this high-impact essential practice at varying levels of enactment. We also introduced and demonstrated a variety of instructional resources that they could use with students, including the Constructive Conversation Skills Poster and the Conversation Analysis Tool (See Appendix A.) During Studio Time in the first session, design teams worked together to integrate these materials into already existing lessons. In the sessions that followed teams developed new lessons that incorporated these tools.

Another important emphasis in the early sessions was helping each design team understand how to utilize effectively the PLC structure that existed at each site. One aspect of this work was developing Inquiry Questions and using them as the focus during PLC meeting time. Tools we used during this component appear in Appendix B.

Graphic organizers were also developed for use by teams in collecting evidence of the effectiveness of their science learning activities with embedded essential practices that were implemented in classrooms and expanded learning programs. Data collected through use of the graphic organizer tool were brought to PLC meetings for discussion, reflection and refinement with fellow team members. An example of a graphic organizer appears in Appendix C.

During the final session, we shared *Introducing Robotics with Scribbler*, a robotics curriculum that was purchased for teacher and expanded learning staff use. We also demonstrated where the essential instructional practices could be woven into lessons plans for classroom and expanded learning program enactment.

Outcomes

Participant interviews were used to gather information and help us understand how successful our efforts were in supporting and engaging school-based design teams in implementing science learning activities across school and expanded learning contexts. A case study approach was taken with the interviews, which were conducted with the principals, teachers, expanded learning site coordinator, district expanded learning director, and county office of education staff. The interview format was modified slightly to reflect the context and role that each individual had on the team.

When asked to describe what the team had accomplished over the year, all of the interviewees spoke of the team co-designing a series of science lessons that incorporated ALLIES+ strategies. This included participating in joint professional learning workshops about ALLIES+ practices and selecting a science content area and accompanying lessons for expanded learning staff to implement. The principal and expanded learning coordinator highlighted how valuable it was to have joint collaboration and planning time with teachers and expanded learning staff during both the formal professional learning workshops and the PLCs. Team members described how the PLCs provided time to debrief after the lessons to talk about what went well and what should be modified or changed for the next lesson.

Bridging Expanded Learning and Expanded Learning Contexts When asked whether the project had successfully brought classroom teachers and expanded learning staff together all of the individuals interviewed responded that it had definitely opened a dialogue and broken down barriers. For example, the principal said, "It made a huge difference last year... I saw the connections build and saw that expanded learning staff was more comfortable asking teachers questions and teachers were more open to sharing resources with expanded learning staff." In addition, the regional leads described how the project had begun to break down barriers and bring the two communities together. Initially they noticed that expanded learning staff was tentative in meetings with teachers and administrators but they gradually began to feel comfortable and by the end they were participating fully. They also felt that teachers had gained a new understanding of some of the challenges that expanded learning staff face.

Academic Language and Literacy and Science Practices The interviewees were asked if the project led to increased use of academic literacy practices and science content in classrooms or expanded learning contexts. All of the interviewees responded that there was definitely an increase in the use of ALLIES+ practices in the expanded learning setting. The principal spoke of a "huge increase in the expanded learning program use of the ALLIES+ practices." The regional leads observed several science lessons being taught and described how EL staff used specific ALLIES+ strategies (e.g. sentence stems, academic vocabulary, finding evidence to support claims). The expanded learning coordinator described how the team had learned both science content and new ways to incorporate ALLIES+ strategies in to lessons. Likewise, the expanded learning director noted, "I saw science and math teachers provide their expertise to expanded learning program staff and then later saw it implemented in the expanded learning program. It was pretty amazing."

The principal and expanded learning coordinator described student presentations that were the culminating activity of the science unit taught by the expanded learning staff. Students presented on the STEM unit to a panel of teachers and the principal. The regional leads felt this was a powerful activity that engaged students as well as the entire ALLIES+ expanded learning team.

Successes One of the primary successes mentioned by interviewees was a shift in how teachers and expanded learning staff communicate and interact. All of the interviewees described how important it was that the entire team (teachers, expanded learning staff, principal) consistently attended all of the workshops and PLC's together. This resulted in the team gaining momentum and setting realistic and actionable goals. Another success related to this was that the principal was able to carve out common planning time for the team to meet. All of the interviewees mentioned how valuable it was to have the principal be an active and engaged member of the team. This, in turn, resulted in significant steps towards bridging the gap between regular day and expanded learning programs.

Another success identified in the interviews was that expanded learning staff gained confidence in their ability to implement both the ALLIES+ strategies and the science content. As the expanded learning director described this shift, "It was great to see the expanded learning program staff taking the activities to the next level, and seeing the pride they had when they implemented the activities." Several interviewees noted that this increase in confidence resulted in students being more engaged and interested in the content as well. In a related comment, one of the COE staff noted the success of the project in giving English learners more opportunities to speak and use academic language – something that they do not typically get enough time to do during the school day.

Continued Supports Needed All of the interviewees noted the importance of continuing to give teachers and expanded learning staff the dedicated time to meet and collaborate. Specifically this involves both financial support and the organizational structure to set aside joint planning time. Expanded learning staff and teachers need additional training in the ALLIES+ strategies and science content in order to effectively collaborate and teach the content. The interviewees mentioned that the team needs to have access to high quality resources and materials that will engage students. This requires giving expanded learning staff the time to properly prepare the materials for students so that they are ready to teach the lesson in the most effective way.

Conclusions

Findings from this study suggest that professional development models that are responsive to the needs and interests of the participating educators, schools, expanded learning programs and districts hold great promise for authentic and generative teacher knowledge development. Specifically, models of professional development designed around the key, research-based practices of effective professional development, can positively impact teacher knowledge and practice. As such, the following features characterize our professional development model:

- *Situated in Practice:* Teams of educators from schools and expanded learning programs came to the professional development sessions and worked collaboratively on science curriculum and artifacts of practice from their contexts. Between sessions and meetings they implemented new lessons and activities in their settings and then they came back together to reflect on implementation and refine these products.

- *Focus on Student Learning:* The professional development sessions were all designed to focus on student learning (i.e., academic-language, science learning, and grade level concepts).
- *Model Instructional Strategies:* The professional development team modeled instructional strategies throughout the professional development sessions. In addition, teachers and expanded learning staff modeled various instructional strategies for each other.
- *Engage Educators in Active Learning:* The design studio components of the professional development meant that teams were active participants in the professional development sessions.
- *Build Professional Learning Communities:* Many activities in the professional development sessions were designed to build learning communities, both among the teams of teachers from each school, expanded learning staff, and among teachers and expanded learning staff.
- *Integrate with Other Aspects of School Change:* This initiative was developed in response to the district's emphasis on NGSS and CCSS. The professional development team met with the district leaders to elicit their goals for the professional development program and to understand the bigger strategic goals for the district. The professional development team then worked to design the professional development sessions such that they aligned with district goals.
- *Sustainable:* The professional development program was offered over an extended period of time consisting of activities that were ongoing and sustainable over time, and that provided the opportunity for participants to engage in cycles of experimentation and reflection. In addition, district instructional leaders participated in the professional development sessions so that they would have the knowledge and skills needed to sustain the work beyond this project.

Learning how to use ALLIES+ practices across educational contexts requires expert instruction, explicit modeling, and ongoing support. Learning to integrate these practices into an existing schema for teaching students in support of STEM learning requires time to practice and collaborate with colleagues. This professional development model, designed around the key principles for building instructional capacity, provided time for teachers and expanded learning staff to learn how to use the practices in support of academic language and science learning through explicit modeling, individual and collaborative experimentation, and expert and peer mentoring. The professional development providers' ability to determine and respond to the needs of design teams, by balancing modeling with appropriate support, were the critical components in what participants reported were authentic and generative learning experiences that promise to impact positively student academic language and their understanding of science concepts.

Appendices

Appendix A

Conversation Analysis Tool

The following scoring tool is meant to help you reflect on two key dimensions of effective interactions. You will fill in the online version of this tool when you complete Assignments 1.1 and 3.1. You can use this for practice and notes.

DIMENSION 1: Turns build on previous turns to build up an idea

4 Half or more of the turns build on previous turns to effectively build up a clear and complete idea

3 Half or more of the turns build on previous turns to adequately build up an idea, which may be incomplete or lack clarity.

2 Few turns build on previous turns to build up an idea.

1 Turns are not used to build up an idea.

DIMENSION 2: Turns focus on the knowledge or skills of the lesson's objectives

4 Half or more of the turns effectively focus on the lesson's objectives and show depth or fostering of the intended learning.

3 Half or more of the turns sufficiently focus on the lesson's objectives, but this focus may be superficial or lack clarity.

2 Few turns focus on the lesson's objectives.

1 Turns do not focus on the lesson's objectives.

Dimension 1: Turns build on previous turns to build up an idea	*Score* 4-3-2-1
Rationale for score	
Dimension 2: Turns focus on the knowledge or skills of the lesson's objectives	*Score* 4-3-2-1
Rationale for score	

Appendix B

Process for Developing Inquiry Questions

What is it that your students struggle with the most?

What do many of your students experience difficulty doing? (*Brainstorm on post-its.*)

What ideas go together? (*Prioritize.*)

What evidence do we need?

How are we going to teach it? (*Fill in Inquiry Question chart.*)

Write it as an inquiry question: *How can I develop my students' __________evidenced by ________by using_____________?*

Question	Response
What do we want students to know and be able to do?	
How will we know? What evidence will we use to determine they met the learning target?	
What strategies or instructional practices will we use to help our students reach the learning target?	

Appendix C

School A: How can we improve student's ability to communicate using scientific, academic language?
Evidenced by written and spoken conversations: presentations with peers, teachers, and classroom guests and parents using Kate Kinsella's 4Ls for academic conversations; PBL; ALLIES frames and survival.

Which strategies were your focus? Sentence stems; Think pair share; Anchor charts

What does the evidence tell us? What students are doing well; What areas are students having difficulty? Patterns? Trends?	Next Steps:
Inquiry Question:	Evidence:
Timeframe:	Next PLC:

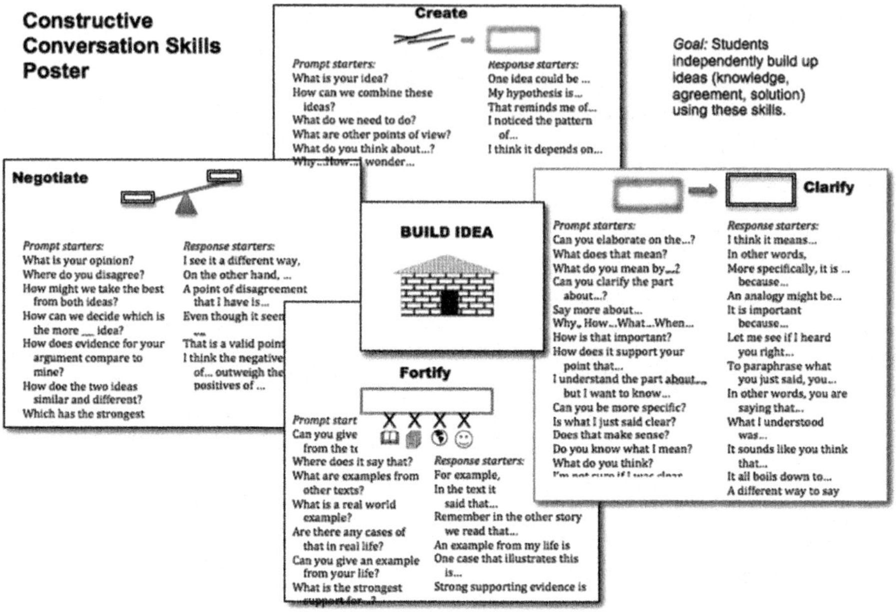

Source: Zwiers, J., O'Hara, S., & Pritchard, R. (2014). *Common Core Standards in diverse classrooms: Essential practices for developing academic language and disciplinary literacy.* **Portland, ME: Stenhouse**

References

Achinstein, B., O'Hara, S., Pritchard, R., & Zwiers, J. (2012). Strategic mentoring for new teachers of English learners. *Journal of Communication and Education, 11*(10), 20–23.

Anstrom, K., DiCerbo, P., Butler, F., Katz, A., Millet, J., & River, C. (2010). *A review of the literature on academic English: Implications for K-12 English language learners*. Washington, DC: The George Washington University Center for Equity and Excellence in Education.

August, D., Artzi, L., & Mazrum, J. (2010). *Improving science and vocabulary learning of English language learners*. Austin: CREATE.

Baker, S., Lesaux, N., Jayanthi, M., Dimino, J., Proctor, C. P., Morris, J., Gersten, R., Haymond, K., Kieffer, M. J., Linan-Thompson, S., & Newman-Gonchar, R. (2014). *Teaching academic content and literacy to English learners in elementary and middle school* (NCEE 2014–4012). Washington, DC: National Center for Education Evaluation and Regional Assistance (NCEE), Institute of Education Sciences, U.S. Department of Education. Retrieved from: http://ies.ed.gov/ncee/wwc/publications_reviews.aspx

Brisk, M. E., & Proctor, C. P. (2012). Challenges and supports for English language learners. In K. Hakuta & M. Santos (Eds.), *Understanding language: Language, literacy, and learning in the content areas* (pp. 115–122). Palo Alto: Stanford University.

Casteel, C. J., & Ballantyne, K. G. (Eds.). (2010). *Professional development in action: Improving teaching of English learners*. Washington, DC: National Clearinghouse for English Language Acquisition.

Cook, H. G., Boals, T., & Lundberg, T. (2011). Academic achievement for English learners: What can we reasonably expect? *Phi Delta Kappan, 93*(3), 66–69. Retrieved from http://intl.kappanmagazine.org

Echevarria, J., Richards-Tutor, C., Chinn, V., & Ratleff, P. (2011). Did they get it? The role of fidelity in teaching English learners. *Journal of Adolescent & Adult Literacy, 54*(6), 425–434. doi:10.1598/JAAL.54.6.4.

Echevarria, J., Vogt, M. E., & Short, D. (2012). *Making content comprehensible for English learners: The SIOP model* (4th ed.). Boston: Allyn & Bacon.

Fogo, B. (2011). Making and measuring the California history standards. *Phi Delta Kappan, 92*(8), 62–67.

Francis, D., Rivera, M., Lesaux, N., Kieffer, M., & Rivera, H. (2006). *Practical guidelines for the education of English language learners: Research-based recommendations for instruction and academic interventions*. Portsmouth: RMC Research Corporation, Center on Instruction.

Genesee, F., Lindholm-Leary, K., Saunders, W., & Christian, D. (2006). *Educating English language learners: A synthesis of research evidence*. New York: Cambridge University Press.

Jaquith, A. (2013). Instructional capacity: How to build it right. *Educational Leadership, 71*(2), 56–60.

Jiménez, R. T., David, S., Fagan, K., Risko, V. J., Pacheco, M., Pray, L., & Gonzales, M. (2015). Using translation to drive conceptual development for students becoming literate in English as an additional language. *Research in the Teaching of English, 49*(3), 248.

Kibler, A., Walqui, A., Bunch, G., & 1. (2015). Transformational opportunities: Langauge and literacy instruction for English language learners in the common core Era in the United States. *TESOL Journal, 6*, 9–35. doi:10.10002/tesj.133.

Lee, O., & Llosa, L. (2015). What are the language demands for science in NGSS? How do they relate to the literacy standards for science in the CCSS? In G. Valdes, K. Menken, & M. Castro (Eds.), *Common core, bilingual and English language learners: A resource for all educators*. Philadelphia: Caslon Publishing.

Moschkovich, J. (2012). Mathematics, the common core, and language. *Teachers College Record, 96*(3), 418–431.

National Governors Association Center for Best Practices & Council of Chief State School Officers. (2010). *Common core state standards*. Washington, DC: National Governors Association Center for Best Practices/Council of Chief State School Officers.

NGSS Lead States. (2013). *Next generation science standards: For states, by states*. Washington, DC: The National Academies Press.

O'Hara, S., Zwiers, J., & Pritchard, R. (2014). Cutting to the common core: Changing the playing field, part 1. *Language Magazine: The Journal of Communication & Education, 13*(5), 24–27.

O'Hara, S., Pritchard, R., & Zwiers, J. (in press). Academic Language and Literacy in Every Subject (ALLIES): A capacity building approach to supporting teachers in grades 4–8. In P. Protor, A. Boardman, & E. Hiebert (Eds.), *English learners and emergent Bilingualism in the common core era*. New York: Guilford Publishing.

Van Lier, L., & Walqui, A. (2012). Language and the common core state standards. *Commissioned Papers on Language and Literacy Issues in the Common Core State Standards and Next Generation Science Standards, 94*, 44.

Windschitl, M., Thompson, J., & Braaten, M. (2013). Developing a theory of ambitious early-career teacher practice. *American Education Research Journal*. doi:10.3102/0002831213476334.

Wong Fillmore, L., & Fillmore, C. (2012). *What does text complexity mean for English learners and language minority students?* Paper presented at the 2012 Understanding Language Summit, Understanding Language Initiative, Stanford. Retrieved from: http://ell.stanford.edu/sites/default/files/pdf/academic-papers/06-LWF%20CJF%20Text%20Complexity%20FINAL_0.pdf

Zwiers, J., O'Hara, S., & Pritchard, R. (2014). *Common core Standards in diverse classrooms: Essential practices for developing academic language and disciplinary literacy*. Portland: Stenhouse.

Chapter 12
A Design-Based Model of Teacher Professional Learning in the LISELL-B Project

Cory A. Buxton, Martha Allexsaht-Snider, Yainitza Hernández Rodríguez, Rouhollah Aghasaleh, Lourdes Cardozo-Gaibisso, and Mehtap Kirmaci

Introduction

The Language-rich Inquiry Science with English Language Learners through Biotechnology (LISELL-B) project is the latest iteration of an ongoing design-based research project to develop, implement, and refine a teacher professional learning framework and a pedagogical model for teaching science with emergent bilingual learners[1]. The LISELL-B framework and model support middle school and high school science and ESOL teachers, their emergent bilingual students, and those students' families in gaining proficiency with science and engineering investigation practices and with the academic language of science. Together, these skills are essential for attaining academic success and accessing college and career pathways. The project uses science, and particularly biotechnology, as a context for developing the problem solving and academic communication skills that emergent bilingual learners need to progress along STEM academic and occupational pathways. The project is a collaborative partnership among the research team, approximately 50 teachers, 4000 of their students, and 100 Latino families (focal families) in 10 schools (5 middle schools and 5 high schools) in two Georgia school districts. These districts are part of what Wortham et al. (2002) have called the new Latino diaspora, the region of the Southeastern U.S. with rapidly changing demographics

[1] Students who are learning English as an additional language are referred to by a range of labels, including English language learners (ELLs), English learners (ELs), and Limited English proficient (LEP). We prefer to use the term emergent bilingual learners to highlight the fact that these students have a home language that is a resource that should be maintained and strengthened while they are learning English.

C.A. Buxton (✉) • M. Allexsaht-Snider • Y.H. Rodríguez • R. Aghasaleh • L. Cardozo-Gaibisso • M. Kirmaci
University of Georgia, Athens, GA 30602, USA
e-mail: buxton@uga.edu; marthaas@uga.edu; y.her.lisell@gmail.com; aghasale@uga.edu; lourdes.cardozoga25@uga.edu; mehtapt2009@gmail.com

A.W. Oliveira, M.H. Weinburgh (eds.), *Science Teacher Preparation in Content-Based Second Language Acquisition*, ASTE Series in Science Education,
DOI 10.1007/978-3-319-43516-9_12

driven largely by Latino immigration from México, Central America, and South America.

In the Southeast, schools and teachers have historically had little interaction or experience with teaching emergent bilingual learners, but are now confronted with classes that may be more than one-half first- or second-generation immigrant students. While in more longstanding Latino communities many students start school speaking English as their first language, in the new Latino diaspora, these newcomer populations are typically Spanish dominant. With these cultural and linguistic shifts in mind, it is not surprising that in the past 10 years, educational policies in the Southeast, which have largely promoted English-only models of instruction, have failed to make use of students' home languages as an instructional support (Garcia and Kleifgen 2010). It is within this context that we developed the LISELL-B pedagogical model and professional learning framework that we describe in the next two sections. We note that small numbers of emergent bilingual students in our project schools spoke a home language other than Spanish, and we discuss our efforts to support these students as well.

Design and Evolution of the LISELL-B Pedagogical Model

Current understandings of how and where people learn science (National Research Council 2009, 2011) support the idea that useful science knowledge includes a blend of practices, core conceptual ideas, and communication skills that are developed in a broad range of life-wide and life-long learning contexts. Additional insights from sociocultural and sociolinguistic research highlight the challenges, resources, and support structures that must be considered so that all learners can be successful with science both in and beyond school (e.g., Buxton et al. 2015; Parsons 2008; Rosebery and Warren 2008). In the case of the science and engineering practices in the Next Generation Science Standards (NGSS Lead States 2013) research is just beginning to address the unique academic needs and resources of bilingual learners (Lee et al. 2013). Indeed, the language of the NGSS practices (e.g., planning and carrying out investigations; obtaining, evaluating and communicating information) is quite broad, and many students, but especially emergent bilingual learners, will require additional specificity and clarity if they are to take ownership of these practices.

There are a number of emergent projects and initiatives focused on making these language demands more explicit. For example, the NGSS includes several appendices with information relevant to language demands: Appendix D presents cases, including a focus on English learners, meant to demonstrate how all students can meet the goals of NGSS, while Appendix F provides lists of specific skills needed to engage in the science and engineering practices, with embedded language demands. In addition, the Council of Chief State School Officers (Pimentel et al. 2012) has published a framework for the development of English language proficiency standards meant to outline the language requirements and strategies that

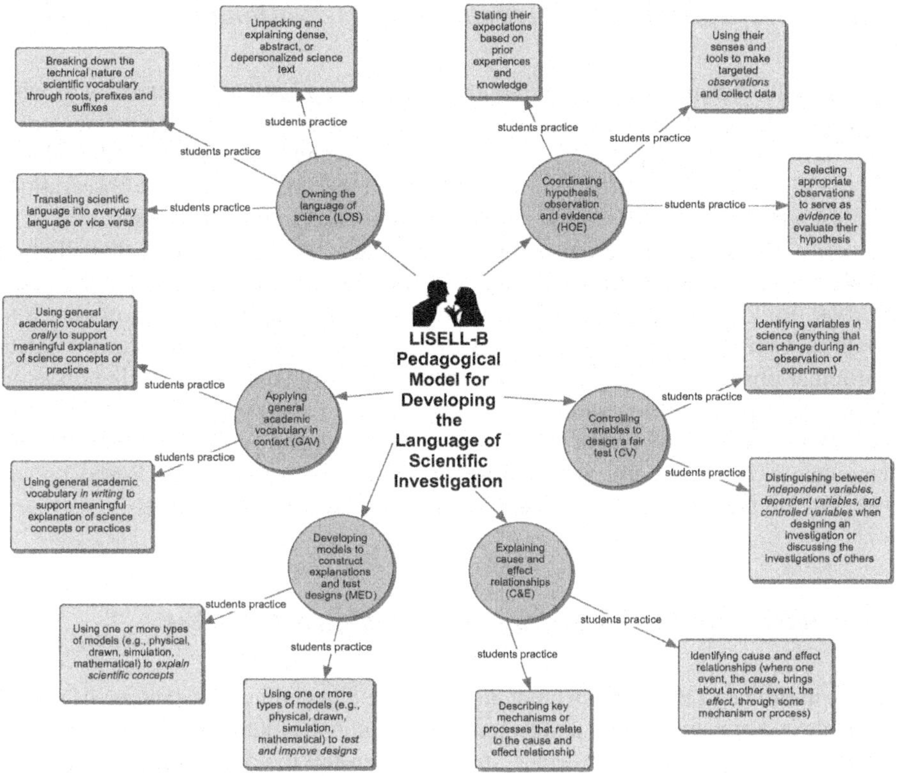

Fig. 12.1 LISELL-B pedagogical model

support the explicit disciplinary knowledge and skills relevant for NGSS as well as the Common Core. Finally, we note the work of the Understanding Language Initiative and the resulting ELPA21 Consortium that has undertaken development and assessment of English language proficiency standards aligned with NGSS science and engineering practices (Linquanti and Hakuta 2012).

Building on and testing assumptions of these initiatives, we used the LISELL-B pedagogical model to identify key features of the language of scientific investigation – those language skills and practices that are needed to engage in, make sense of, and communicate meaningfully before, during, and after participation in scientific investigations. These practices become increasingly important as students transition from elementary school to secondary school, where many students are systematically exposed to the language of science for the first time. Students are asked to contextualize and interpret their experiences of the natural world through a language that may often sound quite foreign (Halliday 2004). To this end, we have developed six *language of science investigation practices* that constitute the pedagogical model for LISELL-B (Fig. 12.1).

Practice 1: Coordinating Hypothesis, Observation and Evidence

By the time students enter middle school, most have a basic understanding of what a hypothesis is and can typically generate examples. However, students may struggle to explain how hypotheses can be evaluated using observations and evidence (McNeill and Krajcik 2011). Students may have the limited idea that observations are just what we see with our eyes, rather than involving all of our senses, as well as the use of tools and technologies (e.g., microscopes, balances, probes) that allow us to observe and measure things that we could not detect with our bare human senses. Further, middle school students often struggle to use observations as evidence to evaluate their hypotheses (Buxton et al. 2013). Students need to learn how to select particular observations that can serve as evidence, while discounting other observations as not pertinent to evaluating the hypothesis. To engage in this practice, students must use both receptive language skills (listening and reading) and productive language skills (speaking and writing) in appropriate ways (e.g., Based on my observation that _____, the evidence supports my hypothesis because _____.).

Practice 2: Controlling Variables to Design a Fair Test

Many middle and high school students also have a general understanding of what a variable is, but may struggle to differentiate between the mathematical and scientific usages common in school. Further, students often fail to conceptualize the need to manipulate and control variables in precise ways as part of experimental science (Lorch et al. 2010). While we acknowledge that an investigative approach involving the manipulation of variables is only one form of scientific inquiry, this LISELL-B language of scientific investigation practice emphasizes the value of accurate scientific communication when designing a fair test. Students learn to communicate about the process of manipulating one variable, determining the effect of this manipulation on another variable, and attempting to control any other possible variables that could affect the process or outcome of an investigation and guarantee a fair test. Students need explicit practice if they are to take ownership of language that expresses the complex interrelationships among variables (e.g., The dependent variable I will observe as an outcome of my investigation will be ____ because ____.).

Practice 3: Explaining Cause and Effect Relationships

Cause and effect relationships are at the heart of scientific explanation, but can take time and practice for young adolescents to develop and articulate (Kuhn 2005). Practicing cause and effect explanations helps students to better understand scientific investigations and processes as they learn to identify the actions, reactions, events, or conditions, linked to particular science content, that lead to or create specific consequences (Kuhn et al. 2000). This LISELL-B language of scientific investigation practice is also foundational to the practice of evidence-based argumentation that is a central feature of the NGSS. Indeed, much of scientific argumentation focuses on crafting convincing explanations of cause and effect relationships grounded in science content (e.g., When ___ happened, then the effect was ___ because ____.).

Practice 4: Using Models to Construct Explanations and Test Designs

Scientists and engineers rely on a wide variety of models, including mental models, physical models, conceptual models, and mathematical or computational models, to make sense of, test, and refine their ideas. Science students can also benefit from the process of developing models as well as from studying models that have been developed by others. Models bring together and unify other LISELL-B practices because most models (especially conceptual models) demonstrate the relationships among variables, and serve to explain causal relationships linked to important science content (McNeill and Krajcik 2011; Schwarz et al. 2009). In addition to constructing explanations, scientists and engineers also use models to test designs. These design tests can build conceptual understanding while also serving as important linguistic scaffolds that students can build on individually and collectively to support collaborative sense-making (e.g., My model shows the relationships between ____ and ____.).

Practice 5: Using General Academic Vocabulary in Context

General-purpose (non-discipline specific) academic vocabulary is often overlooked in school, because teachers, textbook authors, and test developers all routinely make the incorrect assumption that most students are already proficient with this vocabulary (Snow et al. 2009). General-purpose academic words (e.g., indicate, feature) are common in written academic texts, assessments, standards, and teacher talk, but are rare in students' oral, conversational language, making it nearly impossible for many middle school students to fully comprehend the meanings of the content-specific texts they encounter in science (Snow 2010). Coxhead (2000) codified and compiled an academic word list of the most commonly used general academic vocabulary words across the disciplines. In the LISELL-B project, we began with the Coxhead academic word list, but have made continual modifications based on feedback from project teachers and from our own experiences working with students. We specifically highlighted the numerous general academic vocabulary words that are cognates in English and Spanish as one way to promote a focus on using home language resources as an asset for science learning. We use our revised word list to support general academic vocabulary development in meaningful context in science classrooms, embedding this vocabulary in lessons and building bilingual academic word walls.

Practice 6: Owning the Language of Science

Science uses unique language structures to communicate ideas in particular ways. These structures include a reliance on grammatical metaphor, technical vocabulary, dense clausal packing, and rheme to theme structure, among other linguistic features (Halliday 2004). Scientists have developed these language patterns for useful reasons, such as to support the accurate and concise communication of scientific thinking, and to make scientific claims sound authoritative and definitive. These same language structures, however, also tend to confuse students and make science seem more complex and harder to understand than it actually is. Explicit deconstruction of the language of science can help all students, and especially bilingual learners, to better understand and communicate scientific ideas (Fang and Wei 2010). The LISELL-B practice of owning the language of science focuses explicitly on supporting students in adopting this specialized language to better understand and communicate scientific ideas, in addition to building credibility for their own scientific thinking. For example, students practice two-way rewriting, translating academic science language into everyday language and vice versa.

When taken together, these language of science investigation practices help all students, and especially emergent bilingual learners, to adopt the specialized language critical for success in academic science by allowing students to both decode scientific texts and to communicate their own scientific ideas clearly. The LISELL-B pedagogical model was enacted by teachers based on their engagement in our professional learning activities.

Design and Evolution of the LISELL-B Professional Learning Framework

We developed the LISELL-B professional learning framework through an iterative co-design process, to provide multiple ways for teachers to explore the language of science investigation practices that compose the LISELL-B pedagogical model, to assist project staff in modifying and adapting the practices to make them more meaningful for teachers and students, and to consider how these practices might be

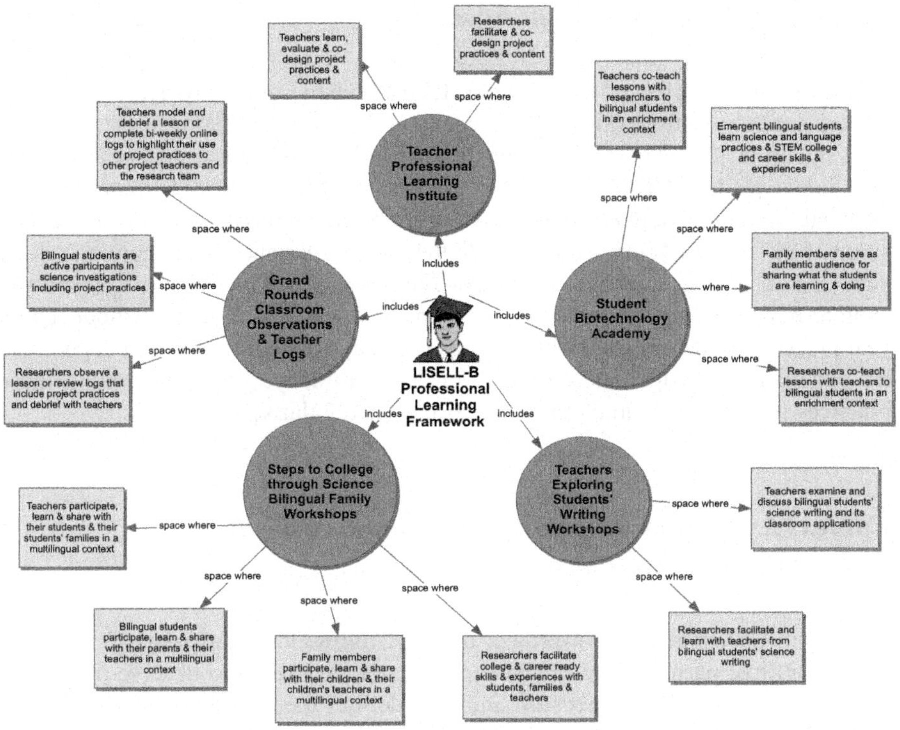

Fig. 12.2 LISELL-B professional learning framework

integrated into daily science teaching repertoires. We designed five professional learning contexts in which different stakeholders come together to do the work of bringing the LISELL-B pedagogical model to life (Fig. 12.2).

Context 1: Summer Teacher Institute

An annual 4-day summer workshop on a university campus serves as a setting for negotiating common understandings of the LISELL-B pedagogical model and for co-developing classroom materials to support enactment of the model during the school year. The summer institute emphasizes both the theoretical ideas behind our pedagogical model and the practical collaboration and planning needed to support enactment of the practices. Teachers also visit research labs, often with a focus on biotechnology, and speak with STEM researchers about their work and about their career trajectories.

Context 2: Summer Student Academy

Based on our experiences in the initial LISELL exploratory project, we added a student academy in the LISELL-B project that follows the teacher institute. Teachers in the institute nominate students from their schools who become the participants in the student academy. The dual goals of the student academy are: (1) to provide science enrichment to students who are emergent bilingual learners, while previewing some of the specific science content that they will experience in the coming school year, and (2) to give teachers a professional learning space in which to dynamically explore new teaching practices learned during the teacher institute without school year pressures and constraints. In this space, as they collaborate with colleagues and try out ideas with students, teachers develop confidence about integrating those practices into their teaching repertoire during the regular school year.

Context 3: "Grand Rounds" Classroom Observations and Online Teacher Logs

Classroom observations with project teachers follow a "grand rounds" model in which multiple participating teachers in a school are invited to observe one teacher's lesson along with project staff and then debrief the lesson together in a mini-workshop format. Teachers have the opportunity to see their colleagues implementing the project practices and then engage in dialogue about the lesson and about student engagement with the practices. This process especially helps teachers new to the project to gain confidence about integrating the project practices in their own teaching. Project teachers also complete an online log every 2 weeks, in which they report on the language of science investigation practices that they have been enacting in their classes. They provide details about the materials and communication structures their students used while engaging in the practices. Periodically, teachers review and discuss their own log data, as well as log data for the other teachers working in their grade level or science discipline. This process allows teachers to self-evaluate the practices they have and have not been using in their classroom and to think about what changes or additions they wish to make in their teaching.

Context 4: "Steps to College Through Science" Bilingual Family Workshops

During these workshops, project teachers come together with groups of bilingual learners from their schools (again nominated by the teachers) and those students' families in a series of Saturday workshops. Participants rotate through three bilingual stations: (1) a session in which LISELL-B staff lead science investigations to model the project's pedagogical practices; (2) a session in which LISELL-B staff facilitate activities exploring academic language, and in which guest speakers, including successful college students and former project participants, discuss the role of family support for academic success; and (3) a session facilitated by the host university, college, technical school, or other institution to share information, activities, and current student testimonials, that often includes a visit to a research lab where families, students, and teachers experience the work in research labs and gain knowledge about current innovative research. Each workshop ends with a shared meal and informal conversations among students, families, teachers, and researchers.

Context 5: Teacher Workshops for Exploring Students' Writing

In these Saturday workshops, teachers work with project staff to examine and learn from students' written responses on the LISELL-B constructed response assessments as well as on other science writing samples. In this professional learning context, project teachers spend mornings working with project staff to score and discuss students' written responses on LISELL-B assessment items. They consider how native English speakers and bilingual learners make and express meaning through their written responses. In the afternoons, teachers utilize the strengths and limitations they saw in their students' written responses to discuss how to design lessons and investigations that support their students in thinking, doing, talking, and writing science together.

Implementing and Improving the Effectiveness of the LISELL-B Pedagogical Model and Professional Learning Framework Through Multiple Iterations

In the remainder of this chapter, we draw upon data from the multiple professional learning contexts of the LISELL-B project to consider two substantive lessons we have learned from our work to support emergent bilingual learners' (and all students') engagement in robust science and language practices. Through a collaborative design-based implementation research design (Penuel et al. 2011), we have worked with participating teachers to plan, implement, and revise multiple iterations of our pedagogical model, our professional learning framework, and the tools and resources that support their implementation over a 7-year period. This evolution of our work highlights the multiple ways in which teachers, students, and the research team came to understand and make use of the LISELL-B language of science investigation practices.

As one example of this design-based collaboration, we collectively developed an approach to curricular adaptation that we came to refer to as LISELLizing. LISELLizing involved teachers and researchers both in bringing science lessons to the table, and then working together to design scaffolds and additional materials to explicitly support the integration of the language of science investigation practices into the lessons. Fundamental to this approach was the idea that there is not one correct way to adapt a lesson for emergent bilingual learners, and that the template we designed was not a "script" for teachers to follow. Instead, the LISELLizing template outlined a range of components that can be added to a lesson to support the goals of language-rich science investigation. We encouraged teachers to keep in mind their knowledge of their students and how students were learning as they adapted their lessons. The LISELLizing process involved the integration of three components. The first component was an explicit focus on the language of science, through the inclusion of science concept cards, lab role cards, general academic vocabulary cards, and language frames (elaborated more fully below). The second component of the LISELLizing process was the development of a LISELL language booster as a high interest way to engage students in both the language and the content of the lesson (also elaborated more fully below). The third component of the LISELLizing process was the focus on one of the LISELL-B language of science investigation practices (described earlier) through the inclusion of an explicit writing scaffold we refer to as a LISELL lab notes template. The evolving process of LISELLizing can be seen as a response to the lessons learned over multiple years of project implementation.

Lesson 1 – Emergent Bilingual Learners Can Benefit from Reading, Thinking, Talking, Writing, and Doing Science Together, Regardless of Their English Proficiency Level, Given Adequate Linguistic Supports, Structured Tasks, and Meaningful Purposes

One guiding principle of the LISELL-B project has been that all students can and must be engaged in rigorous but intentionally scaffolded science learning, regardless of their English language proficiency or their prior educational experiences. As the number of immigrant students in our project schools continued to increase, this challenge became a growing concern for teachers and administrators. Historically, emergent bilingual learners have been excluded from grade-appropriate content instruction during a period of time (up to several years) in which those students were given intensive English language instruction (Harper and Jong 2004). The result of such policies, of course, was that emergent bilinguals fell further and further behind in terms of content learning as well as the content specific academic language that develops through content learning (Calderón et al. 2011). While this model of educating emergent bilinguals still exists in some places (Iddings et al.

2012), one upshot of the standards and accountability movement has been the inclusion of emergent bilinguals (as well as many other students with special learning needs) into general education content courses, but often without the necessary support structures to make this transition effective (Goldenberg 2008). We describe the evolution of two aspects of the LISELL-B project that highlight how we improved the support for students to do and communicate about science together, namely, the role of language boosters and the role of language–rich science investigation kits.

Evolution and Role of Language Boosters From the early phases of our work, we were committed to providing emergent bilingual learners with scaffolds that would support their reading, writing, thinking, and talking about science together in small groups or with partners. Our first attempt to facilitate this work was through the development of LISELL *launching paragraphs*. Launching paragraphs were short (1–2 paragraph) high interest texts written by the research team and by our initial cohort of teachers. Launching paragraphs were designed to be read by students and then discussed with a partner. Two or three discussion questions were provided that related the reading to one of the LISELL language of science investigation practices (such as explaining cause and effect relationships). General academic vocabulary words were also included and highlighted in the reading. While the launching paragraphs were somewhat successful in prompting students to think, talk, and write together, they were generally viewed as stand-alone activities to be done at the start or end of a class period, and were not directly linked to the science topic or standard being taught, nor connected to a particular science investigation.

After several iterations, we now refer to our short readings as *language boosters*. They are meant to accomplish the same general goal as the earlier launching paragraphs, with a modified format and more explicit connections to lessons and standards. Language boosters are still short, high interest science texts with two or three discussion questions that ask pairs of students to read, think, talk, and write together about science, they still systematically incorporate general academic vocabulary words in meaningful context, and they are still created both by members of the research team and by teachers in the project. However, language boosters are now connected thematically to a relevant content area standard and are usually attached as the introduction to a specific science investigation. They also include engaging images to support understanding of the focus of the text. Thus, for example, a physical science investigation we developed to explore potential and kinetic energy using bouncing balls begins with a language booster about how the chemist Norman Stingley invented the Superball (see Fig. 12.3), and an engineering investigation on building earthquake resistant structures begins with a language booster that compares the damage from the 1995 Kobe, Japan, earthquake with that of the 2015 earthquake in Nepal. These shifts have made LISELL-B language boosters more useful both for teachers and for students as they explicitly connect use of the language of science with engagement in doing science together. While most of the language boosters are in English only, emergent bilingual students still engage meaningfully with these texts, supported by their short length, their engaging images, concept and general academic vocabulary cards explicating key concepts,

LISELL-B Language Boosters
Explaining Cause and Effect Relationships

Inventing the Superball

Norman Stingley was a chemist who worked in a rubber factory in the 1960s. His job was to make stronger rubber tubes and belts for cars and machines by inventing new formulas for synthetic (artificial) rubber. Norman Stingley also liked balls – he liked balls a lot. He used to bounce a ball while he thought about how to make stronger rubber tubes and belts. One day he had an idea. He could use the machines that compressed the synthetic rubber to make a stronger and bouncier rubber ball. He reasoned that if he could squeeze the rubber into a denser ball, the **effect** would be a much higher bounce. He worked in the lab at night and came up with a **formula** for a ball that could bounce three times higher than a normal rubber ball. He named his new ball the Superball.

He took his invention to Wham-O, the toy company that also made the Frisbee and the Hula Hoop in the 1960s. They thought the Superball was a great new toy and started **producing** them right away. Superballs can now be found all over the world thanks to Norman Stingley's cause and effect thinking and his love of rubber balls.

Work with a partner to talk about and then write the following:
1) Write a sentence that describes in your own words how Norman Stingley used *cause* and ***effect*** thinking to invent the Superball.
2) Write a sentence that describes a new kind of ball that you would like to invent. What would be special about the ball? What would cause this special way that your ball **functions** and what would the effect be?

LISELL# 23 Mini language booster

Fig. 12.3 Sample LISELL-B language booster

and the structured partner format of reading independently and then talking and writing collaboratively. We also developed a series of bilingual language boosters in Spanish and English for use in our steps to college through science bilingual family workshops, which some teachers have subsequently used in their classrooms.

Evolution and Role of Language-Rich Science Investigation Kits Another way that we scaffolded students' language use from an early point in the project was through the development of resources to support both comprehension and communication as students do science together. We began by developing a series of what we referred to as *LISELL model lessons*. These lessons were science investigations designed for use in our teacher summer institutes and our bilingual family workshops. We occasionally had teachers request to use these same activities in their classrooms during the school year, and we would supply the lessons and the materials needed to teach them. We did not, however, consider the LISELL project to be a curriculum development project, but rather, as providing a framework that could be applied to whatever lessons that teachers already taught. We sometimes used the metaphor of a transparency on an overhead projector, in which teachers could use the LISELL model as an overlay to support their existing science investigations.

As the project developed through multiple iterations, however, we began to see these model lessons, with their scaffolded handouts that structured both the language of science and the practices of our pedagogical model, as an important way to support teachers in implementing the LISELL-B practices. Working collaboratively with teachers in the summer institute and academy, we have now developed approximately 40 LISELL-B language-rich science investigation kits that are aligned with the state science curriculum while providing additional support for students to own the language of science (Appendix lists the current LISELL-B Earth Science Kit topics). The language-rich science investigation kits include language boosters, lab materials, and scaffolded handouts for conducting language-rich science investigations. The kits also include science concept cards, general academic vocabulary cards, and lab role cards. Concept cards highlight a small number (5–6) of key science concepts that are needed to communicate about the investigation. Each card includes the concept in English and in Spanish, as well as relevant images, a student-friendly definition and the concept used in a sentence. Some teachers have students use the cards in their lab groups, while other teachers use the cards to make a concept word wall. General academic vocabulary cards follow a similar format, addressing more general vocabulary incorporated in the language booster and scaffolded handouts associated with the investigation. Lab role cards are directions for students to use in lab groups to play one of four roles (principal investigator, translator, lab technician, or data manager). Each role card gives the student a set of tasks to perform and a set of questions to ask the group that focus on the language of science investigation practices. While most of the language-rich science investigations are in English only, emergent bilinguals are provided with multiple resources to support their engagement, including the concept cards for supporting key ideas and vocabulary, the lab role cards for supporting full participation of all students, and the hands on nature and high interest topics of the investigations themselves. As with the language boosters, we are gradually developing English-Spanish bilingual versions of the kit materials.

Lesson 2 – Monolingual, English-Speaking Teachers Can Still Help Emergent Bilingual Leaners to Use Their Home Language as an Academic Resource for Science Learning While Simultaneously Supporting Students' Acquisition of Academic English

A second fundamental premise of the LISELL-B project is that teaching and learning practices for emergent bilingual learners should enhance the maintenance, development, and use of students' home language(s) as well as English, since all "emergent bilingual learners are developing a complex linguistic multicompetence" (Garcia and Kleifgen 2010, p. 45). From the beginning of the project, we wished to work with teachers to rethink the place that home languages have in the science classroom and to promote translanguaging practices (Garcia and Wei 2014), in which students' multiple languages are viewed as fluid tools for meaning-making, coexisting in ways that can enrich and enhance the learning process. The varied professional learning contexts that we constructed all supported teachers in gaining awareness of how students' home languages can become powerful academic resources for science learning. We considered how school policies and practices often relegate home languages to social interactions, thus reducing their potential to enrich academic learning. Instead of viewing students' multiple emerging languages as separate codes for constructing meaning, we wished to support teachers in adopting the translanguaging perspective that a collective body of linguistic resources can and should be used fluidly and as needed by students independently, in groups, and in interaction with teachers. Several aspects of the LISELL-B project grew and evolved in accordance with this belief, including the role played by the steps to college through science bilingual family workshops and the role of the LISELL-B bilingual constructed response assessments.

Role of the Steps to College Through Science Bilingual Family Workshops During the LISELL-B steps to college through science bilingual family workshops, project teachers (mostly monolingual English speakers), emergent bilingual students (some Spanish dominant and some English dominant), the students' family members (mostly Spanish dominant), and the research team (mostly bilingual or multilingual, but not necessarily in Spanish) all worked together as co-teachers and co-learners. As noted earlier, the workshop format involves groups of families and teachers rotating through three types of activities followed by the whole group coming together for lunch. As initially conceptualized, these workshops were meant to give parents and students a way to interact together around the context of science in the hope that the shared conversations and activities would increase awareness of and enthusiasm for science learning, science related careers, and the opportunities that good academic performance can provide. We saw the role of teachers being present in these workshops largely as a way for teachers to support and connect with their students and their students' families. Because most of the parents in these workshops were more comfortable in Spanish, and because we wanted to support the students in maintaining and developing academic language skills in Spanish, we conducted many of the family workshop activities in a mix of Spanish and English without using simultaneous translation.

As we continued to offer and modify the bilingual family workshops over time, we came to conceptualize the role of teachers in these workshops as more significant than we had originally envisioned. Teachers told us repeatedly during interviews, focus groups, and informal conversations, that their participation caused them to fundamentally shift their thinking about their students' capabilities in science, their assumptions about their students' families, and their ideas about the role of home language in supporting scientific thinking and communicating. Teachers often found themselves working with families during science investigations in which the teachers understood the science content being explored but struggled to participate in the science talk that was largely taking place in a language they did not understand. Even with the supports that were provided, such as bilingual handouts for the investigations, LISELL-B staff keeping running notes bilingually on chart paper in the rooms, and the presence of their students who were usually at least moderately bilingual, the teachers still commented about how exhausting it was to stay focused on learning in a second language. This experience led to greater empathy for their emergent bilingual students and, at least in some cases, a new commitment to supporting their students' use of their home language as well as English as a way to facilitate science learning and communicating. A group of teachers even organized themselves into a Spanish language study class, working with a member of our project team who was an experienced Spanish teacher, as a way to support increased bilingualism in their classrooms. We also note that recently, a group of Burmese-Karen families have begun to attend our family workshops, pushing us to consider new multilingual models of workshop facilitation moving forward.

Role of Bilingual Constructed Response Assessments Another goal of the LISELL-B project was to develop more robust ways of understanding and measuring emergent bilingual students' developing scientific thinking and communicating as expressed through the language of science. To this end, we developed a bilingual constructed response assessment that asks students to write about their understandings of the LISELL-B science practices in the context of science investigation scenarios (Buxton et al. 2014). While there is some evidence for the value of studying bilingual learners' writing as a way to support their science learning (e.g., Gunel et al. 2007), there is less evidence about how best to support emergent bilingual students' science writing on either formative or summative assessments as part of that process. Our analysis of emergent bilinguals' written responses on our assessment points to areas where students demonstrated substantive proficiency and growth using key aspects of the language of science. This was particularly true for students who made use of both the Spanish language and English language resources provided on our assessment. At the same time, the majority of the students showed that they still had a long way to go in order to master the language of science for the purpose of explaining their scientific thinking in written form.

Similar to the bilingual family workshops, we initially conceptualized the bilingual constructed response assessments as a way to help us understand how emergent bilingual students were making sense of their science learning while also providing experiences for the students that could help them understand what it means to think and communicate scientifically. Our initial iteration of the project did not consider how engaging teachers with their students' multilingual science writing could serve as a valuable professional learning opportunity to support teachers in understanding and improving these students' science learning experiences. After the first few iterations of our professional learning framework, we added the *teachers exploring students' writing workshops* as a component of our professional learning. We found, that when teachers were given time and a structure for looking at and talking about their emergent bilingual students' writing about science, they developed a positive appreciation for the emergent science ideas and scientific language usage that their students demonstrated. Teachers commented that on most science assessments used today, the focus is on correct and incorrect responses with little opportunity for students to explain their thinking and therefore, little opportunity for teachers to see their students' emergent science understanding. Because our assessment was bilingual and we tracked the language choices that students made in terms of reading the questions and writing their answers, teachers were provided with another concrete way to see that emergent bilingual students benefit from the opportunity to engage with science in multiple languages where they can choose from moment to moment which language best serves their needs.

While the vast majority of the emergent bilingual students we work with speak Spanish as their home language, and the bilingual resources and spaces we designed highlight Spanish and English, we intend for the LISELL-B pedagogical practices to be beneficial to other linguistic minority groups as well. Smaller linguistic groups in our schools are typically even more marginalized and have less access to academic resources in their home languages than do Spanish speakers. Even when bilingual materials are not available, students who share a home language can be encouraged to use that language as well as English as they work to make and communicate meaning about science. When well intentioned teachers insist on English only in their science classes, they are unknowingly handicapping emergent bilingual students by removing critical tools for scientific sense-making. The bilingual experiences that teachers in the LISELL-B project gained through participation in the family workshops and in workshops exploring students' writing helped these teachers to better understand both the fluidity and the utility of emergent bilingualism.

Discussion of Lessons Learned (Outcomes)

As we noted earlier, current science reform documents make the implicit assumption either that students already possess the language skills needed to fully engage with disciplinary core ideas and science and engineering practices, or that these language skills can be acquired implicitly through the process of participating in robust science learning. Our findings question these assumptions, pointing to the need for the language of science to be explicitly taught and practiced as an integral part of the process of engaging in science learning. In order to be culturally and linguistically responsible science educators, we all need systematic professional learning opportunities to think about, plan, and practice ways of teaching and supporting the language of science. Such explicit preparation has been central to the LISELL-B project from the outset. As the project continued to expand through multiple iterations, we have collaborated with teachers, students, family members, and additional researchers to modify some aspects of the project (such as the language boosters) and to add some new aspects (such as the teachers exploring students' writing workshops). Each of these refinements has helped us to better understand the interplay of science learning and language learning as well as the interplay between first language use and second language use for academic purposes.

In today's high stakes teaching and learning contexts, we cannot wait until emergent bilingual learners have acquired an intermediate level of English proficiency before beginning to engage them in challenging, grade appropriate, academic content. Instead, we must use students' existing linguistic repertoires, while simultaneously helping all students build strength with the language of science, with academic English, and with academic language skills in their home language. We have seen students draw upon their home language as needed while connecting their emergent science knowledge with their everyday experiences both in the United States and from their home countries. As we have shown throughout this chapter, the legitimation and support of students' home language(s) can be empowering for students while simultaneously helping teachers get a more accurate sense of what students do and don't know and what they can and can't communicate.

Finally, we note that the LISELL-B project has demonstrated to us the power of design-based iterative models of professional learning. With each new iteration of the project, we learn from our previous experience and make modifications that continually improve our pedagogical model and professional learning framework. It is also important to point out that the most powerful and useful modifications we have made were co-designed by researchers and teachers working together to think about how best to improve students' language-rich science learning opportunities. Examples of this co-design, such as our work with teachers to LISELLize science lessons, can be seen throughout the chapter. As the LISELL-B project continues to evolve, experienced teachers, as well as experienced students and families, play increasingly important roles in its evolution. Ongoing project iterations include expansions into schools with more multilingualism beyond Spanish and English to better understand how to support science learning in these contexts, efforts to build more systematic sustainability into our model once federal funding ends, and efforts to connect the work of the LISELL-B project with related projects, both nationally and internationally. Building such networks seems critical if we hope to improve the science learning experiences of emergent bilinguals on a large scale and for the long term.

Acknowledgement This material is based upon work supported by the National Science Foundation under Grants No. DRL-1316398 and DRL-1019236.

Appendix

LISELL-B Earth Science Kits Aligned with GA Unit Topics

Inside the earth	Rocks & minerals	Earth's changing surface	Water in earth's processes	Weather & climate	The universe & solar system
LISELL-B kits	LISELL-B kits	LISELL-B kits	LISELL-B kits	LISELL-B kits	LISELL-B kits
Great fossil find	Crayon rock cycle	Modeling constructive & destructive forces	Probing the ocean floor	Characteristics of air	Moon phases glasses
					Shifting shadows
		Earthquake resistant buildings	Bottled water & tap water	Making & using weather instruments	Expanding universe
			Hot and cold water interactions	Heat absorption and wind	
				Atmospheric composition	

Plus general nature of science kits:
Technical writing with Zoobs
Nature of science cards
Hypothesis box
Measuring mass, volume, and time
Estimation with Gummi Bears

References

Buxton, C., Allexsaht-Snider, M., Suriel, R., Kayumova, S., & Choi, Y. (2013). Using educative assessments to support science teaching for middle school English language learners. *Journal of Science Teacher Education, 24*(2), 347–366.

Buxton, C., Allexsaht-Snider, M., Aghasaleh, R., Kayumova, S., Kim, S., Choi, Y., & Cohen, A. (2014). Potential benefits of bilingual constructed responses science assessments for emergent bilingual learners. *Double Helix, (2)*1. http://qudoublehelixjournal.org/index.php/dh/article/view/31/156

Buxton, C., Salinas, A., Mahotiere, M., Lee, O., & Secada, W. G. (2015). Fourth grade English learners' scientific reasoning complexity, inquiry practices, and content knowledge in home, school, and play contexts. *Teachers College Record, 117*(2), 1–36.

Calderón, M., Slavin, R., & Sánchez, M. (2011). Effective instruction for English learners. *The Future of Children, 21*(1), 103–127.

Coxhead, A. (2000). A new academic word list. *TESOL Quarterly, 34*(2), 213–238.

Fang, Z., & Wei, Y. (2010). Improving middle school students' science literacy through reading infusion. *The Journal of Educational Research, 103*(4), 262–273.

García, O., & Kleifgen, J. A. (2010). *Educating emergent bilinguals: Policies, programs, and practices for English language learners*. New York, NY: Teachers College Press.

García, O., & Wei, L. (2014). *Translanguaging: Language, bilingualism and education*. Basingstoke, UK: Palgrave.

Goldenberg, C. (2008). Teaching English language learners. *American Educator*, 8–43.

Gunel, M., Hand, B., & Prain, V. (2007). Writing for learning in science: A secondary analysis of six studies. *International Journal of Science and Mathematics Education, 5*(4), 615–637.

Halliday, M. A. K. (2004). *The language of science*. London, UK: Continuum.

Harper, C., & Jong, E. (2004). Misconceptions about teaching English-language learners. *Journal of Adolescent & Adult Literacy, 48*(2), 152–162.

Iddings, A. C. D., Combs, M. C., & Moll, L. (2012). In the arid zone drying out educational resources for English language learners through policy and practice. *Urban Education, 47*(2), 495–514.

Kuhn, D. (2005). *Education for thinking*. Cambridge, MA: Harvard University Press.

Kuhn, D., Black, J., Keselman, A., & Kaplan, D. (2000). The development of cognitive skills to support inquiry learning. *Cognition and Instruction, 18*(4), 495–523.

Lead States, N. G. S. S. (2013). *Next generation science standards: For states, by states*. Washington, DC: The National Academies Press.

Lee, O., Quinn, H., & Valdés, G. (2013). Science and language for English language learners in relation to next generation science standards and with implications for common core state standards for English language arts and mathematics. *Educational Researcher, 42*(4), 223–233.

Linquanti, R., & Hakuta, K. (2012). *How next-generation standards can foster success for California's English learners*. West Ed Policy Brief. http://www.wested.org/online_pubs/resource1264.pdf

Lorch, R. F., Jr., Lorch, E. P., Calderhead, W. J., Dunlap, E. E., Hodell, E. C., & Freer, B. D. (2010). Learning the control of variables strategy in higher and lower achieving classrooms: Contributions of explicit instruction and experimentation. *Journal of Educational Psychology, 102*(1), 90.

McNeill, K. L., & Krajcik, J. (2011). *Claim, evidence and reasoning: Supporting grade 5–8 students in constructing scientific explanations*. New York, NY: Pearson, Allyn & Bacon.

National Research Council. (2009). *Learning science in informal environments: People, places and pursuits*. Washington, DC: National Academy Press.

National Research Council. (2011). *A framework for K-12 science education: Practices, crosscutting themes, and core ideas*. Washington, DC: National Academy Press.

Parsons, E. C. (2008). Learning contexts, black cultural ethos, and the science achievement of African American students in an urban middle school. *Journal of Research in Science Teaching, 45*(6), 665–683.

Penuel, W. R., Fishman, B. J., Cheng, B. H., & Sabelli, N. (2011). Organizing research and development at the intersection of learning, implementation, and design. *Educational Researcher, 40*(7), 331–337.

Pimentel, S., Castro, M., Cook, G., Kibler, A., Lee, O., Pook, D., Stack, L., Valdés, G., & Walqui, A. (2012). *Framework for English language proficiency development standards corresponding to the common core state standards and the next generation science standards*. Washington, DC: CCSSO.

Rosebery, A. S., & Warren, B. (Eds.). (2008). *Teaching science to bilingual learners: Building on students' strengths*. Arlington, VA: National Science Teachers Association.

Schwarz, C. V., Reiser, B. J., Davis, E. A., Kenyon, L., Acher, A., Fortus, D., Shwartz, Y., Hug, B., & Krajcik, J. (2009). Developing a learning progression for scientific modeling: Making scientific modeling accessible and meaningful for learners. *Journal of Research in Science Teaching, 46*, 632–654.

Snow, C. E. (2010). Academic language and the challenge of reading for learning. *Science, 328*(5977), 450–452.

Snow, C. E., Lawrence, J. F., & White, C. (2009). Generating knowledge of academic language among urban middle school students. *Journal of Research on Educational Effectiveness, 2*(4), 325–344.

Wortham, S. E. F., Murillo, E. G., Jr., & Hamann, E. T. (Eds.). (2002). *Education in the new Latino diaspora: Policy and the politics of identity* (Vol. 2). Oxford, UK: Greenwood.

Chapter 13
Science Teachers as Architects: Building and Supporting Science-Learning Environments with Emergent Bilingual Students

Max Vázquez Domínguez, Martha Allexsaht-Snider, and Amanda Latimer

Introduction

Science teachers play a crucial role in fulfilling societal goals for a better educated and more prepared citizenry (Gordon and DeBard 2014); however, they face a range of challenges. Among the most common challenges are: policy reforms that narrow the curriculum toward what is most easily assessed (David 2011); the changes in ethnic, linguistic, and socioeconomic demographics in the classroom (United States Census Bureau 2014); and concerns about how emergent bilingual students[1] perform on standardized assessments (National Center for Education Statistics 2014). In this light, as teacher educators in the Language-rich Inquiry Science with English Language Learners through Biotechnology (LISELL-B[2]) project, we have collaborated closely with 50 middle and high school science teachers and teachers of English to speakers of other languages (ESOL) across 10 schools with 4000 students over the past 2 years. In this collaboration, Max Vázquez Domínguez, a Mexican educator now studying teacher education, Martha Allexsaht-Snider, a teacher educator with extensive experience in México and working with Latino/a families in the U.S., and Amanda Latimer, an industrial scientist in the area of biotechnology now studying teacher education, functioned as part of a larger team of

[1] In this chapter, we use the term *emergent bilingual students* to emphasize the positive attribute of having a home language that is a resource that can and should be used, maintained, and strengthened in school and in science learning contexts. It refers to a range of students, including those who are new to English and literate in their home language as well as those who report receptive skills in home languages and a language other than English spoken at home but identify themselves as lacking literacy and oral skills in that home language.

[2] The work has been supported by the National Science Foundation (award # DRL-1316398).

M.V. Domínguez (✉) • M. Allexsaht-Snider • A. Latimer
University of Georgia, Athens, GA 30602, USA
e-mail: maxvaz@uga.edu; marthaas@uga.edu; amanda.latimer@uga.edu

A.W. Oliveira, M.H. Weinburgh (eds.), *Science Teacher Preparation in Content-Based Second Language Acquisition*, ASTE Series in Science Education,
DOI 10.1007/978-3-319-43516-9_13

teacher educators, researchers, and teachers. The team of researchers, science teachers, and ESOL teachers co-constructed science materials and learning environments that integrated science and emergent bilingual students' language and cultural practices inside and outside school settings. The LISELL-B project's pedagogical practices are embedded in five teacher professional learning contexts: (a) A summer teacher institute, (b) a student summer academy, (c) 'grand rounds' classroom observations, (d) bilingual 'Steps to college through science' family workshops, and (e) teachers' exploration of students' writing workshops. In this chapter we will discuss how the *Teacher Institute* and the *Student Summer Academy* were designed as interactive learning environments for science teachers and their emergent bilingual students and how these two contexts helped teachers meet the challenges of preparing their ethnically and linguistically diverse students to reach high levels of science learning in classrooms situated in the current demanding policy milieu related to standards, assessments, and teacher evaluation.

Context

Teacher Institute

The LISELL-B project included middle schools and high schools where rapidly growing populations of emergent bilingual students, mostly of Latino/a descent, range from 30 to 50 % of the total school population. In collaboration with two school districts, we identified participating schools and then invited science and ESOL teachers to participate in the project. Teacher participation in the institute and in the student summer academy was voluntary; a total of 29 teachers participated. Of these 29 teachers, 24 were women and 27 were white (one was Latina and one male teacher was from China). These teachers then participated in the process of recruitment of the emergent bilingual students who attended the student summer academy. In this way we brought together students and teachers who had worked together in classrooms during the previous school year.

Teachers in the institute, that included a total of 28 h distributed across 4 days, took part daily in five major activities: learning LISELL-B science investigation and language practices, collaborative planning for science teaching in the student academy, reflection and goal setting, science and engineering laboratory visits, and Spanish lessons.

For 22 h, teachers from the different schools worked in the teacher institute in four teams: Earth Science, middle school Life Science, high school Life Science, and Physical Science groups, depending on what they taught in their home school. In collaboration with teacher educators and in preparation for the upcoming summer student academy, teachers designed science investigations that incorporated one of the six LISELL-B *language of science investigation practices* (coordinating hypothesis, observation and evidence; controlling variables to design a fair test;

explaining cause and effect relationships; applying general academic vocabulary in context; developing models to construct explanations and test designs; and owning the language of science).

Student Summer Academy

The 8-day student summer academy was planned in conjunction with science teachers, ESOL teachers, and teacher educators during the teacher institute. The emergent bilingual population who attended the academy included 90 emergent Latino/a students and 8 Karen (an ethnic group from Myanmar/Burma) students. From a total of 98 students (38 high school students and 60 middle school students) who attended the student summer academy, 43 were girls (25 from middle school) and 55 were boys (35 from middle school). An average of 70 students attended the summer academy each day, spending 4 days on each of two different college campuses, which are located 1 h away from each other. As we have noted, students in the summer academy came from eight schools with large and growing Latino/a populations, as well as small numbers of other groups speaking languages other than English. Between 70 and 80 % of the total student population in these schools qualifies for free or reduced price lunch, and the National Assessment of Educational Progress (NAEP) scores for the participant schools is slightly below the national average.

The emergent bilingual students who participated in the student academy engaged in five types of activities distributed across 58 h of the academy. These activities were: (a) Teacher-directed sessions incorporating LISELL-B science investigation and language practices (where students were grouped based on their grade level and science interests), (b) problem posing projects, (c) science and engineering laboratory visits, (d) problem posing presentations with families, and (e) a soccer tournament (in these other activities students were in mixed groups across grade level and school).

The Model of Professional Learning

The LISELL-B project team developed the professional learning model for the teacher institute linked to the student summer academy in recognition of the current contexts of the participating schools as well as broader aspects of the changing needs of students and teachers, both in U.S. schools and internationally. The general characteristics evident in the educational landscape for Latino/a and many other immigrant students are: (a) international assessments showing that Latino/a students continue to underperform in science (Organisation for Economic Co-operation and Development 2013); (b) teacher workforce demographics showing that the majority of U.S. teachers (almost 77 %) are White, non-Hispanic (U.S. Department of Education 2004); and (c) research showing that most teachers feel unprepared to address different ethnicities in their classrooms as well as children from low socio-economic status (Milner and Laughter 2015), especially in science classrooms where many teachers consider science as culture-free (Lee and Buxton 2011). As we imagined possibilities for teacher learning, we considered what we had learned in our previous LISELL project and current LISELL-B project about how these trends were both constraining and opening opportunities for us as teacher educators and for the science and ESOL teachers and the emergent bilingual students in the schools where they taught (for more details on the program implementation see Buxton et al. 2015; Buxton et al. (this volume); and, "LISELL-B" 2016).

In this light, we designed the *teacher institute* with two main goals: (1) to support teachers in learning how to engage emergent bilingual students in science by bringing language and cultural practices and LISELL-B science investigation practices into science learning environments, and (2) to increase and foster teacher experimentation using different teaching practices designed during the teacher institute to build learning environments with emergent bilingual students that could be adapted into teacher routines and activities in their own science classrooms during the regular school year. The *student summer academy* was designed to accomplish three main goals: (1) to foster teacher-teacher, teacher-student, and teacher educator-teacher-student collaboration; (2) to enact, adapt, and adopt the LISELL-B science investigation and language practices; and (3) to use the spaces of college campuses to construct learning environments that could be transformed into learning territories building on emergent bilingual students' language and cultural practices and fostering robust science learning[3].

Below, we begin with a description and analysis of the activities during the teacher institute followed by explication of the activities of the student summer academy. Due to space limitations, we will only include selected samples of the resources used during the institute and the academy, as well as samples of the teachers' and students' work. Both the teacher institute and the student summer academy can be conceptualized using an architectural metaphor incorporating learning environments that involved both *material* and *expressive* components (De Landa 2006). For De Landa (2006, 2010), material components are the physical elements (e.g., bodies, science equipment, soccer balls) and expressive components are both the linguistic (i.e., language and symbols) and nonlinguistic elements (e.g., gestures, postures) in an environment. We used this architectural framework to consider what Didakis and Phillips (2013) (following Heidegger) have described as the difference between a house (which is inhabited) and a home (which is dwelt in), in conceptualizing the design of teacher and student learning contexts in the LISELL-B project. According to Heidegger, for a house to become a 'home,' it needs to go through a

[3] We use the term *space* to refer only to a physical setting; we use the terms *environment*, *context*, and *territory* to refer to the combination of a physical setting and the corresponding social elements (e.g., interactions between people, institutions, and larger social organizations). We further differentiate between territory and environment or context by claiming that the creation of a territory requires a different (and more intentional) relation between the space and the social settings than is the case for a context or environment.

process of additive interactions that serve to build an experience between the individual and the space until the architectural space "becomes an extension of the inhabitant, absorbing preferences, customs and rituals" (Didakis and Phillips 2013, p. 308). We found this metaphor useful in conceptualizing our work in that we actively encouraged the consideration and incorporation of emergent bilingual students' language and cultural practices in both the teacher institute and summer academy contexts, seeking to foster a sense of teachers and students together dwelling in these purposefully constructed and linked environments.

The daily interactions that took place in the summer academy on university campuses among the emergent bilingual students, their science and ESOL teachers, and teacher educators typically started with an experience that supported the emergent bilingual students in contemplating the possibility of dwelling within these spaces as future university students. In order to reinforce our goal for the science teachers to think about how the emergent bilingual students might dwell in university spaces, we constructed learning environments that allowed science teachers and emergent bilingual students to explore science learning using the varied spaces around them. Thus, we provided teachers and students with the opportunity to turn indoor and outdoor spaces, i.e., classrooms, science labs, and soccer fields, in these university contexts into learning territories.

In the following example, we explain how we came to see the conceptual difference between a learning environment and a learning territory. First, we consider the relationship between dwelling and constructing (Heidegger 1975) as these concepts connect a learning process to a physical space. It is also important to know that a space, an environment, or a context becomes a territory only when an individual inhabits it and adds an expressive component to that space. The function of an expressive component is achieved when using and possessing the space so it has a signature of the possessor (Bogue 2007). In general, expressivity consists of two elements, the nonlinguistic part and the linguistic one (De Landa 2006; Deleuze and Guattari 1987). In the case of science learning environments, these aspects of expressivity can take the form of meaningful written and oral communication such as reporting observational data, stating hypotheses, and explaining cause and effect relationships (linguistic elements); as well as in the creation of charts, diagrams, sketches, or models (nonlinguistic elements) to support scientific thinking and meaning-making.

To better support science learning and communicating in the student academy, we (both teacher educators and science teachers) also needed to consider the material component of how we worked with emergent bilingual students in the university classroom settings. Material and expressive components come in mixtures and are always embedded in processes (De Landa 2006). For example, a written science lab report, when linked to a science investigation, includes both expressive and material elements. We wished to question and explore how expressive components were reflected in the learning spaces of the LISELL-B student academy.

To continue with the example of a written science lab report, common in most science classroom settings, students are typically asked to respond to a previous activity they have completed by explaining their ideas or findings. In our classroom observations of LISELL-B teachers during the regular school year, we have typically seen teachers explain what needs to be included in the lab report to the whole class, then have students work individually or in pairs doing their written lab reports, and then either (a) the teacher gathers the reports to grade them, or (b) asks the students for answers to compare with the rest of the class. We wish to distinguish between viewing a written report as part of a learning environment, and viewing this activity as part of a learning territory in which students learn to express, construct, and dwell within the science classroom. In the first case, where the science teacher gathers the lab reports to grade them later, the student's expressive understanding of the lesson is not maintained in relation to the space around her, but has been shifted to a different relation in which the teacher now has a primary role and the space has a secondary role in regard to the student's learning process. In the second scenario, where the students compare their work with their peers, the social interaction is an additional linguistic element of support for students' learning process; however, the process still does not support students *dwelling* in the science learning space. For some students these learning practices are enough to support scientific meaning making, but for other students creating an additional relationship to the space might be necessary to support science learning. In the student academy, we collaborated with teachers to consider what these additional relationships to the learning space might entail.

In order to improve the science learning process as well as to territorialize the learning environments in the institute and academy, we consider how the production of artifacts and the use of space may or may not allow emergent bilingual students to transform a space into a learning territory, and by extension, provide students with sufficient opportunity to dwell in these spaces. There cannot be a learning territory without a learning environment, but there may be a learning environment without a learning territory. Purposeful use of the space and creation of ways for students to communicate about their work and relate to it can facilitate the construction of a learning territory, which also depends on the teachers' promotion of the student-space interaction. We argue that by including cultural practices in the learning process, we can help emergent bilingual students to dwell within their science classrooms. For instance, including emergent bilingual students' home language in the written activities and in materials displayed throughout the science classroom can help link the students' learning process to the space and turn it to a learning territory by emphasizing the students' expressivity using language familiar to them. When activities that strengthen a learning territory are presented with sufficient frequency, they support students in domesticating the science investigation activities and the space around them. As Didakis and Phillips (2013) suggest, domestication is a cumulative process of interactions between the space and the student that builds "meaning, affection and emotion" (p. 308). Science teachers can promote the domestication of science classrooms by increasing the space-student interaction through science activities using the emergent bilingual students' cultural resources, including home languages.

We conceptualized the process of dwelling in these spaces as only possible if the inhabitants (teacher and students) domesticated them. We considered that incorporating cultural practices, such as home languages (García and Kleifgen

2010) and family and community practices (González et al. 2005; Gutiérrez and Rogoff 2003), that have proven to promote learning with emergent bilingual learners, could facilitate this process of domestication for LISELL-B participants. The cultural practices we incorporated in the LISELL-B project were based on activities that our Latino/a students practiced regularly; for example, the use of Spanish language and connections to soccer as a communal practice. This theoretical framework that we have described led us to two driving questions:

1. What are the lessons learned from utilizing an architectural perspective to make sense of science and ESOL teachers' work in an outside-of-school learning environment for emergent bilingual students?
2. How can those lessons be adapted by science and ESOL teacher educators designing dynamic learning opportunities for current and future teachers committed to language-rich science learning for all of their students, including their emergent bilingual students?

Implementation of the Teacher Institute and Student Summer Academy

How could we start this process of constructing supportive teacher professional learning contexts promoting language-rich science learning environments for middle school and high school emergent bilinguals? After observing a number of our project teachers' science classrooms in the past 2 years, we found that many teachers are already using, to varying extents, some of the practices needed to build productive learning environments for their emergent bilinguals. We have also noticed that teachers' use of these practices often focuses more on the content of the lesson than on the students' cultural practices and strengths. If science is not culture-free, as Lee and Buxton (2011) suggested, then science content along with the science standards can be linked to students' cultural practices. To this end, we begin by describing the science investigation practices and resources developed in the teacher institute, that were later utilized in the student summer academy, that included the emergent bilingual students' language and cultural practices. Figure 13.1 shows the relationship between the activities and goals of the teacher institute and the student summer academy in which science teachers and ESOL teachers worked together to use the resources and practices necessary to create learning territories.

Institute for Teachers

Daily sessions at the ***Institute*** provided teachers with opportunities for learning about the LISELL-B model for developing the language of scientific investigation practices and how this model and its practices support national and state science standards; Buxton et al. (this volume) describe these practices in detail. For our

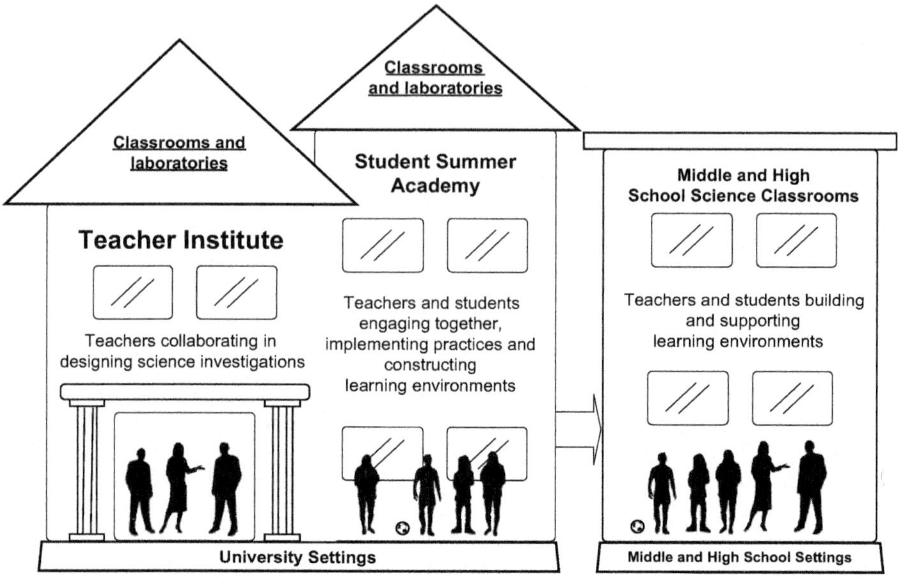

Fig. 13.1 Constructing possibilities of learning territories for emergent bilingual students and their teachers

purposes, it is enough to say that these practices help teachers to engage all students, and particularly their emergent bilingual students, in developing the process of thinking scientifically and communicating scientific ideas using the language of science and taking advantage of students' home language resources. Teachers also learned how their current teaching in the science classroom could be adapted to this model to better meet the needs of their students while still addressing curricular standards.

Working in Spanish As a first step, teachers took part as students in a science investigation activity where directions and explanations were in Spanish with bilingual language supports such as concept cards (explained below), general academic vocabulary cards, and talk moves/language frame guides that provided relevant sentence stems to be used orally and in writing to articulate understandings using the language of science (illustrated in Fig. 13.3). The goal of this activity was to help teachers empathize with their emergent bilingual students while considering how to balance existing science curriculum requirements with students' linguistic needs and the supports they could make available in their classrooms. The teachers in the Institute were asked to write their ideas and answers in Spanish as they collaborated with other science and ESOL teachers during the soccer and science investigation[4]

[4]These investigations were co-developed with a science teacher who was also the soccer coach at one of our participant schools. Most students on the soccer team were emergent bilingual students of Latino descent.

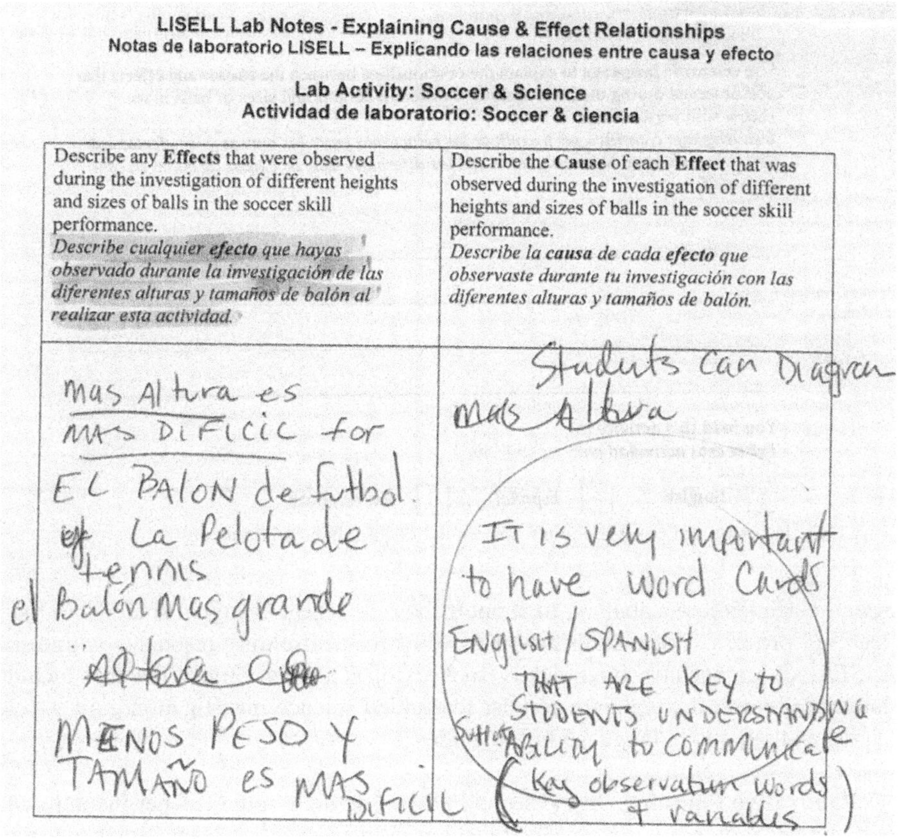

LISELL Lab Notes - Explaining Cause & Effect Relationships
Notas de laboratorio LISELL – Explicando las relaciones entre causa y efecto

Lab Activity: Soccer & Science
Actividad de laboratorio: Soccer & ciencia

Describe any **Effects** that were observed during the investigation of different heights and sizes of balls in the soccer skill performance. *Describe cualquier **efecto** que hayas observado durante la investigación de las diferentes alturas y tamaños de balón al realizar esta actividad.*	Describe the **Cause** of each **Effect** that was observed during the investigation of different heights and sizes of balls in the soccer skill performance. *Describe la **causa** de cada **efecto** que observaste durante tu investigación con las diferentes alturas y tamaños de balón.*
mas Altura es MAS DIFICIL for EL BALON de futbol. y La Pelota de tennis el Balón mas grande ~~Altura e~~ MENOS PESO Y TAMAÑO es MAS DIFICIL	Students can Diagram mas Altura IT IS very important to have word cards English/SPANISH THAT ARE KEY TO STUDENTS UNDERSTANDING Support Ability to communicate (Key observation words + variables)

Fig. 13.2 Teacher's second page of the soccer with science activity

titled 'Explaining cause and effect-Ronaldinho controlling the soccer ball' science investigation as shown in Fig. 13.2.

In this activity, teachers read an investigation guide and looked for clues as they highlighted the Spanish text that helped them understand and carry out the investigation and record their ideas. Figure 13.2 also shows how a White non-Hispanic female science teacher wrote a note on the second column of the table she filled out as part of the investigation that reads, "It is very important to have word cards English/Spanish that are key to students' understanding," and "support ability to communicate," as she was finishing recording her notes in Spanish about the investigation. As they experienced learning in a second language, similar to the situation in which many emergent bilinguals are placed, teachers recognized the need to use cultural practices such as home languages in their classrooms. They also saw the utility of specific resources, such as *concept cards* with important concepts written in both English and Spanish, explanatory graphics, and carefully crafted student-friendly definitions, and *general academic vocabulary cards* with non-discipline

Language frames for Explaining Coordinating Hypothesis, Observation, and Evidence

Teacher Language Frames

- Based on your knowledge of science and other life experiences, what do you think will happen? Why?
- What observations could you make to test that idea?
- What evidence would support your hypothesis?
- What evidence would disprove your hypothesis?
- How can you use your observations as evidence to judge your hypothesis?
- Based on your observations and evidence, can you make a new hypothesis?

Student Language Frames

- Based on my experience with _____, my hypothesis _____.
- I can observe/measure ____ to find out ____.
- Evidence that would support my hypothesis is _____.
- Evidence that would disprove my hypothesis is _____.
- Based on my observations that ____, the evidence supports/does not support my hypothesis because _____.
- Based on the evidence from my observations, a new hypothesis I have is _____.

Explicando la Coordinación entre la Hipótesis, Observación, y la Evidencia usando marcos de lenguaje

Marcos de lenguaje para los maestros y maestras

- Basándose en tus conocimientos de ciencias y otras experiencias de vida, ¿qué crees que va a pasar? ¿Por qué?
- ¿Qué observaciones podrías hacer para probar esa idea?
- ¿Qué evidencia apoyaría tu hipótesis?
- ¿Qué evidencia refutaría tu hipótesis?
- ¿Qué te dicen tus observaciones acerca de tu hipótesis?
- ¿Cómo puedes utilizar tus observaciones como evidencia para juzgar tu hipótesis?
- Basándose en tus observaciones y la evidencia, ¿puedes hacer una nueva hipótesis?

Marcos de lenguaje para los Estudiantes

- Basándose en mi experiencia con _______ mi hipótesis es _______.
- Yo puedo observar/medir _______ para encontrar que _______.
- La evidencia que apoya mi hipótesis es _______.
- La evidencia que refutaría mi hipótesis es _______.
- Basándose en mis observaciones que son _______, la evidencia apoya/refuta mi hipótesis porque _______.
- Basándose en la evidencia de mis observaciones, la nueva hipótesis que tengo es _______.

Fig. 13.3 Example of language frames posters in both English and Spanish

specific academic vocabulary, to support their emergent bilingual students in their learning process. These cards are just two of the scaffolding resources within the LISELL-B science investigation kits (see Buxton et al. this volume) that also include language booster texts, lab materials, scaffolded science investigation guides, role cards, and talk move/language frame guides.

Collaborative Planning In the second type of session in the teacher institute, *the collaborative planning for science teaching in the student academy*, science teachers from different schools who taught the same grade level and ESOL teachers gathered together to brainstorm ideas about adapting their current practices and lesson plans and constructing new ones using the LISELL-B pedagogical model. This exercise helped teachers to incorporate the LISELL-B language of science investigation practices into their planning, and to share their experiences working with emergent bilingual students in the science classroom while considering students' academic strengths, cultural and language practices as resources, and challenges. During the institute, teachers co-designed science investigation activities planned for the student academy using classroom resources such as posters that described language frames for communicating their scientific ideas through the six LISELL-B language of science investigation practices (coordinating hypothesis, observations, and evidence; controlling variables to design a fair test; explaining cause and effect relationships; owning the language of science; using models to construct explanations and test designs; see Fig. 13.3 for examples). Two additional posters were designed to highlight classroom norms for supporting a culture of science investigation and talk moves for promoting productive science discourse.

Reflection The third activity in the teacher institute, *reflection and goal setting*, provided opportunities for teachers to think about the knowledge, practices, and resources they were learning about; and then consider how these could be included in spaces such as the student summer academy and their regular science classrooms. Through this reflective process, the teachers were coming to see that the science learning spaces that they control can evolve into learning territories if they become an extension of their emergent bilingual students' learning processes, that is, if the classroom captures the affectivity, emotions, and interests of the students who inhabit it (Didakis and Phillips 2013).

Visits The *science and engineering laboratory visits*, the fourth activity in the teacher institute, supported the teachers in thinking about the connections among the LISELL-B pedagogical model, the middle and high school science curriculum, their emergent bilingual students' pursuit of college and career pathways, and the student summer academy. These teacher laboratory visits also provided previews for the laboratory visits that the students in the summer academy would participate in; the specific goals for the student lab visits are described in the next section regarding the Student Summer Academy.

Using L2 The fifth activity in the teacher institute, daily *Spanish language lessons*, functioned as a language support and as an informal bonding experience for both the science teachers and the ESOL teachers. Here, teachers, according to their personal experience and needs, learned basic Spanish words and phrases with the goal of communicating with and building stronger relationships with their emergent bilingual students and their students' families. LISELL-B teachers and teacher educators, drawing on experiences in the Summer Academy with small numbers of students speaking home languages other than Spanish (such as the Karen language) are considering ways to adapt project materials and practices to support emergent bilingual students in multilingual and multiethnic settings.[5]

The activities of the teacher institute were co-designed and put into practice by a group of teacher educators and teachers who emphasized collaboration and group work. Not only did this collaboration between science and ESOL teachers and teacher educators enhance the richness of each activity by incorporating different perspectives, but it was also an opportunity to share and learn from each other's experiences working with emergent bilingual students as lessons were collaboratively planned. As a consequence, the design of the student summer academy was based on enriching science content knowledge, supporting curricular requirements, and validating emergent bilingual students' linguistic and cultural practices connected to the local context, at the same time as it incorporated the LISELL-B pedagogical practices. The student summer academy offered teachers a more flexible space to try new activities and approaches when compared to that of their regular school year science classrooms.

[5] Recent research about teaching science with Karen refugee group (Harper 2015) is serving as a resource.

Student Academy

After the teacher institute finished, the ***student academy*** took place in the same university facilities where teachers implemented science investigation activities based on the LISELL-B pedagogical framework using bilingual resources, physical objects, and the indoor and outdoor spaces. The academy provided teachers with an interactive professional learning context, and at the same time engaged students in language-rich science learning environments.

Investigations In the student academy, one central activity was the *teacher-directed science investigation sessions*, designed to put into practice the language of science investigations that the teachers planned during the teacher institute. An additional purpose was to evaluate the materials, directions, resources, and procedures they had developed so they could be further modified for classroom use. Many of these activities were developed into science investigation kits for teachers to check out and use in their own classrooms during the school year.

Problem Posing In the *problem posing project activity*, each student brainstormed three possible research/problem interests and posted them on the board (see Fig. 13.4). After collecting and categorizing all the research/problem interests, students chose one to work on with one or two other students during the summer academy. Students worked collaboratively, with support from teachers and mentors and using the LISELL-B language of science investigation practices, to set guiding questions, search for information about possible relevant causes and effects and variables, elaborate on findings, and construct presentations to share with families.

Visits The goal of the *science and engineering laboratory visits* was to increase the emergent bilingual students' perspectives on science and engineering practices and professions as they participated in guided visits to different university laboratory facilities. These visits also helped students and teachers to connect current science

Fig. 13.4 Students' problem posing project ideas

research with science learning taking place in the summer academy. The group visited research labs in the departments of kinesiology, public health, and animal sciences; the college of engineering; and a microbiology research center.

Community Building 1 After lunch, students went to the athletic fields to play soccer for 40 min a day. Latino/a and Karen students and boys and girls played soccer in mixed teams and some of them played soccer matches, while others just kicked the ball around the field. On the final day of the student academy we had *a soccer tournament* in which students, teachers, and teacher educators participated. The ambiance in the tournament, incorporating an important informal cultural practice for local Latino/a and other immigrant communities, further encouraged the bonding process between teachers and students in the student summer academy.

Community Building 2 At the end of each week, students had the chance to *present their problem posing projects to their families* before a shared dinner with teachers, students and families. In university classrooms, students used the language of science as they explained to the assembled families, peers, and teachers their research/problem topic of interest, relevant causes and effects and variables they had identified, and ideas about next steps in their inquiry.

Teacher Experience of Building Science Environments for Emergent Bilingual Students

A learning territory is in the process of construction when participants take active part in and collaboratively build the activities that occur in a particular space. As Heidegger (1975) suggests, we can dwell in these spaces only when we construct them and when we domesticate them. This architectural perspective allowed us, as teachers and teacher educators, to identify, conceptualize, and enact the processes incorporating material and expressive elements that became relevant for constructing learning territories where science and ESOL teachers supported emergent bilingual students in learning language-rich science investigation practices. This architectural perspective emphasizes the role of the physical space in which the learning occurs. Put simply, the location is a necessity because no learning happens without being situated in a location. However, the relation we have with a location depends on whether we engage it as a learning resource or as something distinct from or extraneous to the learning process.

During the teacher institute and the student summer academy, three important processes were emphasized that supported the creation of learning territories. The first process was the importance of including students' home language as a resource for their learning, to which one science teacher reflected,

> I think that the LISELL[-B] program brings that ELL perspective that I am not familiar with. I didn't grow up, I didn't move to any state, I didn't move anyways... or I don't know anybody and I've never been to an area where they don't speak my language. [...] So I've

> never been in a situation where I cannot bridge a way of communicating and I think this program really brings this perspective that allows me to see how to bridge that gap and what strategies I can use to help those students, that are ELL students, to feel more comfortable in the classroom.

The second process in the teacher institute that was important for creating learning territories emphasized the use of science inquiry along with the structure of the LISELL-B pedagogical model to design science investigations for teachers' own classrooms, to which one of our teacher participant shared,

> I think the [LISELL-B] practices are really great because they provide that inquiry-based learning to the kids… As a teacher, to highlight those practices within the units has been very helpful, so I know that when I do a lab that has variables, and then I make sure to highlight those practices, which are now incorporated throughout our lessons. Cause and effect is a topic that I am now more intentional when I use those words, so being through this program has allowed me to think more about my planning and how I am going to phrase the lesson and use the wording in the lesson, make sure I highlight, for example, the cause and the effect nicely, because sometimes they get those two switched easily.

The third process we stressed in the institute for creating learning territories was the use of the particular learning space as a key element to the learning process, especially for the emergent bilingual students, with the purpose of supporting students in domesticating the space they inhabit. In this light, another teacher shared that, "the communication component is the most important for me. The sentence starters that you showed in the posters are very useful for me. I think I can use them in my classroom." Another example is the general academic vocabulary and concept cards as mentioned by the science teacher in Fig. 13.2. This shows teachers thinking about how to prepare the classroom learning space to be ready to interact with and be used by students, but this material element has to be systematically included and utilized in the activities in the class so that students domesticate their use and the location where the material elements are being provided. Another science teacher added,

> I think that participating in the LISELL-B project has helped with pushing a focus on the vocabulary that it didn't necessarily happen before, so, understanding the importance of breaking down vocabulary into the root words, showing students how to interpret phrases or modeling that, using the Spanish-English interpretations, and focusing on particular vocabulary words instead of the majority of it, because in science we have a very large vocabulary to be able to focus on maybe six of the most [important], I think it has really encouraged me in my classroom to take those tools and skills and make sure that I am always thinking with my ESOL hat on or think about my ELLs that I don't necessarily… I didn't have that on before I entered into this program.

With these three processes, science and ESOL teachers had the chance to collaborate in learning, designing, and enacting the LISELL-B language of science investigation practices with other middle and high school teachers and teacher educators and to plan for integrating those practices into their own classrooms. When teachers design and put into practice the different elements of the LISELL-B pedagogical model, they go through a similar process to that of the students when they build a learning environment, one that will allow them to support all students and especially emergent bilingual students in their learning process, a necessary step to building a learning territory. As one teacher mentioned, "So what we are doing is translating this very dense language not just only for newcomers but for everybody. And to me all the LISELL-B [activities] are very sound educational practices for all, regardless of the language ability. It is good teaching," and he continued, "It's just like the soccer activity in Spanish... I get that now."

Results and Implications

Our research with the professional learning components of the LISELL-B project confirms and expands the assertion that educators committed to equity for all students do not and cannot work in isolation (Allexsaht-Snider and Hart 2001). It also shows that science teachers can play a critical role in designing and implementing the use of science investigation activities, group work, bilingual materials, and resources for supporting their emergent bilingual students. This approach serves to structure the classroom setting to promote science thinking, the use of the language of science in communication about science learning through investigations, and science collaboration with peers. For instance, the group of students working on the political conflict topic wrote that, "The cause is people that are open minded and kind and understanding" and as the "effect is there will be more peace and less political conflict." the group of students researching about the relation between exercise and learning, after visiting a kinesiology exercise research lab and considering variables related to their investigation, wrote that, "it is beneficial because when you exercise your brain gets healthier and so does your body;" another student wrote that, "it's good because people that are fit have good brains and have a lot of energy." In an investigation coordinating hypothesis, observation, and evidence and using indicators to test for the presence of macromolecules in food from many restaurants and fast food places where students work and eat, a student answered that,

> My hypothesis is the victim's last meal was at Waffle House. I use results indicators to test [for] the presence of Lipids, proteins, and carbohydrates. Indicator's test for starches and lipids are positive. Indicator's test for sugar, glucose, and protein are negative. The victim's last meal was at Waffle House because of the presence of starches and lipids. The evidence supports my hypothesis.

Reflecting on student writing such as excerpted above, science teachers saw the utility of the LISELL-B pedagogical model with their emergent bilingual students in the summer academy, viewing the model as aiding them in building a bridge that would also provide continuity in applying the practices in their classrooms during the upcoming school year. For instance, a 6th grade earth science teacher shared that,

> I think these investigations are really beneficial for these students because they have to be engaged and do things in the classroom [...] so we are constantly taking what we learned and using these investigation practices in our classroom, whether it is the language boosters, or practices like cause and effect and the variables-practices that we put together within the activities that we do in the classrooms.

Moreover, this model helped both the emergent bilingual students and the science teachers to connect their science knowledge, interests, and practices to the laboratory visits and industrial practices in science. In regard to how teachers think about connecting and enacting these practices in higher education institutions where the student academy took place, one participant noted,

> We are seeing them [the emergent bilingual student population] staying and hanging on because they know that, one, they are intelligent, and two, they see their peers going to college. This program we all are involved in, it is one of the best things that shows them that there are many colleges and I think these kids realize that there are more opportunities available to them and there is a lot of opportunities in science, technology. And a lot of our kids are good at, I mean, we do a lot of labs and a lot of our ELLs are really good at doing things, making things… nonverbal communication. And I think as they learn the language of science they realize that, man, I am good at this, I can do this, I can make a living in this. I think that opens up more opportunities.

In order to understand and work toward change in the current educational landscape, the use of an architectural perspective can provide the tools for science teachers who are looking to balance their students' educational needs with curricular demands by strengthening connections between secondary and postsecondary educational institutions and other resources in the community.

When emergent bilingual students have purposeful experiences that involve visiting a higher education setting such as a university or a science or engineering laboratory under the guidance of science teachers and university educators, they can make meaningful connections between the content and science investigation practices explored in class to the practices they observe in the laboratories. We recommend that science teachers continue to build stronger connections to post-secondary institutions and organize purposeful visits for all their students, but in particular for their immigrant and emergent bilingual students, to higher education settings and laboratories. An architectural perspective highlights how such experiences benefit science teachers by supporting the propagation of inclusive science learning environments for all students, while also offering a richer experience to continue learning about industrial and other career sectors where employees with STEM expertise are in demand.

When science teachers actively participate in programs where they bring and use their knowledge, ideas, and experience to design and implement new educational practices in dynamic, interactive spaces, the chances of teachers adopting and adapting these practices in their school classrooms increase. This finding is supported by the current high demand for the LISELL-B science investigation kits in the school classrooms. At the same time, when emergent bilingual students experience these science practices using their language and cultural practices as learning supports, the chances of students incorporating new knowledge also increase. The purposeful university and laboratory visits also reinforce the possibilities of students dwelling in these higher education spaces while linking science learning to career practices. That is, the purposeful visits link an idea, an experience, and a practice to a space where students see those experiences and practices being used and valued.

Challenges still exist in creating and sustaining initiatives that support science teachers in better meeting the needs of emergent bilingual students. These challenges include: coherent integration of out-of-school experiences with the regular science classroom settings; collaborating in the implementation and elaboration of science investigation kits for classroom use; engaging and supporting families in science learning and college preparation; connecting middle and high school science education to college and science careers; and promoting the use of both English and Spanish to support powerful science learning for emergent bilingual students who speak Spanish. More research and development work is needed, in a wide range of contexts across the United States and in other countries educating multilingual populations. We hope that what we have learned along with the teachers and students in the LISELL-B project can help others to construct viable and sustainable models for supporting current and future science teachers towards these goals.

References

Allexsaht-Snider, M., & Hart, L. E. (2001). 'Mathematics for all': How do we get there? *Theory Into Practice, 40*(2), 93–101.

Bogue, R. (2007). *Deleuze's way: Essays in transverse ethics and aesthetics*. Burlington: Ashgate.

Buxton, C., Allexsaht-Snider, M., Kayumova, S., Aghasaleh, R., Choi, Y., & Cohen, A. (2015). Teacher agency and professional learning: Rethinking fidelity of implementation as multiplicities of enactment. *Journal of Research in Science Teaching, 52*(4), 489–502.

Buxton, C., Hernandez R. Y., Cardozo G. L., & Kimarci, M. (this volume).

David, J. L. (2011). Research says .../high-stakes testing narrows the curriculum. *Educational Leadership, 16*(6), 78–80.

De Landa, M. (2006). *A new philosophy of society*. New York: Continuum.

De Landa, M. (2010). Material(ism) for architects: A conversation with Manuel De Landa. *Cluster.* Retrieved from http://www.cluster.eu/2010/10/08/materialism-for-architects-a-conversation-with-manuel-delanda/

Deleuze, G., & Guattari, F. (1987). *A thousand plateaus: Capitalism and schizophrenia* (B. Massumi, Trans.). Minneapolis: University of Minnesota Press.

Didakis, S., & Phillips, M. (2013). Objects of affect: The domestication of ubiquity. *Technoetic Arts: A Journal of Speculative Research, 11*(3), 307–317.

García, O., & Kleifgen, J. A. (2010). *Educating emergent bilinguals: Policies, programs, and practices for English language learners*. New York: Teachers College Press.

González, N., Moll, L., & Amanti, C. (2005). Introduction: Theorizing practices. In N. Gonzalez, L. C. Moll, & C. Amanti (Eds.), *Funds of knowledge* (pp. 1–28). New York: Routledge.

Gordon, C., & DeBard, A. (2014). *CEOs say skill gap threatens U.S. economic future*. Retrieved from Washington, DC: http://changetheequation.org/press/ceos-say-skills-gap-threatens-us-economic-future

Gutiérrez, K. D., & Rogoff, B. (2003). Cultural ways of learning: Individual traits or repertoires of practice. *Educational Researcher, 32*(5), 19–25.

Harper, S. (2015). Keystone characteristics that support cultural resilience in Karen refugee parents. *Cultural Studies of Science Education, 2*(10), 1–32.

Heidegger, M. (1975). *Poetry, language, thought*. New York: Harper & Row.

Lee, O., & Buxton, C. (2011). Engaging culturally and linguistically diverse students in learning science. *Theory Into Practice, 50*(4), 277–284.

LISELL-B: Language-rich Inquiry Science with English Language Learners through Biotechnology. (2016, February 19). Retrieved from https://coe.uga.edu/research/grants/lisell-b

Milner, H. R., & Laughter, J. C. (2015). But good intentions are not enough: Preparing teachers to center race and poverty. *The Urban Review, 47*(2), 341–363.

National Center for Education Statistics. (2014). *Racial/Ethnic enrollment in public schools*. Retrieved from http://nces.ed.gov/programs/coe/indicator_cge.asp

Organisation for Economic Co-operation and Development. (2013). *PISA 2012 results* Retrieved from http://nces.ed.gov/surveys/pisa/pisa2012/pisa2012highlights_4e_1.asp. Science literacy: Student race/ethnicity. Retrieved January 27, 2014.

U.S. Department of Education. (2004). *Percentage distribution of public school teachers by race/ethnicity, percentage minority, and state: 2003–04*. Retrieved from https://nces.ed.gov/surveys/sass/tables/state_2004_18.asp

United States Census Bureau. (2014). *As the nation ages, seven states become younger, Census Bureau reports*. Retrieved from http://www.census.gov/newsroom/releases/archives/population/cb14-118.html

Chapter 14
An Alternative Approach to Educating Secondary Science and Mathematics Teachers: Meeting the Needs of Culturally and Linguistically Diverse Youth

Cecilia M. Hernandez, Jamie S. Baker, Christine M. Reyes, and Lida J. Uribe-Flórez

Introduction

Despite efforts of the education community to attract and retain high quality secondary Science, Technology, Engineering, and Mathematics (STEM) teachers nationwide, there is still a teacher shortage; particularly in culturally, linguistically, and economically diverse communities. The U.S. Department of Education, Office of Innovation and Improvement (2004) revealed that rural school districts face serious issues related to the recruitment and retention of new and experienced teachers. In light of more rigorous standards, a beginning teacher must possess the content knowledge, pedagogical skills, and dispositions for teaching in high-need schools. However many states are falling short of providing adequate teachers due to increasing student enrollment, retirement, attrition of novice educators, and low production of secondary STEM graduates from traditional teacher preparation programs. In response, many states and institutions of higher education have partnered to create alternative routes to certification. Over time "alternative" has acquired a number of connotations to mean sub-standard support for individuals teaching on waiver to highly structured and well-designed programs. This chapter seeks to identify how one high-quality sponsored program, *Aggie Prep*, develops practicing secondary

C.M. Hernandez (✉) • J.S. Baker
New Mexico State University, Las Cruces, NM 88003, USA
e-mail: cecimh@nmsu.edu; jabaker@nmsu.edu

C.M. Reyes
University of Texas at El Paso, El Paso, TX 79902, USA
e-mail: drcmr2013@gmail.com

L.J. Uribe-Flórez
Boise State University, Boise, ID 83725, USA
e-mail: Lidauribeflorez@boisestate.edu

A.W. Oliveira, M.H. Weinburgh (eds.), *Science Teacher Preparation in Content-Based Second Language Acquisition*, ASTE Series in Science Education,
DOI 10.1007/978-3-319-43516-9_14

science and mathematics alternative certification candidates' understanding of relationships between standards-based conceptual mastery and culturally responsive pedagogy that meet the needs of diverse youth. Using a standards-based approach, we present literature on language acquisition in the content areas (science and mathematics), describe the teacher preparation model, and highlight effective professional development experiences supported by best practice that address the need to (1) make science and language learning accessible; (2) elevate content rigor; and (3) build collaborative learning environments with attention to integrated content, language and literacy practices. We sought to accomplish all of these goals using active science and mathematics activities that provided opportunities to gain language proficiency and construct knowledge in the content area.

Standards-Based Professional Development Approach

"In order to improve the educational outcomes of culturally and linguistically diverse students in general, research indicates that reform is needed in teacher education to more adequately prepare teachers to meet the needs of *all* students" (Hernandez et al. 2013, p. 803). In order to address this issue from a science education perspective, work to create a new framework for K-12 science education began with a focus on three key dimensions:

1. Scientific and engineering practices;
2. Crosscutting concepts that unify the study of science and engineering through their common application across fields; and
3. Core ideas in four disciplinary areas: physical sciences; life sciences; earth and space sciences; and engineering, technology, and applications of science (NRC 2012, p. 2).

The framework aided the development of a new set of science standards released in 2013, the Next Generation Science Standards (NGSS 2013) based on identified understandings of teaching and learning, particularly in science education. In their report for the Office of Science Education National Institute of Health, Bybee et al. (2006) focused on the statement that, "The sustained use of an effective, research-based instructional model can help students learn fundamental concepts in science and other domains" (p. 1). The report described the importance of teaching science through the use of instructional models that makes connections for students and illustrates, "learning results from an interaction between what information is encountered and how the student processes that information based on perceived notions and extant personal knowledge" (p. 15). For linguistically diverse students, these standards value the opportunity to merge the context of science, mathematics, and funds of knowledge (González et al. 2005) youth bring to the classroom into cohesive learning experiences.

Prior to the changes in science education curriculum, the Common Core State Standards (CCSS 2010) were introduced and implemented in the majority of states across the country. By the end of 2012, 46 states, three U.S. Territories and the District of Columbia voluntarily adopted the CCSS. The English language arts reading and writing standards for literacy in science and technical subjects require students to have regular practice with complex texts and their academic language; reading, writing, and speaking grounded in evidence from texts, both literary and informational; and building knowledge through content-rich nonfiction (2010). In order to meet the demand of the literacy standards anchored in a focus on career and college readiness, students in science must engage with technical and non-technical text, summarize complex concepts and processes, and cite textual evidence from a range of sources to support analysis and findings. Even though several states have withdrawn their support amid the political controversy surrounding the standards, the key shifts initially called for by CCSS resonate nationwide and argue that subject matter content is an authentic context for learning language and increasing content rigor. Coupled with the "prominent focus of NGSS on productive language use via the identification of language-intensive science and engineering practices has opened up new possibilities for all science teachers to consider the role of language in science and engineering instruction" (Tolbert et al. 2014, p. 68). Secondary science and mathematics teachers who traditionally identify themselves as teachers of content, rather than teachers of language, often need support re-conceptualizing their role in fostering a content, language, and literacy-rich environment. Valdés et al. (2014) add that the era of new standards specifically pose two key challenges in the expertise of ESL teachers: (1) language practices required by the new standards and (2) inclusion of ELLs in standards-aligned instruction.

Similarly, Second Language Acquisition (SLA) literature indicates there are fundamental stages in acquiring both the first and second language. Cummins (1979) proposed that the first and second languages are interdependent on a cognitive level, in what he refers to as the Common Underlying Proficiency (CUP). As a student develops a second language, his/her knowledge of a first language in phonology, morphology, semantics, syntax and pragmatics are used to transfer to the new language. Krashen (1982) developed five hypotheses regarding SLA, suggesting there is a natural order for the progression of language development. It is a predictable sequence in which learners progress from listening to speaking and then to reading and writing in the new language. This complex process occurs over time and is aided by direct language instruction in a context rich environment that includes content rigor. The Center for Research on Educational Diversity and Excellence (CREDE 2015), established by the US Department of Education, suggests five standards for effective pedagogy and learning within a grounded sociocultural framework that is language rich, contextualized and content specific at all grade levels for culturally and linguistically diverse learners. These standards of practice value content rigor and support principles of effective pedagogy incorporated in *Aggie Prep*.

Teacher Preparation Model – *Aggie Prep*

Recruiting individuals with strong backgrounds in science and mathematics has the potential to bring rich personal and professional experience to the secondary classroom. Chambers (2002) found career changers to be more cognizant of diversity and more likely to adopt methodologies that were more student-centered and helped to make connections between content in the classroom and the outside world. These individuals tend to carry a belief that they can impact students' aspirations to pursue careers in science. However, they also feel challenged to reach lower-achieving students who demonstrate little interest in science and mathematics. For this reason, the program chose a competitive portfolio and interview application process to identify individuals with dispositions and skills the model could support in terms of serving students' needs in the field as well as the process of forming a teaching identity. The ideal participants are those with strong content backgrounds, a desire to serve in diverse communities, and basic proficiency in online learning environments. Initially the program sought to select up to 15 science and mathematics teachers however, after the selection process was completed we selected seven teachers (four science and three mathematics teachers) who met the selection criteria. Once members of a cohort are identified, they participate in a range of collaborative experiences that prepare them to be responsive to the unique language and cultural diversity of the southwest border communities. In this document, we are sharing our experiences with this first cohort.

Teacher Preparation Program Features

Since participants work in rural communities with many language learners, it was imperative to design the program to support novice teachers with understanding language acquisition through their content area instruction. Several key programmatic features define *Aggie Prep* (Fig. 14.1). Like other alternative pathways, participants work as the full time teacher of record while completing the 12-month program that leads to initial teacher certification. *Aggie Prep* expands pedagogical content knowledge through an online learning environment that supports peer and faculty collaboration. Each competitively selected participant is matched with a master teacher for site-based mentorship and co-teaching experiences. They are also provided monthly professional development and field coaching by a highly qualified university clinical educator (CE). The program faculty made field visits to observe the participant, model best practices and support culturally responsive pedagogy (Rychly and Graves 2012). The reflective capacity of each participant to engage students through data-driven instruction is developed by regularly required videotaping and analysis of classroom instructional practices. The support for each member of the cohort continues in the form of field coaching and expanded professional development in his/her second year of teaching. Lastly, upon successful

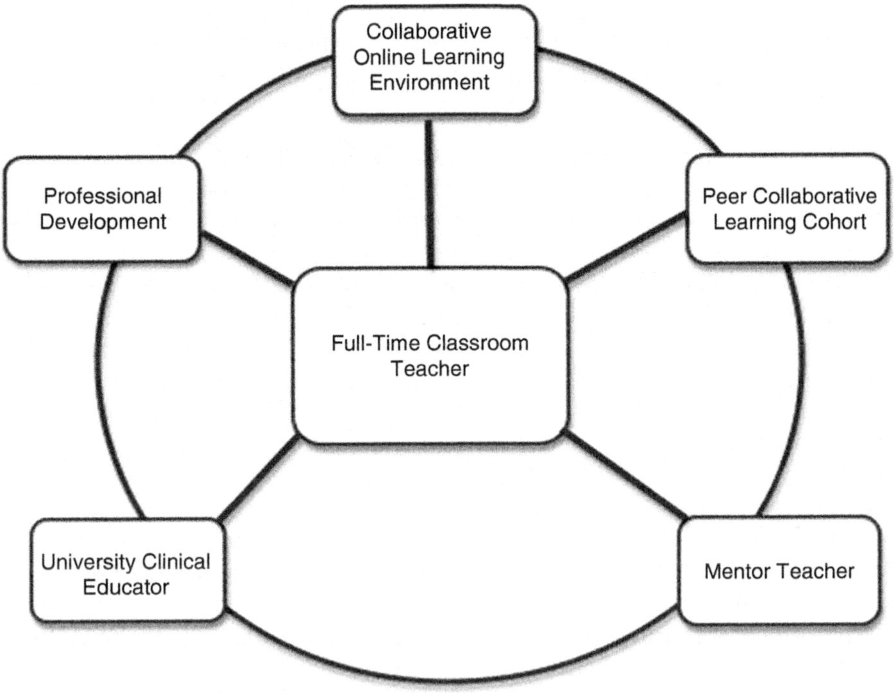

Fig. 14.1 Aggie Prep program key features

completion of *Aggie Prep*, participants with a 3.0 GPA or higher are eligible to take an additional fifteen credits of coursework at the university toward earning a master's degree in curriculum and instruction.

Program Implementation

Program Overview As faculty of a nationally accredited NCATE/CAEP teacher education institution, we understand the value of utilizing frameworks and standards that guide program development that are responsive to the local needs of our community and globally competitive. Throughout each *Aggie Prep* candidate's program of study (admission to completion), growth and progress is monitored through a systematic and ongoing performance assessment process that includes, but is not limited to: key assignments, observation of teaching in the field with feedback, mentor teacher assessment of field performance, self-reflection, and licensure exams. In the Content Rigor section below, we provide a description of some of the key assignments used to evaluate course completion. University faculty, clinical educators, and building mentors all use a departmental observation form to ensure all candidates meet expectations, and a debriefing session takes place after each observation. Our courses, curriculum, observations, and assessment instruments are aligned with the InTASC Model Core Teaching Standards (2013), New Mexico Secondary Entry-Level Teacher Competencies, and New Mexico Teacher Evaluation criteria.

Participants in the cohort are content experts, however, none of them had experience teaching in a public school environment prior to beginning their first year in the alternative licensure program. The program includes online courses and professional development opportunities that integrate pedagogical content rigor and collaboration between science and mathematics education faculty. Participants are required to enroll in six three-credit courses over the span of two semesters and one summer session (Appendix A). In addition to coursework, the program includes two face-to-face professional development sessions led by *Aggie Prep* faculty and attendance at a state STEM conference. It was important for the initial professional development session to include face-to-face interactions that established clear expectations and a sense of community since participants were selected from districts across the state.

The online coursework in *Aggie Prep* is not viewed separately from site-based mentoring, field observation and coaching. Rather, the program faculty, clinical educators, and mentors communicate monthly through conference calls to evaluate the progress of the participants and discuss their needs. The faculty responded to the feedback by adjusting the schedule of topics (Appendix B) for the weekly interactions via our university's course management system (Canvas™) to address specific classroom interactions that were of immediate concern. Figure 14.2 presents the course homepage, where participants have access to information about the course, syllabus, and grading. The timeliness of communication created opportunities for the *Aggie Prep* faculty to collaborate, design, implement and model active learning inquiry based lessons in the participants' classrooms during site visits.

Build Collaborative Learning Environments In order to build a collaborative learning environment, all courses in *Aggie Prep* are designed with the unique context of the participants in mind. For example, the STEM methods course was co-taught combining teaching strategies for Science and Mathematics classrooms. Through the use of our Canvas courses students have access to the discussion boards, announcements, key assignments, document files and modules that indicate the course activities for each week (Fig. 14.2). The Methods course was organized using modules (Fig. 14.3), which includes readings, discussion board, and assignments. The science and mathematics methods instructors met over the course of a semester before the program began to align their syllabi. The instructors grew professionally and learned throughout the course design and implementation process. They combined best practices from their areas of expertise to create a progression of experiences that would support better understanding of the content and how to choose appropriate instructional activities to engage all learners. This was an important component considering participants were all teaching in bilingual/bicultural communities in the Southwest.

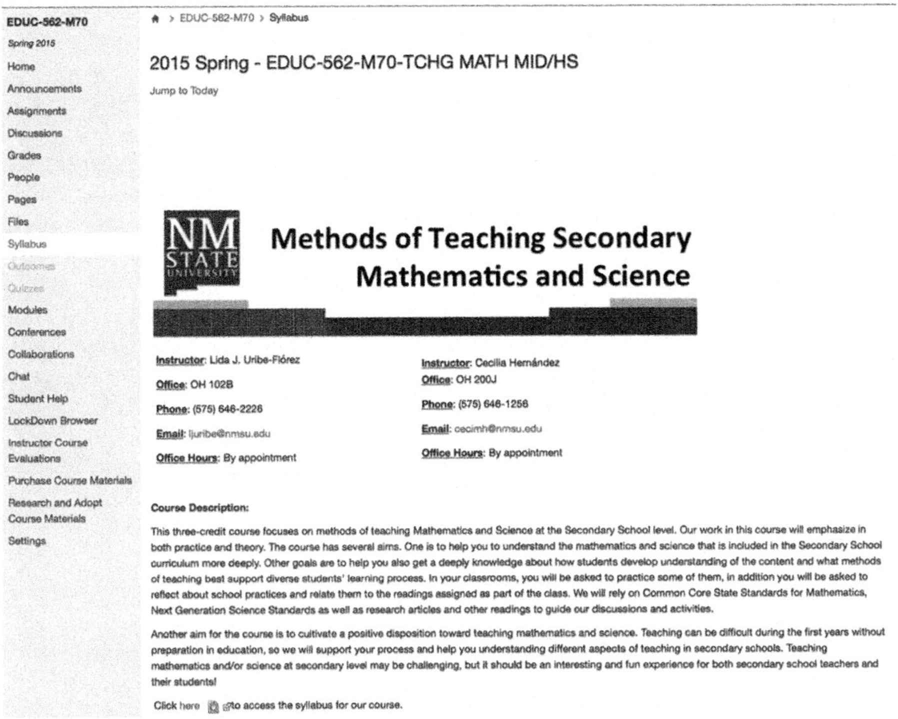

Fig. 14.2 Course management homepage

Our goals for the course emphasized both theory and practice. Throughout the semester we sought to help deepen the participants' understanding of secondary science and mathematics curriculum. We also wanted to assist our participants' in expanding their knowledge about how students develop understanding of the content and what methods of teaching best support diverse students' learning processes. Each week, participants had specific readings, assignments, and ongoing discussions to complete via the course management system (Fig. 14.3). We wanted to create a virtual collaborative learning environment so all students were required to interact with their peers for each weekly discussion. For example, one of the discussion prompts stated:

> For this discussion, we would like to include your understanding and concerns regarding the Common Core State Standards for Mathematics (for mathematical practices and content) as well as the NM State Science Standards. Remember to post your comments by noon on Sunday, and reply and/or comment on at least two of your classmates' responses by Sunday evening.

In addition to the weekly discussion topics, all participants were required to "meet" synchronously online as a cohort once a week. These synchronous meetings were facilitated through the university's course management system which through Adobe Connect ™. Each participant was able to interact with their instructors and

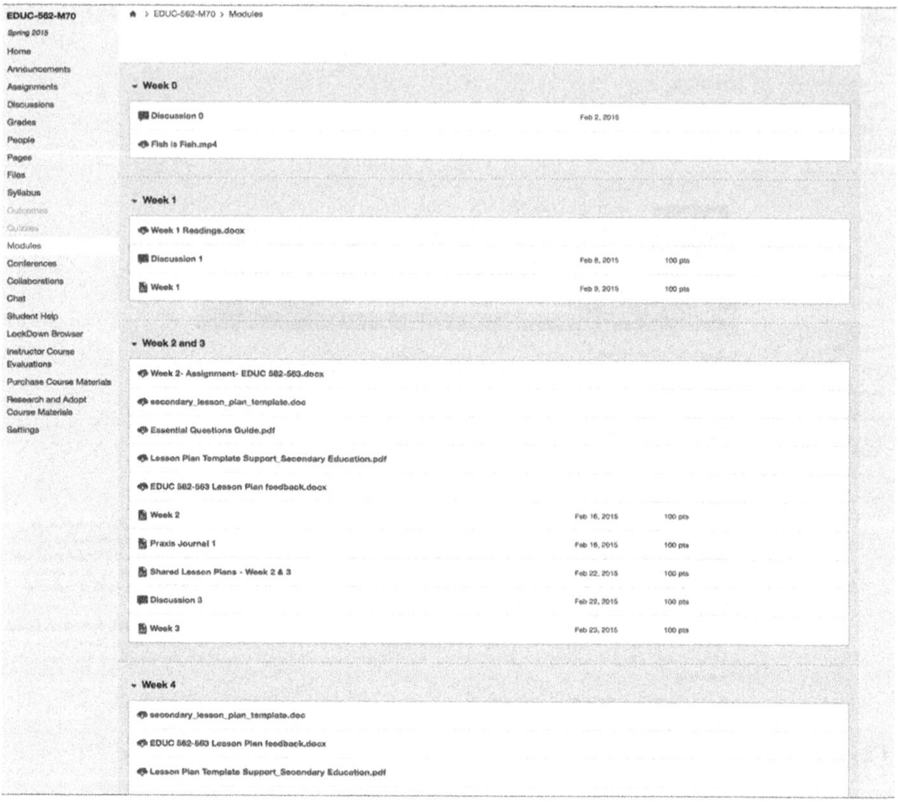

Fig. 14.3 Science and mathematics methods course modules

colleagues in real time to ask questions and get advice on how to implement the different methods introduced throughout the course. The main idea of the weekly meetings was to foster the community building and reflection process within the online environment (Duemer et al. 2002). Ultimately each participant could regularly utilize the expertise of the co-instructors and their peers to reflect and build competencies necessary for student success.

Elevate Content Rigor One feature of the course was to instruct participants in writing and delivering lesson plans based on the 5E Model (Bybee et al. 2006). This model is based on several learning cycle models that emphasize a hands-on, inquiry-based approach to learning science, and is a structure that can be used in planning active learning mathematics activities (pp. 4–10). Additionally, participants were introduced to the *Secondary Education Template* (Appendix C) that was developed by the secondary education department, and was structured with diverse students in mind. Within the lesson description, the participants use the 5E Model (Bybee et al. 2006) to address the learning objectives, and the structure to provide scaffolding opportunities for language learners and other diverse student needs. Several of the class discussions centered on how to address the needs of the language learners in the classroom while maintaining high expectations for science and mathematics knowledge construction.

In order to assist the beginning teachers in making the content relevant and grade level appropriate, we adapted the high cognitive demand mathematics task based on Stein et al. (2009) for STEM areas. The goal of this project was for the participants to learn how to evaluate, design, and implement activities that promote high order thinking in secondary science/mathematics students. We used the adapted rubric from Stein et al. (2009), to discuss and evaluate tasks created by participants to ensure high cognitive demand was present. Each participant was to select an ELL or bilingual secondary school student in a science/mathematics classroom who was not a relative or friend. They were asked to meet with this student three times during the semester and focus on working on science/mathematics tasks they designed. The goal for each activity they designed was to elicit high order thinking in the content area. The three meetings were required so that participants could evaluate the implementation, and create a new activity for the student that reached a higher level of thinking each time they met. The participants were given the opportunity to design each task and then discuss it during one of our weekly meetings before implementing it with the student. As a learning community, peers and instructors, could assist in determining the cognitive demand level, and modify if needed, before working with their chosen student. Following each meeting with their student, they were asked to reflect upon the process in order to help them write the next task. After all tasks were completed, each participant wrote a reflection on the whole process, which was submitted during finals week.

Plan with the Needs of Language Learners in Mind Once participants became more familiar with the lesson planning process and how to create active learning opportunities, the next step was to teach them how to achieve more depth in planning and teaching with the needs of language learners in mind. This process supports what Lee and Buxton (2013) identify as the types of strategies that effective teachers use to communicate ideas using multiple modes of representation. This concept was introduced during a face-to-face professional development workshop. Participants were placed in a situation of being language learners, so they could identify how their students may feel and identify strategies utilized to work with language learners. In this case, one of the *Aggie Prep* instructors led a mathematics activity fully in Spanish, and modeled strategies suggested in the literature (i.e., Cummins and Swain 1986; Lesh 1979; Uribe-Flórez 2014). The main idea was the use of multiple representations, context and grouping to support language learners in their mathematics learning process. The classroom visits by university faculty, clinical educators and building mentors also reinforced these strategies.

When teaching science and mathematics content to linguistically diverse students it is important to have clear content and language objectives so as to increase academic vocabulary development. Increasing vocabulary through the content methods adds to a student's word knowledge. A student's maximum level of reading comprehension is determined by his or her knowledge of words. Word knowledge allows students to comprehend text. In science and mathematics it is important to help students understand word development. As many science and mathematics content words are from Latin and Greek roots, word meaning can be explicitly taught to improve comprehension. Knowing a word means knowing it in all of the following dimensions:

- The ability to define a word (e.g., photosynthesis or sum)
- The ability to recognize when to use that word (e.g. mol or cm)
- Knowledge of its multiple meanings (e.g., control and table or mode and median)
- The ability to decode and spell that word (e.g. metamorphosis or denominator

Teaching key-words within the content assists students in comprehending texts, learning the content in the texts, and increases academic success in formative assessments. Words are taught through direct instruction of word meanings as well as through discussions about words (including prefixes, suffixes, and roots) – all combined with reading in the content (Beck et al. 2002). The *Aggie Prep* participants practiced planning for English Language Learners in a Content Area Literacy course. For instance, students identified appropriate content-based informational text, read with an eye for where students might struggle, and developed targeted before, during, and after comprehension and vocabulary strategies to support learners. Specifically, students practiced identifying appropriate text using the article of the week strategy (Gallagher 2009).

Tovani (2004) suggests, "Accessible text help students make a connection between school subjects and the real-world because it helps them experience reading that is done in the real world" (p. 39). It was an aim in the *Aggie Prep* program to teach each participant not to limit students' ability to think about content because the textbook is too difficult or sparse. Teachers who avoid over-reliance on the textbook create multiple sources of text at various reading levels, select from a range of options based on interest and background, and individualize instruction with more small-group work and less large-group lecture (Allington 2011). Each participant was invited to select an upcoming topic of study and develop a text set that became the basis for planning a full unit. Each participant found seven varied examples of text (print and non-print) that related to their unit of study. An annotated bibliography was written for each text so the beginning teacher could think through how to use the text with specific reading strategies and anticipate places where learners might struggle (see Daniels and Zemelman 2004). For instance, one student explored text related to Genetics of Inheritance: Alleles, Mutations, GMOs, and evolution. While another identified Earth Science related text for a 9th grade Integrated Science class.

After the text set was complete, the participant created five sequential lesson plans connected to the unit that incorporated at least three of the text sources. The unit articulated the intention guiding the selection of sequence, structure of information, and strategies used to form a cohesive, standards-based learning experience for students. The unit plan was presented to the class. The cohort chose to create a shared Dropbox™ since many high quality resources were identified in the process of searching for age, level, and content-appropriate text. Several participants worked in locations with limited access to technology and materials, so this process helped broaden the planning process and available resources beyond the textbook. As the field coaches observed some of the lessons in practice, they were able to reinforce teaching strategies that cannot fully be addressed through planning alone such as speaking slowly and audibly, assisting with one-to-one instruction when appropriate, and using visual cues as much as possible. Participants also filmed their teaching, uploaded it to the course management system and reflected on the effectiveness of their instruction.

Outcomes

As previously detailed, the overall goal for *Aggie Prep* was to build the participants' ability to reflect on the nature of science and mathematics, how adolescents learn, and implement instructional best practices of science and mathematics that accommodate the language and learning needs of students in their classrooms. These aims were best accomplished through the university faculty, clinical educator, building mentor, and participant collaboration and engagement in the learning process. Regular communication supported each team member's ability to support the daily work of each participant. The context for each participant was initially gathered using a competency-based pre-assessment tool (Appendix D). The faculty facilitated a discussion with all of the participants and their mentors at a face-to-face orientation at the onset of the program with all seven participants, four lead faculty, two clinical educators, and five mentors in attendance. One of the purposes of the orientation program was to build community and a sense of how to support each individual alternative licensure candidate.

The information gleaned from the pre-assessment showed the majority of participants felt confident in the areas of content knowledge and professionalism. They felt less comfortable with lesson planning, differentiating instruction, and managing student behavior. One student expressed concern that resolving professional conflict was very different than her experiences working in the environmental engineering industry. Every member of the *Aggie Prep* team received a copy of this document. It was revisited throughout the program to ensure the field coaching and participants' increasing levels of self-awareness were purpose driven and context specific.

The goals were met as illustrated by the anecdotal evidence gathered throughout the semester via CE and mentor conference calls, face-to-face meetings, and during professional development sessions. At the beginning of the program, the CE remarked the majority of students did not understand the lesson plan template and they were struggling with classroom management and how to create active learning opportunities in their science and mathematics classrooms. However, by the end of the semester, several of the mentors noted the participants were providing much more detailed lessons with active learning strategies in place that addressed multiple learning styles aimed at meeting the needs of diverse learners. One mentor stated, "I have seen a huge difference in [her] confidence since January. I have been in her classroom all year. Her growth since January has been amazing (personal communication)." Another mentor who completed an alternative licensure program herself said, "If I had this opportunity in an alternative program, I would have been better off. I really think it is a positive thing to have for alternative teachers (personal communication)."

As part of the design, coursework was flexible and responsive to the need for participants to allow their experiences to serve as the context for learning and developing strategies for responding to learners in the environment. Overall, students felt that the flexibility of the courses and the ability to give and receive feedback was an important aspect of the program, and that, "Instructors have great feedback and comments that helped [me] to improve my teaching and planning." Student feedback regarding the benefits of the *Aggie Prep* online courses were positive. One student in particular stated,

> I really enjoyed having the weekly meetings. It really helped me stay on track for this class and also see how others were doing in the class. I was really able to get some good ideas from the assignments that we performed and it helped me structure some of my lessons. Cohort collaborations and partner work encouraged me to stay on track and turn in assignments on time since someone else's grade was also on the line. Thank you for providing the readings online, that helped greatly!

While other students appreciated the emphasis on teaching strategies and stated, "I learned many new techniques and strategies that will enable me to teach better;" "The instructors know what work you have in other classes. It has challenged me to be a better teacher;" and "Aggie prep program is wonderful for new, alternative licensure teachers!"

Summary and Implications

Attracting and retaining high quality secondary STEM teachers who are content experts and who are prepared to teach in culturally and linguistically diverse classrooms has become an important tool in addressing the teacher shortage. In order to find highly qualified content experts, it has become necessary to look outside of a traditional teacher education program. As discussed in this chapter, the authors described a high quality alternative licensure program aimed at preparing high quality content experts in the pedagogical content skills they need in order to be successful in the secondary classroom. The curriculum of this online program emphasized a standards based active learning environment and bilingual/bicultural education. Within this context, the authors also chose to focus on language development through science and mathematics instruction. Language learning is an essential component of the educational process, especially for those students whose first language is not English, however, the process of becoming literate should not and cannot be limited to language arts. The potential for increasing language acquisition through meaningful science and mathematics activities while providing experiences to learn the content can be beneficial to all students regardless of their language proficiency. This successful program illustrates the potential for an alternative teacher preparation program to reach and support novice teachers placed in rural areas with limited assistance and resources.

Appendices

Appendix A: Aggie Prep Coursework

1st Semester	2nd Semester	Summer I
EDUC 563/563: Secondary STEM Methods	EDUC 518: Technology and Pedagogy	RDG 514: Content Area Literacy
EDUC 505: Classroom Management	SPED 500: Special Education for General Educators	EDUC 515: Multicultural Education
NES Essential Academic	NES Secondary Content	NES Prof. Knowledge

Appendix B: STEM Methods Course Outline

Session	Topic
Week 1	Introductory Activities
Week 2	Lesson Planning 101
Week 3	Lesson Planning using the Standards
Week 4	Learning Mathematics and Science in Secondary Classrooms
Week 5	Teaching Mathematics and Science in Secondary Classrooms
Week 6	High Cognitive Demand Tasks
Week 7	Teaching Mathematics and Science in Secondary Classrooms
Week 8	*Assessment*
Week 9	Mathematics, Science & Technology
Week 10	Teaching Diverse Learners
Week 11	Motivation in Mathematics and Science Secondary Classrooms
Week 12	Incorporating the Practices of Mathematics and Science in Secondary Classrooms
Week 13	Shared Lesson Demonstrations
Week 14	Break
Week 15	Shared Lesson Demonstrations
Week 16	Final Reflection

Appendix C: Secondary Education Lesson Plan Template

<table>
<tr><td>Grade:</td><td>Subject:</td><td>Date:</td><td>Prepared By:</td></tr>
<tr><td colspan="4">Overview: (essential question, connection to larger unit, concept, etc.)</td></tr>
<tr><td colspan="4">Standards & Benchmarks (Common Core State Standards, and/or other state content standards listed in their entirety)</td></tr>
<tr><td colspan="4">Measurable Objectives:</td></tr>
<tr><td colspan="4">The student will be able to…</td></tr>
<tr><td colspan="2" rowspan="2">Description of the Lesson (Write/script each step in the order that you will proceed through the lesson)</td><td>Anticipated Time</td><td>Questions for Consideration (Self & Student)</td></tr>
<tr><td></td><td>(Write/script questions you want to ask students during this part of the lesson and questions that emerge as you are teaching the lesson.)</td></tr>
<tr><td colspan="4">Materials:</td></tr>
<tr><td colspan="4">Assignment(s):</td></tr>
<tr><td colspan="4">Summative Assessment: (May not be specific to the lesson, but will ultimately measure desired student outcome on overarching concept.)</td></tr>
</table>

Appendix D: Aggie Prep Orientation to Education Pre-Assessment

Student: Please reflect and respond to each prompt below. Circle the number that best reflects your perception at this time. When appropriate, please provide a response to the short-answer prompts.

To what degree do you feel you perform each competency?
(1) Needs improvement (2) Developing (3) Effective (4) Highly Effective

Domain 1: Preparation & Planning (Learner Development, Content Knowledge, Planning for Instruction)

Making adequate time to prepare instructional materials	1	2	3	4
Developing daily and long-term lesson plans	1	2	3	4
Using state standards and benchmarks to create lessons	1	2	3	4
Planning my lessons to achieve optimum learning by students	1	2	3	4
Writing measurable objectives	1	2	3	4
Including knowledge of students' backgrounds, cultures, skills, academic language, interests, and special needs into planning and practice	1	2	3	4
Understanding the central concepts, tools of inquiry, and structures of the discipline I am teaching	1	2	3	4

Comments/Goals/Needs

Domain 2: Creating an Environment for Learning (Learning Differences, Learning Environments)

Organizing class activities to sustain attention and interest of students	1	2	3	4
Establishing an environment that utilizes feedback in a positive manner	1	2	3	4
Challenging unmotivated students	1	2	3	4
Challenging students to use higher order thinking skills	1	2	3	4
Adhering to classroom schedules activities	1	2	3	4
Developing and/or implementing classroom management procedures	1	2	3	4
Creating interactions with students that are polite & respectful	1	2	3	4
Relating to and providing for English Language Learners	1	2	3	4
Identifying stereotypes and bias in curriculum materials & adapt instruction accordingly	1	2	3	4
Providing for the individual needs of students	1	2	3	4
Relating to and providing for students with exceptionalities	1	2	3	4
Participating in the IEP process	1	2	3	4
Comments/Goals/Needs				

Domain 3: Teaching for Learning (Application of Content, Assessments, Instructional Strategies)

Employing a range of vocabulary to ensure learning expectations are comprehensible to all students.	1	2	3	4
Integrating content area concepts across all areas of instruction	1	2	3	4
Incorporating a variety of pedagogical strategies into practice	1	2	3	4
Promoting creativity and independent thinking	1	2	3	4
Accessing a range of technology to support instruction	1	2	3	4
Incorporating technology into instruction to support student learning	1	2	3	4
Allowing sufficient time for students to answer through active engagement with peers and teacher	1	2	3	4
Connecting assessment to stated goals and objectives	1	2	3	4
Developing valid and varied evaluation tools to measure student outcomes	1	2	3	4
Incorporating formative assessment into instruction	1	2	3	4
Using effective questioning techniques to assess student knowledge	1	2	3	4
Adjusting teaching style to meet the learning styles of all students in the classroom	1	2	3	4
Engaging learners in applying methods of inquiry and standards of evidence used in the discipline.	1	2	3	4

Stimulating learner reflection on prior content knowledge and link new concepts to familiar concepts	1 2 3 4
Recognizing learner misconceptions that lead to errors	1 2 3 4
Comments/Goals/Needs	

Domain 4: Professionalism (Professional Learning & Ethical Practice, Leadership & Collaboration)

Implementing record keeping procedures (school and district)	1 2 3 4
Maintaining accurate and up-to-date records	1 2 3 4
Communicating with families in a culturally appropriate manner	1 2 3 4
Communicating frequently with families	1 2 3 4
Actively participate in professional community, and in school/district events and projects.	1 2 3 4
Maintain positive and productive relationships with colleagues	1 2 3 4
Accepting opportunities for professional development after an individual assessment of need	1 2 3 4
Promoting a safe environment for students when monitoring students and activities	1 2 3 4
Comments/Goals/Needs	

1. Do you have a required lesson planning template? Explain.
2. How comfortable are you using TeachScape? Explain.
3. Do you have a planning period? If yes, how do you typically use this time? Explain.
4. Do you work with an inclusion teacher? If yes, how?
5. How do you typically give feedback to students?
6. What do you feel are your content area strengths?
7. What do you feel are your content area weaknesses?
8. What is your understanding of differentiated instruction?
9. Are you currently satisfied with how your class operates on a daily basis? Explain.

References

Allington, R. (2011). *What really matters for struggling readers: Designing research-based programs* (3rd ed.). Boston: Pearson.

Beck, I., McKeown, M., & Kucan, L (2002). *Bringing words to life: Robust vocabulary instruction*. Retrieved from http://www.colorincolorado.org

Bybee, R. W., Taylor, J. A., Gardner, A., VanScotter, P., Powell, J. C., Westbrook, A., & Landes, N. (2006). *The BSCS 5E instructional model: Origins and effectiveness*. Colorado Springs: Author.

Chambers, D. (2002). The real world and the classroom: Second career teachers. *The Clearinghouse, 75*(4), 212–217.

CREDE. (2015). *Center for research on education, diversity and excellence*. www.manoa.hawaii.edu/coe

Cummins, J. (1979). *Cognitive/academic language proficiency, linguistic interdependence, the optimum age question and some other matters.* Working Papers on Bilingualism, 19, 121–129.

Cummins, J., & Swain, M. (1986). *Bilingualism and education: Aspects of theory, research and practice*. New York: Longman Inc.

Daniels, H., & Zemelman, S. (2004). *Subjects matter: Every teacher's guide to content area reading*. Portsmouth: Heinemann.

Duemer, L., Fontenot, D., Gumfory, K., Kallus, M., Larsen, J., Schafer, S., & Shaw, B. C. (2002). The use of synchronous discussion groups to enhance community formation and professional identity development. *The Journal of Interactive Online Learning, 1*(2), 1–9.

Gallagher, K. (2009). *Readicide: How schools are killing reading and what you can do about it.* Portland: Stenhouse Publishers.

González, N., Moll, L., & Amanti, C. (Eds.). (2005). *Funds of knowledge: Theorizing practices in households, communities, and classrooms*. Mahwah: Lawrence Erlbaum Associates.

Hernandez, C. M., Morales, A., & Shroyer, M. G. (2013). The development of a model of culturally responsive science and mathematics teaching. *Cultural Studies in Science Education, 8*, 803–829.

Krashen, S. (1982). *Principles and practice in second language acquisition*. University of Southern California.

Lee, O., & Buxton, C. A. (2013). Integrating science and English proficiency for English language learners. *Theory Into Practice, 52*, 36–42.

Lesh, R. (1979). Mathematical learning disabilities: Considerations for identification, diagnosis, and remediation. In R. Lesh, D. Mierkiewicz, & M. Kantowski (Eds.), *Applied mathematical problem solving* (pp. 111–180). Columbus: ERIC.

National Governors Association Center for Best Practices & Council of Chief State School Officers. (2010). *Common core state standards*. Washington, DC: Authors.

National Research Council [NRC]. (2012). *A framework for K-12 science education: Practices, crosscutting concepts, and core ideas*. Washington, DC: National Academy Press.

NGSS Lead States. (2013). *Next generation science standards: For states, by states*. Washington, DC: The National Academies Press.

Rychly, E., & Graves, L. (2012). Teacher characteristics for culturally responsive pedagogy. *Multicultural Perspectives, 14*(1), 44–49.

Stein, M. K., Smith, M. S., Hennigsen, M. A., & Silver, E. A. (2009). *Implementing standard-based mathematics instruction: A case for professional development*. New York: Teachers College.

The Council of Chief State School Officers. (2013, April). Interstate Teacher Assessment and Support Consortium In TASC model core teaching standards and learning progressions for teachers 1.0: A resource for ongoing teacher development. Washington, DC: Author.

Tovani, C. (2004). *Do I really have to teach reading? Content comprehension, grades 6–12*. Portland: Stenhouse Publishers.

Tolbert, S., Stoddart, T., Lyon, E. G., & Solís, J. (2014). The next generation science standards, common core state standards, and English learners: Using the SSTELLA framework to prepare secondary science teachers. *Issues in Teacher Education, 23*(1), 65.

U.S. Department of Education, Office of Innovation and Improvement. (2004). *Alternative routes to teacher certification*. Retrieved November 2, 2015, from http://www.ed.gov

Uribe-Flórez, L. J. (2014). Manipulative use to support ELL's mathematics learning. In D. F. Berlin & A. L. White (Eds.), *Initiatives in mathematics and science education with global implications* (pp. 33–44). Columbus: International Consortium for Research in Science and Mathematics Education.

Valdés, G., Kibler, A., & Walqui, A. (2014, March). *Changes in the expertise of ESL professionals: Knowledge and action in an era of new standards.* Alexandria: TESOL International Association.

Chapter 15
A Functional Approach to Twenty-First Century Science Literacy

Wei Zhang

Disciplinary literacy development has become increasingly emphasized in recent decades. Studies on new literacies have shifted the research paradigm from the traditional notion of building students' discipline-general literacy skills to the development of multiliteracies, or discipline-specific literacy (e.g., Leu et al. 2009; Shanahan and Shanahan 2008; Tang 2015; Zygouris-Coe 2015). New education standards also emphasize that all students should master disciplinary literacy in order to participate constructively in today's world with its unprecedented technological advancement and scientific discoveries. In the United States, the Common Core State Standards (CCSS) adopted by 43 states lays down not only academic standards for English language arts and mathematics but also literacy standards for other subject areas, including social studies/history, science, and technology subjects (NGA & CCSSO 2010).

While disciplinary literacy develops in tandem with content knowledge, students do not automatically acquire disciplinary literacy as they engage in content knowledge learning. This is particularly true for such disciplines as science where highly specialized language is used to communicate fundamental ideas, key concepts, and core knowledge (Fang 2005; Shanahan 2013; Shanahan and Shanahan 2008). To be literate in science entails both the essential knowledge of science and the literacy skills to read, write, and communicate in scientific language (Norris and Phillips 2003). The teaching of science should include not only hands-on experiential activities to build up scientific concepts, but also literacy instruction on critical reading and writing skills to deepen experiential learning (Fang 2013; Krajcik and Sutherland 2010; Shanahan 2012).

W. Zhang (✉)
The University of Akron, Akron, OH 44325, USA
e-mail: wz23@uakron.edu

A.W. Oliveira, M.H. Weinburgh (eds.), *Science Teacher Preparation in Content-Based Second Language Acquisition*, ASTE Series in Science Education,
DOI 10.1007/978-3-319-43516-9_15

However, there is a general consensus that science texts are too difficult to read and therefore are not read by students or required by teachers. An often noted substitution for asking students to read science textbooks is that science teachers transform the textbook content to lecture notes and assess students on hands-on activity only (Fang 2005, 2013; Shanahan 2012). This practice leads to a problematic dislocation of science literacy instruction from science teaching and learning (American Association for the Advancement of Science 1989). While various educational and societal factors contribute to this situation, the high demand of the language of science itself (Fang 2005) and teachers' insufficient literacy preparation (Shanahan and Shanahan 2008) are two of the major factors. On the one hand, science texts are filled with technical words with condensed meanings and are structured and organized in ways unique to science itself; they require more than general literacy skills for an accurate interpretation and thorough understanding. On the other, science teachers, like other content teachers at secondary schools, usually do not receive training to teach science language and are not required to do so (Barry 1994; Fang 2014; Ross and Frey 2009; Shanahan and Shanahan 2008; Short 2002).

Given the situation, the shifted emphasis on science literacy in science education poses new challenge to science teachers in general, and particularly so to science teachers of English language learners (ELLs). In the past few decades, ELLs have become the fastest-growing student body in public schools (Roseberry-McKibbin and Brice 2015; Uro and Barrio 2013). They need to develop academic language to access content knowledge while still learning English for daily communication. At secondary schools, they are required to use disciplinary literacy skills to learn various subjects without adequate time to build up basic and general literacy skills, in spite that the latter is the foundation for disciplinary literacy development (Shanahan and Shanahan 2008). To meet the needs of ELLs, sheltered instruction has been adopted to support their content knowledge and academic language development (Echevarría et al. 2012; Lightbown and Spada 2013). As a content-based approach to disciplinary literacy, sheltered instruction requires teachers to integrate disciplinary academic language instruction to disciplinary content teaching. In order to do so, content teachers should be equipped with in-depth knowledge in how language works within a content. This chapter describes a content teacher professional development model that incorporates a functional text analysis into sheltered instruction training for in-service content teachers to strengthen their capacity to work with ELLs. In this professional development program, science teachers and language teachers collaborate on designing teaching strategies for effective science literacy instruction in a sheltered setting, which proves to have prepared the teachers well to successfully implement sheltered instruction with ELLs in their classrooms.

Teacher Preparation Model

The in-service teacher professional development program is delivered by a midwestern university funded by a grant from the Department of Education of the United States. Four cohorts of content teachers from six local school districts in the northeastern part of the state are taking courses toward a TESOL (Teaching English to Speakers of Other Languages) Endorsement. STEM (Science, Technology, Engineering, and Mathematics) teachers are prioritized to participate in the grant. Each cohort consists of at least half STEM teachers and teachers of other subjects, including English language arts (ELA), ESL (English as a second language), and social studies. The selected teachers are working with ELLs at grades K-12 in both urban and suburban schools at various capacities, with 10–97 % of their students being designated as ELLs in their schools. All of them are licensed teachers holding at least a Bachelor's degree in education. Most of them have not received prior training in language, linguistics, or methods and techniques to teaching ELLs.

The program is taught by four tenured or tenure-track faculty members at the university. The courses are delivered via distance learning with 30–90 % of the components being offered online to accommodate travel constraints of the teachers and offer more flexibility for time management. The grant is evaluated annually by the education evaluation center in another university. Four types of data are collected for program evaluation: (1) pre-and-post training survey data from the participating teachers; (2) a focus group with the participating teachers; (3) pre-and-post-training class observation data of about two thirds of the participating teachers; and (4) state test scores of ELLs served by the participating teachers.

The TESOL Endorsement is a 22-credit interdisciplinary program with eight courses. These courses are fully aligned with the *TESOL/NCATE Standards for the Recognition of Initial TESOL Programs in P-12 ESL Teacher Education* (Teachers of English to Speakers of Other Languages, Inc. 2010). They cover linguistics, English grammar, second language acquisition theories, language teaching methods, language teaching techniques, second language literacy, and inclusive education. One of these courses is designated to the SIOP (Sheltered Instruction Observation Protocol) model, a content-based instruction model to teach ELLs (Echevarría et al. 2012).

Striking a balance between theory and practice, the TESOL Endorsement program is aimed at developing the grant teachers' knowledge base to teach ELL in content instruction with an emphasis on strategies to teach disciplinary language and disciplinary literacy skills. A major focus of the training is to prepare the grant teachers to effectively use the SIOP model to teach ELLs in an inclusive classroom. The SIOP model is a systematic synthesis of the best practices in language and content instruction to ELLs. It is composed of eight components or thirty features to guide the preparation, delivery, and assessment of each lesson (See Fig. 15.1). It is the only research validated model in content-based instruction (Echevarria et al. 2011; Echevarría et al. 2012; Short et al. 2011, 2012). With adequate training and a high level of implementation of at least 75 % of the time, the model has proven to be effective in ELL instruction (Short et al. 2012). While teaching models generally provide guidance on how (to teach) but not necessarily why (to teach as such), effective implementation of the SIOP model needs to be supported by the other courses in the TESOL Endorsement program to deepen the grant teachers' understanding of the model and to build up their confidence in using the model. In particular, Systemic Functional Linguistics (SFL) is incorporated the TESOL

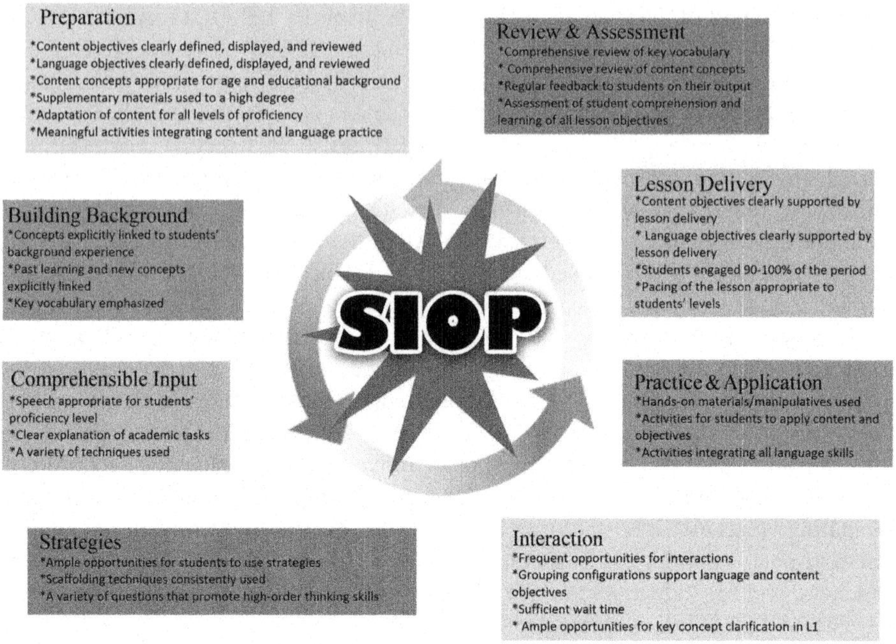

Fig. 15.1 The SIOP model (Adapted from Echevarría et al. 2012)

Endorsement program to strengthen the grant teachers' understanding of the linguistic complexity in order to support ELLs' academic language development in the implementation of the SIOP model.

Functional Linguistics Integration

Systemic Functional linguistics, or SFL, is a genre-based semantic-functional approach to analyzing the linguistic features of different types of texts. It takes as its goal to "construct a grammar for purposes of text analysis: one that would make it possible to say sensible and useful things about any text, spoken or written, in modern English" (Halliday 1994, p. XV). It has been used to unpack the linguistic complexity of content texts for disciplinary literacy instruction (e.g., de Oliveira 2010a, b; Fang 2005, 2013, 2014; Fang and Schleppegrell 2008; Schleppegrell 2004; Zygouris-Coe 2015). In the TESOL Endorsement, SFL is integrated into two courses taught in two consecutive semesters: a foundation course in linguistics and a course on second language acquisition theories and teaching methods. In the first course, the grant teachers learn the principles of SFL text analysis and the unique linguistic features of texts in their respective content areas through guided readings. They then conduct a thorough analysis of a content text to demonstrate their understanding of the reading materials. In the second course, a science teacher and an ELA or ESL teacher work together to illustrate one particular linguistic feature of science texts with examples from science textbooks across grade levels and design activities to target the identified linguistic feature to make science texts more accessible for ELLs. In addition, all teachers write a reflection on the usefulness of the course projects in helping them to teach disciplinary academic language and how they can implement the teaching strategies with their own students. Finally, the SFL integration into the training is evaluated in the pre-and-post-training class observations that are part of the grant evaluation using the SIOP class observation protocol.

From an SFL perspective, science language is hard to understand because it is "simultaneously technical, abstract, dense, and tightly knit" (Fang and Schleppegrell 2008, p. 20). These four features capture the distinctive use of language in science texts: frequent use of technical terms, naming complex processes by means of nominalization, packing dense information into long noun groups and complex sentence structures, and using specific text organization patterns for logic reasoning. In the following section of this chapter, a middle school science text on the water cycle is presented to illustrate the linguistics features of science texts with teaching strategies to teach the linguistic features. It also serves as an example of how to incorporate SFL into the SIOP model to support science teacher to deliver language-based content instruction.

The Water Cycle: Functional Analysis and Teaching Strategies

The text on the water cycle is selected from the seventh grade science textbook *Interactive Science: Science and Technology* published by Pearson Education, Inc. (2015). It is a typical text found in science textbooks (See Fig. 15.2). In this section, the text on the water cycle is analyzed in the SFL framework to illustrate the linguistics features of science texts as being technical, abstract, dense, and tightly knit. Based on the text analysis, teaching strategies and activities for focused instruction are explained to target scientific vocabulary, science word formation, sentence structure, and text organization of science texts.

Science Language Is Technical Science texts are technical due to the large amount of specialized words used in a single text (Fang and Schleppegrell 2008). These words are of two types: (1) words unique to science, such as *photosynthesis*; and (2) common words with scientific meanings, such as *water table*. Focused instruction on the technical words in science texts is necessary to build up students' disciplinary literacy in science, but it is not adequate for ELLs. To read academic texts in English requires a large size of vocabulary of more than 40,000 (Hirsh and Nation 1992; Horst and Cobb 2006; Laufer 1992). Research on word frequency and word families has identified that (1) the first 2000 most frequent words account for about 80 % of all the words in spoken and written English texts; (2) another 570 or so words used

What is the Water Cycle?

Earth has its own built-in water recycling system: the water cycle. The water cycle is the continuous process by which water moves from Earth's surface to the atmosphere and back, driven by energy from the sun and gravity. **In the water cycle, water moves between land, living things, bodies of water on Earth's surface, and the atmosphere.**

Water Evaporates Where does the water in a puddle go when it disappears? It evaporates, becoming water vapor. Evaporation is the process by which molecules at the surface of a liquid absorb enough energy to change a gaseous state. Water constantly evaporates from the surfaces of bodies of water such as oceans and lakes, as well as from soil and your skin. Plants play a role, too, in this step of the water cycle. Plants draw in water from the soil through their roots. Eventually the water is given off through the leaves as water vapor in a process called transpiration.

Condensation Forms Clouds After a water molecule evaporates, warm air can carry the water molecule upward. Air tends to become colder as it rises. Water vapor as it rises. Water vapor condenses more easily at lower temperatures, so some water vapor cools and condenses into liquid water. Droplets of liquid water clump around solid particles in the air, forming clouds.

Water Falls as Precipitation As more water condenses, the water droplets grow larger. Eventually, they become so heavy that they fall back to Earth. Water that falls to Earth as rain, snow, hail, or sleet is called precipitation.

Most precipitation falls directly into the ocean. Of the precipitation that falls on land, most evaporates. A small amount of the remaining water runs off the surface into streams and lakes in a process called runoff, but most of it seeps into groundwater. After a long time, this groundwater may flow down to the ocean and evaporate again.

Precipitation is the source of almost all fresh water on and below the Earth's surface. For millions of years, the total amount of water cycling through the Earth system has remained fairly constant—the rates of evaporation and precipitation are balanced.

Fig. 15.2 Water Cycle Text from the book *Interactive Science: Science and Technology*

most frequently in academic texts across disciplines bring the coverage of a text to about 90 %; and (3) discipline-specific words make up for the rest (Crossley et al. 2013; Hirsh and Nation 1992; Horst and Cobb 2006; Laufer 1992; Nation and Waring 1997; O'Keeffe et al. 2007). Based on the research, vocabulary instruction in science should not limit to the technical words in science only, but also include words in the academic wordlist.

In *The Water Cycle*, the science-specific words are highlighted in yellow, but more academic words are packed in this text. The classic version of *VocabProfilers* on www.lextutor.ca offers a four-category profile of words in a text: K1: the first 1000 most frequent words; K2: the second 1000 most frequent words; AWL: words in the Academic Word List that are used across disciplines; and Off-list: discipline-specific vocabulary (Cobb 2013). Table 15.1 shows the *VocabProfilers* analysis of the words in *The Water Cycle* by token (represented in percentage) and type (listed words). As can be seen in the table, the first 2000 most frequent words do account for about 83 % of all words in the text, but cross-discipline and scientific words account for about 17 % of the words with the majority of them, 12.04 %, being scientific words. Among the scientific words, only *precipitation* and *transpiration*

Table 15.1 VocabProfiler analysis

	Percentage (%)	Cumulated percentage (%)	Words
K1	77.87	77.87	
K2	5.04	82.91	
AWL	5.04	87.95	*Constant constantly cycle cycling energy eventually process role source*
Off-list	12.04	100.00	*Absorb atmosphere clump condensation condense droplets evaporate evaporation gaseous gravity groundwater hail molecule precipitation puddle recycling runoff seeps sleet transpiration vapor*

are unique to science and the rest all have at least a scientific meaning and an ordinary meaning, including *absorb, atmosphere, clump, condensation, condense, droplets, evaporate, evaporation, gaseous, gravity, groundwater, hail, molecule, puddle, recycling, runoff, seep, sleet*, and *vapor.*

Depending on ELLs' proficiency levels, not all of the AWL and Off-list words need to be taught, but the key words that are central to the description of the water cycle, such as *atmosphere, condense/condensation, evaporate/evaporation, precipitation*, and *transpiration* should be emphasized. One strategy to teach science vocabulary is to focus on recognizing the prefixes, suffixes, and word roots. The suffix *–tion* is used most frequently as an ending of nouns. In science, it often signals a word that describes a process, as in *condensation, evaporation, precipitation*, and *transpiration.* Another strategy is to differentiate the everyday meaning and the scientific meaning of science words. For instance, *atmosphere* refers to the whole mass of air surrounding the Earth in science, and generally refers to surrounding, environment, or tone in everyday life. Also note that a lot of science words are with multiple syllables and words with three or more syllable are difficult for ELLs. They are less frequent and are harder to pronounce, but being able to say these words helps ELLs to retain them for academic conversation. One strategy to teach multisyllabic words is to mark the stressed syllables in speech by articulating it with gesture and in writing by certain notations, such as underline, capitalization, or hyphenation (e.g., *at mosphere* or *AT-mosphere*).

Science Language Is Abstract Science texts often read abstract. This can be attributed to the more frequent than usual use of nominalization, a process of turning verbal or adjectival groups into nominal group. Nominalization is used frequently to condense meaning or labeling science processes. These words are a packed with multiple and condensed meanings that require more than a definition for an accurate and thorough understanding. For instance, the word *osmosis* has an everyday meaning, *a gradual absorption of something*, but it describes a process in science with much richer meanings than its scientific definition denotes as shown in Fig. 15.3 (Adapted from Shanahan 2012, p. 166).

Definition	Process where solvent molecules move through a semipermeable membrane from a dilute solution into a more concentrated solution.
General definition	A gradual process of absorption. "I learned through osmosis."
Explanation/ Example of the process	Two parts of a container separated by a semipermeable membrane are filled with water. In one side, there is also a lot of salt. In the other, there is less salt. Water moves from the less salty side to the salty side.
Illustration	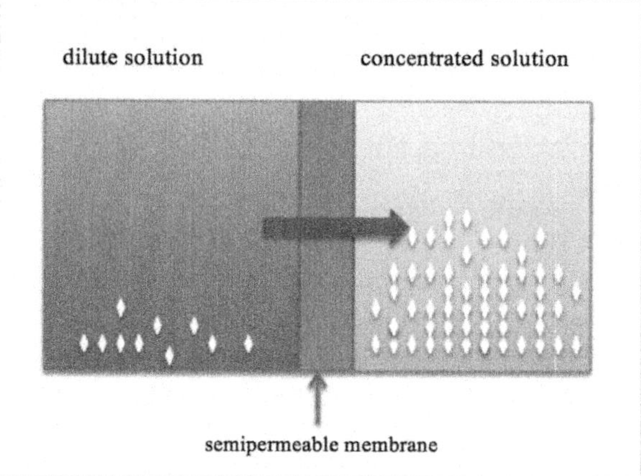
Formula	(V2-V1) ÷ (T2-T1) = Rate of osmosis
Related terms	Solvent, solute, molecule, semipermeable, dilute

Fig. 15.3 Vocabulary notebook example

Table 15.2 Nominalization

Verbal group	Nominalization
It evaporates, becoming water vapor.	*Evaporation is ....*
Eventually the water is given off through the leaves as water vapor in a process called	transpiration.
...so some vapor cools and condenses into liquid water	*Condensation (Forms Clouds)*

In *The Water Cycle*, three instances of nominalization are found. They are all nouns to describe and label processes in the water cycle. While nominalization most often follows the verbal or adjective groups immediately, such as *evaporation* and *transpiration* shown in Table 15.2, *condensation* is used as a heading preceding the verbal group in Paragraph 3, adding to the difficulty of understanding. Dissecting nominalization to the actual process that it describes in a Vocabulary Notebook exercise as illustrated in Fig. 15.3 helps to clarify meaning. Another strategy is to rewrite long sentences with nominalization into shorter ones. For instance, the sentence *Eventually the water is given off through the leaves as water vapor in a process called transpiration* can be broken down to two shorter sentences: *Eventually the water is given off through the leaves as water vapor. This process is called transpiration.*

Table 15.3 Long noun groups

Pre-modifier	Head noun	Post-modifier	Embedded clause
The continuous	*process*		*by which water moves from Earth's surface to the atmosphere and back, driven by energy from the sun and gravity*
The total	*amount*	*of water cycling through the Earth system*	

Table 15.4 Theme-rheme pattern analysis

Theme	Rheme
Earth	*has its own built-in water recycling system: the water cycle.*
The water cycle	*is the continuous process by which water moves from Earth's surface to the atmosphere and back, driven by energy from the sun and gravity.*
In the water cycle	*water moves between land, living things, bodies or water on Earth's surface, and the atmosphere.*

Science Language Is Dense Science texts are high in lexical density. The ratio of content words to the total words in *The Water Cycle* is 0.60 (214 content words/357 words = 0.6), that is, six out of ten words in the text are content words. Apart from lexical density, science texts also contain long sentences that include expanded noun groups with pre-and-post modifiers and embedded clauses as illustrated with two examples from *The Water Cycle* in Table 15.3. Deconstructing long noun groups into its components, expanding simple nouns, and comparing different types of noun groups are effective strategies in helping students to identify the head noun, revealing the meaning of long noun groups.

Science Language Is Tightly Knit Science texts employ a tightly knit structure to describe step-by-step processes and present rigorous reasoning. Dividing sentences into Themes (the departing idea of a clause) and Rhemes (the rest of a clause) and tracking the relationships among them reveals the organization of ideas in a science text. Two unique patterning of Themes and Rhemes are commonly found in science texts: reiterating of Themes and zig-zag patterning. As illustrated by the Theme-Rheme pattern analysis of the first paragraph of *The Water Cycle* in Table 15.4, the Rheme of the first sentence, *the water cycle*, is picked up immediately by the second sentence as its Rhemes, forming a zig-zag pattern, tying the two sentences together. Then the third sentence starts with the same Theme as the second sentence, *the water cycle*, to give more information about it by reiterating the same Theme. As a paragraph become longer, there are more Themes and Rhemes to grasp, but they mostly still fall into these two types of patterning. Tracking clause Themes, creating Themes in text joining activities, and comparing different thematic patterns of science texts are strategies to help students to discover the organization scheme of science texts and therefore gain a deeper understanding of what a science text means.

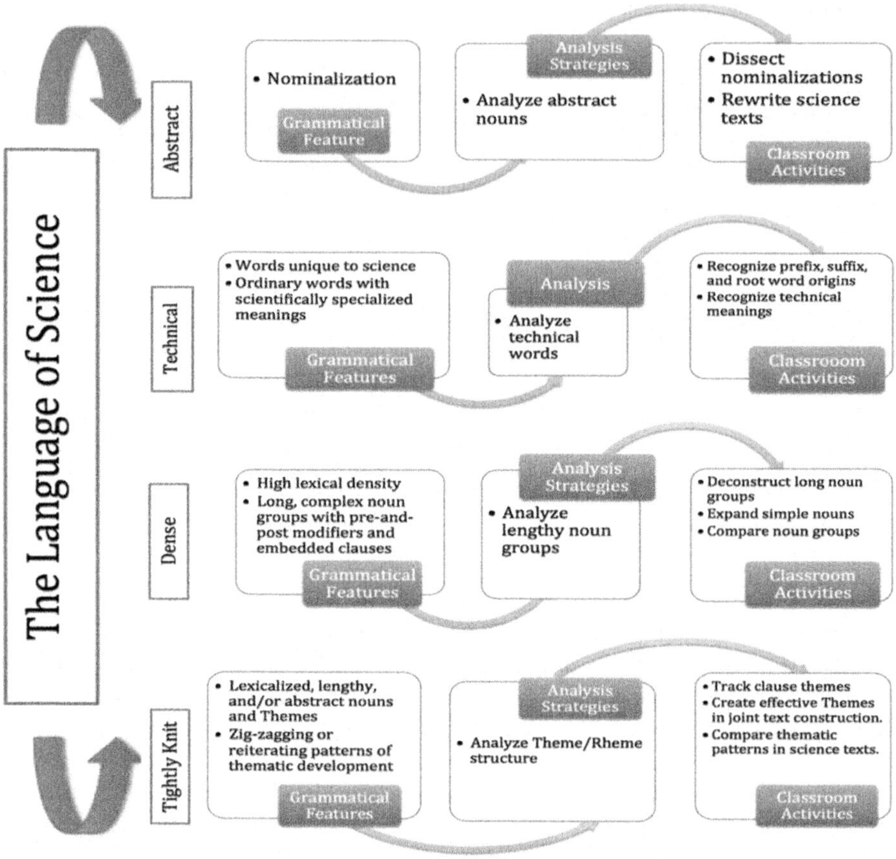

Fig. 15.4 The language of science and teaching strategies

As has demonstrated above, science language has four distinct linguistic features that simultaneously contribute to its text complexity: technicality, abstractness, density, and tightly knit text structure. An SFL approach to science literacy helps science teachers to identify what to teach in terms of language in science classes. Focused instruction on language can be used before students read a science text to provide language scaffolds for text comprehension. It can also be used after students read a science text as review and assessment of text comprehension. Figure 15.4 is a summary of the science text linguistic features and suggested teaching strategies.

Effectiveness of SFL Integration

The effectiveness of the SFL integration into the TESOL Endorsement program is evaluated by a qualitative instrument and a quantitative instrument. The qualitative data are collected from the teachers' self-reflection on the process of analyzing the linguistic features of science texts and designing teaching strategies. The quantitative data are collected as four-point Likert scale ratings on language-related items on the SIOP class observation protocol given to each teacher by two faculty members during pre-training and post-training lass observations. The analysis of the qualitative data follows a category system required of interview data analysis to look for emerging themes (Gall et al. 2007). The analysis of the quantitative data follows a within-subjects, repeated-measures design so that the evaluation of growth can be attributed to the training. The pre- and post-observation scores are matched by teacher so that significant differences will be attributed to growth due to the training instead of differences inherent between teachers.

In the teachers' self-reflection, a general theme that emerges is that the teachers themselves now perceive science texts to be much more accessible for them to teach and for their students to read. A common comment is that the teachers used to think science texts are very difficult to read, not only for their students, but also to themselves; after unpacking the science texts into the four features of science language, science texts becomes much more "reader friendly" and much more "teachable". They also acknowledge that reducing the language without reducing the content is possible by adopting techniques such as vocabulary builders and nominalizations. Two teachers discussed a technique in which they break down sentences into smaller, more easily understood parts using tools such as highlighters. One teacher mentioned a "word wall" in which students add their interpretation of a target word to the wall. Here are their specific statements:

> "I also have tried taking a sentence out of a paragraph that I feel is something that they need to know and we break it down."
>
> "I've got a Word wall in the hallway. I have a big word that says soil because we're starting soil and top soil and some stuff. So the word soil, and I had all the kids around the world find out what the word soil- how to look at it in their language and they put it on a bubble. So it's this big spider web out in the hallway."

In conducting the statistical analysis of the SIOP class observation protocol, it is anticipated that by comparing pre- and post-observation scores growth can be identified through a statistical comparison of scores averaged across the teachers. Therefore, a dependent samples difference score was calculated for each of the item rated to assess if there were significant differences between pre- and post-observations. Among the 30 items that represent 30 features or the eight components of a SIOP lesson, five are identified as directly related to the SFL integration into the training:

2. **Language objectives** clearly defined, displayed, and reviewed with students
9. **Key vocabulary** emphasized (e.g., introduced, written repeated, and highlighted for students to see)
22. Activities integrate **all language skills** (i.e., reading, writing, listening, and speaking)
24. **Language objectives** clearly supported by lesson delivery
27. Comprehensive **review of key vocabulary**

Table 15.5 SIOP observation: language-related items

Item	Number of teachers1	Pre-observation average	Post-observation average2
2. Language objectives defined	12	1.08	3.42*
9. Key vocabulary emphasized	12	1.83	2.67
22. Integrate all language skills	12	1.83	3.08*
24. Language objectives supported	12	1.08	2.83*
27. Review key vocabulary	11	1.18	2.18

1. Only teachers with both pre- and post-observation scores are included in the analysis
2. * $p < .01$, indicates the post-observation is significantly different from pre-observation score

Table 15.5 is the breakdown of the average ratings of the five items on the pre- and post-observations with average differences that are statistically significant marked. The average of all five items has greatly improved with statistically significant results for three of them, which has demonstrated that teacher growth can be attributed to the training they have received.

In summary, the analysis of data collected by both qualitative and quantitative instruments reveal that SFL text analysis, as a functional approach to science literacy is a necessary and useful addition to the training of the science teachers in the TESOL Endorsement program. It deepens the teachers' own understanding of the language of science and prepares them well in identifying the language items to be taught together with the science content and in articulating and implementing the language objectives of a lesson.

Conclusion

This chapter describes a TESOL teacher training program that aims to build up content teachers' capacity to deliver content-based language instruction with a focus on discipline-specific academic language development of ELLs. With the double foci of the SIOP model on content instruction and language teaching, the linguistic analysis tool of SFL is incorporated into the linguistics and language acquisition and teaching methods courses to strengthen science teachers understanding of the linguistic complexity encoded in science texts. This linguistic integration constitutes a solid foundation for them to design teaching strategies to support ELLs' academic language and content knowledge development.

The incorporation of functional linguistics into the TESOL Endorsement program meets the disciplinary literacy standards in twenty-first century science education. Its successful implementation hinges on two factors: (1) the grant teachers' confidence in adopting the SFL text analysis for classroom practice; and (2) individualized guidance to facilitate the grant teachers to complete the SFL content text analysis and teaching strategy design. To build up the grant teachers' confidence in SFL, the principles of SFL and how it can be applied to the science content area and other content areas are explained in an interactive presentation before the grant teachers start the guided readings. As they are completing the guided readings, conducting the content text analysis, and designing the teaching strategies, they are given individual attention in conferences to ensure that they understand the SFL text analysis principles to such a level that they are able to accurately apply the newly gained knowledge into practice. As a new approach to science literacy teaching and learning, SFL holds promise for strengthening the knowledge base of science teachers to ensure effective delivery of sheltered instruction to meet the academic literacy of ELLs, but more empirical research is necessary to validate its usefulness in teacher training and classroom instruction.

References

American Association for the Advancement of Science. (1989). *Science for all Americans: A project 2061 report on literacy goals in science, mathematics*, and *technology*. Washington, DC: Author. Retrieved from http://www.project2061.org/publications/sfaa/online/sfaatoc.htm

Barry, A. L. (1994). The staffing of high school remedial reading programs in the United States since 1920. *Journal of Reading, 38*, 14–22.

Cobb, T. (2013). *VocabProfilers* [Computer software]. Québec: Université du Québec à Montréal.

Crossley, S. A., Cobb, T., & McNamara, D. S. (2013). Comparing count-based and band-based indices of word frequency: Implications for active vocabulary research and pedagogical applications. *System, 41*(4), 965–981.

de Oliveira, L. C. (2010a). Beyond general strategies for English language learners: Language dissection in science. *Electronic Journal of Literacy Through Science, 9*, 1–14.

de Oliveira, L. C. (2010b). Enhancing content instruction for ELLs: Learning about language in science. In D. Sunal, C. Sunal, M. Mantero, & E. Wright (Eds.), *Teaching science with Hispanic ELLs in K-16 classrooms* (pp. 135–150). Charlotte: Information Age.

Echevarria, J., Richards-Tutor, C., Canges, R., & Francis, D. (2011). Using the SIOP model to promote the acquisition of language and science concepts with English learners. *Bilingual Research Journal, 34*, 334–351.

Echevarría, J., Vogt, M. E., & Short, D. J. (2012). *Making content comprehensible for English learners: The SIOP model* (4th ed.). Boston: Pearson.

Fang, Z. (2005). Scientific literacy: A systemic functional linguistics perspective. *Science Education, 89*(2), 335–347. doi:10.1002/sce.20050.

Fang, Z. (2013). Disciplinary literacy in science: Developing science literacy through trade books. *Journal of Adolescent & Adult Literacy, 57*(4), 274–278. doi:10.1002/JAAL.250.

Fang, Z. (2014). Preparing content area teachers for disciplinary literacy instruction. *Journal of Adolescent & Adult Literacy, 57*(6), 444–448. doi:10.1002/JAAL.269.

Fang, Z., & Schleppegrell, M. J. (2008). Technicality and reasoning in science: Beyond vocabulary. In Z. Fang & M. J. Schleppegrell (Eds.), *Reading in secondary content areas: A language-based pedagogy* (pp. 18–38). Michigan: University of Michigan Press.

Gall, M. D., Gall, J. P., & Borg, W. R. (2007). *Educational research: An introduction* (8th ed.). Boston: Pearson Education, Inc.

Halliday, M. A. K. (1994). *An introduction to functional grammar* (2nd ed.). London: Edward Arnold.

Hirsh, D., & Nation, P. (1992). What vocabulary size is needed to read unsimplified texts for pleasure? *Reading in a Foreign Language, 8*(2), 689–696.

Horst, M., & Cobb, T. (2006). Second language vocabulary acquisition. *The Canadian Modern Language Review, 63*(1), 1012.

Krajcik, J. S., & Sutherland, L. M. (2010). Supporting students in developing literacy in science. *Science, 328*, 456–459.
Laufer, B. (1992). How much lexis is necessary for reading comprehension? In P. J. Arnaud & H. Béjoint (Eds.), *Vocabulary and applied linguistics* (pp. 126–132). London: Macmillan.
Leu, D. J., O'Byrne, W. I., Zawilinski, L., McVerry, J. G., & Everett-Cacopardo, H. (2009). Comments on Greenhow, Robelia, and Hughes: Expanding the new literacies conversation. *Educational Researcher, 38*, 264–269.
Lightbown, P. M., & Spada, N. (2013). *How languages are learned* (4th ed.). Oxford: Oxford University Press.
Nation, P., & Waring, R. (1997). Vocabulary size, text coverage, and word lists. In N. Schmitt & M. McCarthy (Eds.), *Vocabulary: Description, acquisition, pedagogy* (pp. 6–19). New York: Cambridge University Press.
NGA & CCSSO (National Governors Association Center for Best Practices & Council of Chief State School Officers). (2010). *Common core state standards for English language arts and literacy in history/social studies, science, and technical subjects*. Washington, DC: Authors. Retrieved from http://www.corestandards.org
Norris, S., & Phillips, L. (2003). How literacy in its fundamental sense is central to scientific literacy. *Science Education, 87*(2), 224–240. doi:10.1002/sce.10066.
O'Keeffe, A., McCarthy, M. J., & Carter, R. A. (2007). *From corpus to classroom: Language use and language teaching*. Cambridge: Cambridge University Press.
Pearson Education, Inc. (2015). *Interactive science: Science and technology*. United States: Authors.
Roseberry-McKibbin, C., & Brice, A. (2015). *Acquiring English as a second language: What's "normal," what's not*. Retrieved from http://www.asha.org/public/speech/development/easl.htm
Ross, D., & Frey, N. (2009). Real-time teaching: Learners need purposeful and systematic instruction. *Journal of Adolescent and Adult Literacy, 53*(1), 75–78.
Schleppegrell, M. J. (2004). *The language of schooling: A functional linguistics perspective*. Mahwah: Lawrence Erlbaum Associates, Inc.
Shanahan, C. (2012). Learning with text in science. In T. L. Jetton & C. Shanahan (Eds.), *Adolescent literacy in the academic disciplines: General principles and practical strategies* (pp. 154–171). New York: The Guilford Press.
Shanahan, C. (2013). What does it take? The challenge of disciplinary literacy. *Journal of Adolescent & Adult Literacy, 57*(2), 93–98.
Shanahan, T., & Shanahan, C. (2008). Teaching disciplinary literacy to adolescents: Rethinking content-area literacy. *Harvard Educational Review, 78*, 40–59.
Short, D. J. (2002). Language learning in a sheltered social studies class. *TESOL Journal, 11*, 18–28.
Short, D. J., Echevarria, J., & Richards-Tutor, C. (2011). Research on academic literacy development in sheltered instruction classrooms. *Language Teaching Research, 15*(3), 363–380.
Short, D. J., Fidelman, C. G., & Louguit, M. (2012). Developing academic language in English language learners through sheltered instruction. *TESOL Quarterly, 48*(2), 334–361.
Tang, K. S. (2015). Reconceptualising science education practice from new literacies research. *Science Education International, 26*(3), 307–324.
Teachers of English to Speakers of Other Languages, Inc. (2010). *TESOL/NCATE standards for the recognition of initial TESOL programs in ESL teacher education (Introduction)*. Retrieved from http://www.tesol.org/s_tesol/bin.asp?CID=219&DID=13040&DOC=FILE.PDF
Uro, G., & Barrio, A. (2013). *English language learners in America's great city schools: Demographics, achievement and staffing*. Washington, DC: Council of the Great City Schools. Retrieved from http://files.eric.ed.gov/fulltext/ED543305.pdf
Zygouris-Coe, V. (2015). *Teaching discipline-specific literacies in grades 6–12: Preparing students for college, career, and workforce demands*. New York: Routledge.

Part III
International Perspectives

The Project

The Inheritances Books are a collaborative project between Ichabod Crane High School's Illustration and ELL students. It is made possible by a grant from the Berkshire Taconic Community Foundation.

The Student-Artist

Michael Tomaso, grade 12, 18 years old.

Chapter 16
Promoting the Integration of Inquiry Based Science and English Learning in Primary Education Through Triadic Partnerships

Mariona Espinet, Laura Valdés-Sánchez, Núria Carrillo Monsó, Laura Farró Gràcia, Roser Martínez Vila, Núria López Rebollal, and Ana Castillon Pascual

Introduction

The Specificity of Situated Multilingual European Contexts in Education

The educational demands that a global society places on most European education systems are high. One of these demands is the need to accept that our society and schools are multilingual contexts and that language diversity is a cultural heritage in need of conservation. The command of at least three languages is considered one of the most important basic competences that every European citizen should acquire through compulsory education (European Commission 2007). However the repertoire of language use in Europe can be seen as divided in two types of multilingualism (Guasch and Nussbaum 2007): a first order multilingualism constituted by the big European languages which are strongly valued and worth learning, and the second order multilingualism constituted by the minority languages present as a consequence of immigration which can be tolerated but have a lower status.

The multilingual context experienced in Catalonia adds a third factor making multilingual education more difficult. In fact, Catalonia is an autonomous region of Spain considering itself a nation without a state. The Catalan social and cultural

M. Espinet (✉) • L. Valdés-Sánchez
Departament de Didàctica de la Matemàtica i de les Ciències Experimentals, Universitat Autònoma de Barcelona, Catalonia, Spain
e-mail: Mariona.Espinet@uab.cat; Lauravaldessanchez@gmail.com

N.C. Monsó • L.F. Gràcia • R.M. Vila • N.L. Rebollal • A.C. Pascual
CESIRE, Center of Educational Innovations, Department of Teaching, Catalan Government, Catalonia, Spain
e-mail: ncarrill@xtec.cat; lfarro@xtec.cat; rmarti13@xtec.cat; nlopez@xtec.cat; acastil4@xtec.cat

A.W. Oliveira, M.H. Weinburgh (eds.), *Science Teacher Preparation in Content-Based Second Language Acquisition*, ASTE Series in Science Education,
DOI 10.1007/978-3-319-43516-9_16

identity is built around the core element of its particular language, Catalan, which has been suppressed throughout the history of the country in several occasions and particularly during the time of General Franco's dictatorship. The revival of the Catalan language only began after Franco's death, in 1975, at the birth of Spanish democracy. Since then, language policies promulgated by the Catalan government have had an important role in ensuring that Catalan is now commonly used in all aspects of daily life especially in education. The region has officially two languages: Catalan and Spanish although the school system has adopted a compulsory language immersion model in Catalan as a tool for social cohesion. The Spanish central government is at present legislating against the use of Catalan in schools. Any attempt to introduce new languages in the education system might be considered a potential danger for Catalan extinction and might add extra tension in the Catalan education system.

A Content and Language Integrated Learning Approach (CLIL) to Primary Science Education

In response to this demand, European educational institutions at all levels are developing new teaching approaches which could be included in the broad umbrella of bilingual education. Cummins, one of the fathers of multilingualism in education, defined bilingual education as the "use of two or more languages of instruction at some point in a student's school career" (Cummins 2008, p.xii). One of the bilingual education approaches recently promoted in Europe has been the "Content and Language Integrated Learning" (CLIL). This approach advocates the need to design learning environments in which both specific content and a specific foreign language can be taught and learned together: "The acronym CLIL is used as a generic term to describe all types of provision in which a second language (a foreign, regional or minority language and/or another official state language) is used to teach certain subjects in the curriculum other than languages lessons themselves" (Eurydice 2006, p. 8).

A CLIL approach to science education implies the teaching of both science content and foreign language in the same classroom. Multilingual science education contexts are very varied with multiple models and structures existing in different European education systems. In this chapter we want to present the case of Catalonia, Spain, which has developed a particular model on multilingualism in education that strongly affects primary science education.

When developing CLIL approaches in primary science classrooms in Catalonia, teachers need to manage the learning of science and the learning of three languages at the same time: Spanish, Catalan and English, the last one being a foreign language for both teachers and students. Primary teachers usually feel unconfident about the mastery of a foreign language such as English and about the way to teach it. This is why the Spanish government framed the English Primary Education

Specialist as a way to increase the provision of foreign language teachers in primary schools. The fact is that at present those who teach primary science under a CLIL approach in Spain and Catalonia are English specialist teachers who feel unconfident about teaching science and more specifically teaching inquiry based science (Martín 2008; Navés and Victori 2010).

Goals of the Program

The consequence of this is that the profile of primary teachers who teach science in Catalan schools is changing rapidly and that the quality of the science taught might be seriously at stake. New professional development programs need to be developed so that we can help primary teachers teach inquiry based science and English by taking into consideration all linguistic resources brought by primary students. The authors of this chapter addressed this challenge by collaborating in the design of a primary teacher development program called *"IBS and English learning in primary education"*. This program started in September 2014 and was the result of 5 years of research on CLIL approaches to science teacher education undertaken at the Universitat Autònoma de Barcelona (Espinet et al. 2012; Ramos and Espinet 2013a, b; Valdés et al. 2013). The primary goal of this program is to support primary science teachers and primary foreign language teachers to develop teaching units that promote the integration of Inquiry Based Science and English learning taking into account the language diversity of the primary classroom. The secondary goal of this program is to create sustainable triadic partnerships which foster educational innovations and research in the integration of primary science and language teaching and learning in Catalonia, Spain.

Teacher Preparation Model

A Triadic Partnership

The teacher preparation model is framed under the construct of *Learning Communities* inspired by the work of Wenger (1998) on communities of practice. In the field of teacher education, this approach implies a view on teacher development which focuses on reflection but extends it towards including a constructivist view of knowledge acquisition, the relevance of context, and the need to switch from individual to group work among participants (LeCornu and Ewing 2008).

A triadic partnership was built around two axes (Fig. 16.1): (a) the educational institution of reference (schools, university, and educational administration), and (b) the discipline of reference (Science and English). The intention was to design a teacher development program in which all partners had the opportunity to experience

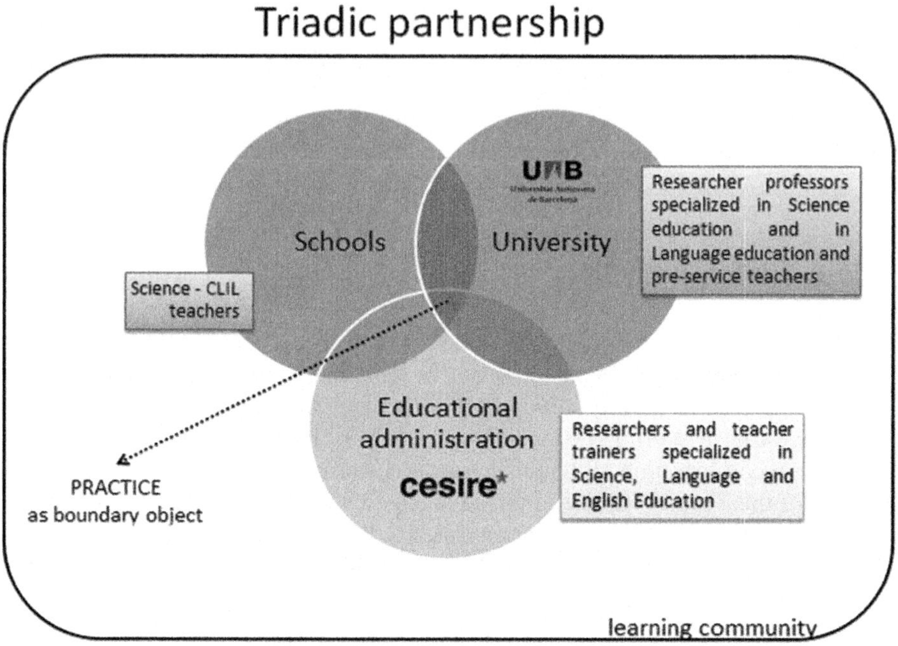

Fig. 16.1 Triadic partnership for teacher preparation including primary schools, UAB (Universitat Autònoma de Barcelona), and CESIRE (Center for Support of Educational Innovation and Research from the Catalan Government)

professional development through their participation in the community. The triadic partnership was managed with two types of activities: (a) staff meetings to promote collaborative research and innovation in Science and English teaching in primary education, and (b) teacher development sessions to promote reflective practice in teaching inquiry based science and English in their own primary schools and with the help of primary student teachers. The institutions directly involved within this triadic partnership were the primary schools, the Autonomous University of Barcelona (UAB), and the Center for Support of Educational Innovation and Research from the Catalan Government (CESIRE) (Fig. 16.1).

Practice as a Boundary Object

Learning communities have a diversity of actors which come from different discourse communities, and thus held different identities. In our program these actors came from different professional institutions as well as different disciplines. Boundary objects are useful tools to help different actors cross the boundaries within communities (Akkerman and Bakker 2011). We took the construct of

Practice as a unifying conceptual tool that would facilitate the building of connections between Science education and English education. This construct acted as a Boundary Object understood as "those objects that both inhabit several intersecting worlds and satisfy the informational requirements of each of them... [They are] both plastic enough to adapt to local needs and the constraints of the several parties employing them, yet robust enough to maintain a common identity across sites. They are weakly structured in common use, and become strongly structured in individual site use" (Star and Griesemer 1989, p. 393).

Conceptual Framework

The conceptual framework supporting our view on the integration of Inquiry based science and English adopts a synergistic view on the integration of content and language (Stoddart et al. 2002). From this point of view the relationship between Science learning and English learning is considered to be reciprocal so that each domain complements and reinforces the other resulting in enhanced learning in both Science and English. Science and English learning are considered to be contextually bounded and thus functional. However, our framework departs from Stoddart et al. in that it introduces the concept of discursive practice instead of the concept of language function in order to describe the characteristics of student and teacher engagement in English teaching and learning. We align with the work of Azevedo et al. (2015) which tried to build a general theory on the relationship between science classrooms activity types and epistemological discourse practices.

The practices of integrated Inquiry based science and English included in our model are the result of two different practices that take place at the same time when teaching science and language in the classroom: Scientific practices and Discourse practices. The scientific practices align with the orientations and conceptualization of the NGSS (Next Generation Science Standards) (NRC National Research Council 2012; Osborne 2014). The Discourse practices included in our model are of an epistemic nature such as identifying, describing, questioning, comparing, explaining, justifying, argumenting or defining. It is assumed that each scientific practice is linked to one or more discourse practices through their enactment. In addition, the framework introduces the idea of scientific and linguistic resources as the necessary tools for the joint enactment of both scientific and discourse practices. The linguistic resources include key scientific vocabulary, specific structures, grammatical cohesion, connectors for discourse coherence, discursive patterns and textual composition process. The scientific resources, instead, include the general concepts necessary to understand natural phenomena from a systemic point of view and correspond to the crosscutting concepts emerging from the NGSS. A conceptual tool representing those practices and their interrelationship was developed to act as a boundary object within the program (Table 16.1).

Table 16.1 Conceptual framework for Inquiry based science and English integration

Practices and resources for Inquiry based science and English integration	
Scientific practices	**Discourse practices**
Asking questions	Identifying
Developing and using models	Describing
Planning and carrying out investigations	Questioning
Analyzing and interpreting data	Comparing
Constructing explanations	Explaining
Communicating	Justifying
	Argumenting
	Defining
Scientific resources	**Linguistic resources**
Specific concepts	Key scientific vocabulary
Patterns	Specific structures
Cause and effect	Grammatical cohesion
Scale, proportion, and quantity	Connectors for discourse coherence
Systems and system models	Discursive patterns: conversation and debate
Energy and matter	Textual composition process: oral and written
Structure and function	
Stability and change	

Implementation of the Model

Program Staff, Participating Teachers and Student Teachers

The staff involved in the program came from two different communities: the School of Education of the Universitat Autònoma de Barcelona (UAB) and the Center for Support of Educational Innovation and Research (CESIRE) from the Catalan Government. The staff involved from the UAB included one science education professor and one doctoral student from the Science Education Department, and one professor from the Language Education Department. The staff involved from the CESIRE included three foreign language teacher educators and two science teacher educators with an extensive classroom teaching experience.

Twenty five primary (6–12) education teachers participated in the program. These teachers were teaching in either private or public urban as well as rural schools. The target profile of the teachers was either science specialists working in collaboration with the foreign language specialist, or foreign language specialist teaching science in English. The criteria used to select the participating teachers included the following: (a) Maximum coverage of the national territory; (b) Diversity of teaching experience; and (c) Known expertise in either foreign language and/or science teaching in public primary schools.

The program included student teachers from the English mediated Primary Education Graduate Program offered at the School of Education (UAB) recruited on a voluntary basis. Each primary school participating in the program hosted one or two student teachers in his or her third year of the Primary Teachers Graduate Program. The function of their participation was to assist the classroom teacher in the implementation of the teaching units and to participate in the teacher development sessions thus acting as teacher assistants.

Development of the Program

The first year of the program was developed through six staff meetings lasting 3 h each, five teacher development sessions lasting two and a half hours each, and implementation of teaching units in each participating school for 3 months. All sessions were tape recorded and reported through the writing of extensive minutes by the doctoral student. All materials were made available to program participants through the moodle electronic platform located within the Catalan government server. The program development was organized around four phases of a teacher development cycle: Exploration, Introduction, Implementation, and Reflection. Table 16.2 reports on the phases and aims of the program development.

Exploration Phase

In the first session both the staff and the teachers shared their experiences on the teaching of Science and English with the purpose of creating a common ground. The World-Café methodology was used as a way to encourage participating teachers to engage in open conversations based on mutual appreciation and collective sharing. They were distributed in one room around small tables to answer one open question and write the answers on a piece of paper. After 20 min, participating teachers freely moved to another table to answer the second questions using the original piece of paper previously set. One person in each table, in our case a staff member, acted as an ambassador ensuring the transmission of the ideas collected in the first round to the second round. Through a World-Café methodology run by the English education specialists, teachers worked in small groups to answer orally and visually the following two questions in two different rounds:

(a) How do you teach Science and English in your primary education classroom and school?
(b) What challenges do you confront when teaching Science and English in your primary education classroom?

One of the most representative challenges highlighted in the discussion of the World-Café visual representations was the need to increase teachers' planning time and effort. Another challenge dealt with the acceptance by all teachers in school that the integration of Science and English learning could be possible. Finally another

Table 16.2 Structure and aims of program development

Teacher development sessions	Program staff meeting
Phase 1: Exploration	
1 session (2.5 h/session)	**3 meetings (3 h/meeting)**
Sharing classroom experience on science and English integrated teaching	Teachers and student teachers' recruitment
	Theoretical reflection and boundary object creation
	Planning of teacher development sessions
Phase 2: Introduction	
2 sessions (2.5 h/session)	**1 meeting (3 h/meeting)**
Understanding science inquiry as the enactment of scientific practices.	Theoretical reflection and boundary object refinement
Understanding English learning as the enactment of discourse practices from a communicative approach	Planning of teacher development sessions
	Preparation of student teacher participation
Phase 3: Design and implementation	
3 months	**3 months**
Design and implementation of teaching units integrating Inquiry based science and English in primary classrooms	On line teacher's assistance
	Monitoring student teacher participation
Phase 4: Reflection	
2 sessions (4 h/session)	**2 meetings (3 h/meeting)**
Communication of teachers' implementations	Analysis of teachers' implementations
Reflection on teachers' implementations	Theoretical reflection on the boundary object
Program evaluation	Planning of the teacher development final session
	Program evaluation and communication

important challenge pointed at the fear teachers had on students' lack of vocabulary to be able to express their scientific ideas using oral English.

The second part of the first session was conducted by the Science education staff members. Teachers were encouraged to work in groups on an inquiry based science activity using the materials offered to them such as the building of a catapult. They were asked to identify questions related to the available materials and phenomena and to answer these questions through and inquiry process. The purpose of doing this activity was to create a common ground where to reflect on the meaning of inquiry based science. Student teachers were invited to participate in this phase although they were excused if they had to attend university course work in the evenings. They were offered a specific preparatory meeting held in the university to inform them about the purposes, methods and expectations of their participation.

Introduction Phase

In the following two sessions participating teachers approached the conceptualization of both Scientific Practice and Discursive Practice through the hands of Science Education and Language Education specialists as well as the support of specific readings. The NGSS approach to scientific practice was presented to them as a tool to interpret their own practice when teaching science. Teachers were confronted with a classroom narrative written by a practicing primary teacher which included transcripts of classroom interactions on water condensation to identify the scientific practices performed by the teacher and students in that specific situation. In addition, participating teachers were offered the opportunity to engage more deeply in a firsthand science inquiry activity on friction as a way to identify some of the scientific practices and to fill the table shown in Fig. 16.2. This table aims at supporting teachers to build a particular inquiry story constituted by a concrete sequence of scientific practices according to the phenomena at hand.

In the second session of the introduction phase the Discourse practices were presented to the teachers and examples were provided to illustrate the most commonly used discourse practices in primary science education such as describing, explaining, justifying and comparing. Participating teachers engaged into a firsthand science inquiry activity on sound as a way to identify the type of discourse practices that were involved in the inquiry activity. The table shown in Fig. 16.3 was given to the teachers as an example to illustrate the discourse practices associated to

Identifying the sequence of Scientific Practices within a Science Inquiry TS	
Scientific Practices	**Description**
Asking questions	
Developing and using models	
Planning and carrying out investigations	
Analyzing and interpreting data	
Constructing explanations	
Communicating	

Fig. 16.2 Tool for helping teachers identify scientific practices in a teaching sequence

<table>
<tr><th colspan="10">Table 1. Scientific and discursive practices. IBS&E Teaching Unit</th></tr>
<tr><th colspan="5">Educational level: Primary Education</th><th colspan="5">Topic: How can we make a good telephone?</th></tr>
<tr><th rowspan="2">Scientific practices</th><th rowspan="2">Inquiry questions</th><th colspan="8">Discourse practices</th></tr>
<tr><th>Identifying</th><th>Describing</th><th>Questioning</th><th>Hypothesizing</th><th>Explaining</th><th>Justifying</th><th>Argumenting</th><th>Defining</th></tr>
<tr><td rowspan="2">Asking questions</td><td rowspan="2">What materials do we have on the table? How can we escribe them? What is the telephone made of?</td><td>X</td><td>X</td><td>X</td><td></td><td></td><td></td><td></td><td></td></tr>
<tr><td colspan="8">We will introduce the telephone as a technological instrument that facilitates human's communication. How do telephones work? How can we know if an object is a telephone? We will introduce a handmade telephone to children and let them play with it so they become familiar with its functioning. It will be important to have different types of telephones so that children can choose which ones to use in different places. Focus on the way it is made of and ask the children: How can we make a good telephone?</td></tr>
<tr><td rowspan="2">Developing and using models</td><td rowspan="2">How can we explain how the sound travels? Can we draw it? What are the important variables to make a good telephone? What would happen if we use longer or shorter cords? Why?</td><td></td><td></td><td>X</td><td>X</td><td>X</td><td></td><td></td><td></td></tr>
<tr><td colspan="8">Through whole group conversation children show the way they have used the telephone to communicate with each other. The teacher uses this opportunity to ask the question: How is it possible that you can hear each other? The fundamental idea is that the sound travels through a medium such as the cord and the transmission depends on the type of material the cord is made of and its length. Children will use a bridging vocabulary. We start thinking how this is possible that we can hear each other so that we start building the first ideas of the model of sound. Children use their ideas to explain the phenomena of sound propagation and to identify some variables they believe are important to make a good telephone. We encourage children to make hypothesis such as: "If we make a telephone with a longer cord, the sound will be worse".</td></tr>
<tr><td rowspan="2">Planning and carrying out investigations</td><td rowspan="2">What do I need to do to test the hypothesis? What materials do we need? What data are we going to collect? How are we going to represent the data?</td><td>X</td><td>X</td><td></td><td>X</td><td></td><td></td><td></td><td></td></tr>
<tr><td colspan="8">Present the materials to children and ask them to select one hypothesis and plan an experiment to test it. The experiment implies making a telephone that will facilitate children the testing of the selected hypothesis. Children will work in small groups. It is important that children start the planning of experiments by being able to justify the hypothesis with their own knowledge. They will need to identify the materials needed, and to describe the necessary actions. During the testing children will need to collect the data and they will need to represent t visually through drawings, graphs, etc.</td></tr>
<tr><td rowspan="2">Analyzing interpreting data</td><td rowspan="2">What has happened?
What does the data in the graph/table tell us? What can we say about our initial hypothesis? Have we conducted a good experiment?</td><td></td><td>X</td><td></td><td>X</td><td>X</td><td></td><td></td><td></td></tr>
<tr><td colspan="8">Children share the results of their recently made telephone to the whole class and try to state if they got evidence to accept their hypothesis. They can share experiments that did not work as expected and provide new hypothesis to test or to improve the experimental designs.</td></tr>
<tr><td rowspan="2">Constructing explanations</td><td rowspan="2">How can you explain what you have found out?</td><td></td><td></td><td></td><td></td><td>X</td><td>X</td><td>X</td><td>X</td></tr>
<tr><td colspan="8">It is the moment to introduce important scientific ideas on sound such as "sound travel as waves", "sounds travel and takes time to do so", "sound travels through materials". We need to encourage children to transfer from the bridging vocabulary to a more scientific one when possible. It is the moment to introduce a definition of wave or sound that can be shared among the children and the community and that it is used to justify the best telephone made. We can use analogies to represent the movement of waves.</td></tr>
<tr><td rowspan="2">Communicating</td><td rowspan="2">How are you going to present your findings?</td><td></td><td>X</td><td></td><td></td><td></td><td></td><td></td><td></td></tr>
<tr><td colspan="8">Children will produce a functional text describing what they have done and learned using any language modality such as oral, visual, artistic etc.</td></tr>
</table>

Fig. 16.3 Tool for planning integrated IBS and English teaching highlighting scientific and discourse practices on sound travelling: "How can we make a good telephone?"

<table>
<tr><td colspan="2">Activity title:</td></tr>
<tr><td colspan="2">Description of the activity:</td></tr>
<tr><td>Scientific practice</td><td>Discursive practice</td></tr>
<tr><td>Scientific demands</td><td>Linguistic demands</td></tr>
<tr><td>Scientific materials</td><td>Linguistic supports</td></tr>
<tr><td colspan="2">Science & English Integrated Learning Objective:</td></tr>
</table>

Fig. 16.4 Tool for planning one integrated inquiry based science and English single activity highlighting the relationship between the scientific and discourse practices, and the scientific and linguistic resources in English

the different scientific practices of a teaching sequence on sound. The difference between the tables shown in Figs. 16.2 and 16.3 is that the later introduces the guiding Inquiry questions and the discourse practices associated to each particular scientific practice. It was suggested to them the use of this table as a tool for planning their own science inquiry teaching sequences.

The discussion unfolded around the ways English use could be promoted when children were actively engaged within the science inquiry activity. The tool presented in Fig. 16.4 was designed to support teachers in the planning of a single activity highlighting on the one hand the relationships between scientific and discourse practices in that particular activity, on the other hand the activation of scientific and linguistic resources necessary for the enactment of those practices. In the case of English, as it is a language that pupils do not master, teachers need to plan the vocabulary and grammatical structures pupils will need. They need language support, not only for the appropriation of subject matter and scientific patterns, but also for the development of classroom activities. This tool was presented to the participating teachers at the end of the second session and they agreed to use it when planning single sessions or activities within a whole unit.

Student teachers were invited to participate in all the teacher development sessions of the Introduction Phase. However, some of them had to attend university course work in the evenings and were unable to join the sessions. In these cases they were offered specific training sessions in the university to be ready to collaborate with the assigned teachers. It is expected in the future that student teachers fully join the teacher development program. Although these students were trained in Inquiry based science education during the course on Didactics of Science, they lacked the pedagogical content knowledge on how to integrate it with English in primary classrooms.

Design and Implementation Phase

The participating teachers and the assigned student teachers engaged into collaborative design of the teaching units aiming at the integration of inquiry based science and English. During the period of 3 months they regularly met once a week in schools to assist each other in the planning and implementation of the teaching units in the teachers' own classrooms. The teaching units were made available to all program staff, participating teachers and student teachers for online revision. Both teachers and student teachers collected teaching materials, observations, and students' work so that the teaching innovation could be documented. Teachers prepared a visual slide presentation for 15 min on the purpose, activities and results of the teaching unit implementation in their classroom with the assistance of student teachers.

Reflection Phase

Finally, two sessions were devoted to teachers' presentation for 15 min of their teaching units' implementation, and a collective reflection followed on these presentations. In addition, during the last session participating teachers were organized in small groups and were suggested to engage into program evaluation for 45 min.

The following list of suggested questions was provided to help them focus the evaluation conversation:

(a) Let's reflect on the group work done
To what extent have the teacher development sessions contributed to the improvement of your classroom teaching?
How do you value the presentations of the participating teacher experiences?
In what ways have both science education and language education specialists' presentations been useful to you? How would you improve their contributions?
What challenges on the Integration of inquiry based science and English are still important in your school and classroom?
(b) Let's reflect on the developed tools
In what ways have you used the developed tool in the planning of your teaching unit?
What strengths and weaknesses can you identify in the tool?
How would you improve the tool so that it is more useful for the planning of integrated inquiry based science and English teaching units?
(c) Let's reflect on the student teacher participation
To what extent has the participation of the student teacher been useful for your classroom and school?
In what ways has the student teacher helped you and the children during the implementation of the teaching unit?
What functions do you think students teachers should have in assisting the implementation of the teaching units?

Outcomes

The outcomes of the program have been elaborated as a consequence of a reflective process among the program staff. Data from the development of the program were collected through the following four types of instruments: (a) minutes of the staff meetings and the professional development sessions written by the doctoral student from the Science Education Department of UAB one week after each event; (b) educational materials written by the teachers (design of the teaching intervention, report at the end of the teaching intervention); and (c) individual evaluation questionnaire at the end of the program, and final evaluation session. The major outcomes can be grouped around the participation of each particular actor within the triadic partnership.

Program Staff

The program staff has found a fertile ground to continue working as a learning community at the intersection between science and foreign language acquisition in primary education. The staff meetings were long and usually intense since we needed to approach each other and understand the interactions and proposals set on the table. Along the collaboration, the program staff felt the strong need to make explicit their own science education and language education approaches in order to evaluate their compatibility. The challenge to be confronted in the near future is how to build a functional didactical framework that integrates a model based inquiry science education approach and a communicative approach to foreign language education. The program staff is determined to support the collaboration and promote its growth in number of people involved, funding, geographical coverage and stability. The use of an electronic platform such as Moodle monitored by the program staff from CESIRE will be a necessary resource to maintain the learning community throughout the project overcoming the time lapses and geographical distances of both staff and participating teachers. Finally, some members of the staff have felt confident enough to participate in the dissemination of this experience in national and international conferences and events (Carrillo et al. 2016) indicating that the experience has been valuable for them and the rest of the program staff.

Primary Teachers

The primary teachers participating in this program have been strongly challenged and have been able to identify their strengths and weaknesses when teaching integrated inquiry based science and English. These teachers were recruited based on their experience in the teaching of CLIL approaches in primary classrooms and they freely accepted to participate in the program. Although the original program call was addressed to both primary science teachers and primary English teachers, the response came mostly from the English specialists. These teachers believed that science could be an optimal context for a functional approach to English learning in the primary classroom and wanted to learn more on how to do this. These teachers have appreciated the systematic reflection on science teaching undertaken along the program development and the exposure to a different way to teach science based on direct experimentation with phenomena. The challenges experienced by the English teachers concentrate on grasping the nature of scientific practices, especially the modelling ones and the shift from experimenting to knowing. One of the teachers mentioned: *"I would like to better define the shift from investigations to the acquisition of true scientific knowledge by children"*. In addition, they have also expressed the need to expand their ability to present the science inquiry activities so that they become more authentic and functional contexts for both language and knowledge development. Their suggestions for improvement have pointed at the need to engage into collective planning of one teaching unit during the teacher development sessions. This collective product would act as an example for all participating teachers when planning their own units in the school.

Prospective Primary Teachers

Student teachers from the English mediated Primary Education Graduate Program at the UAB have experienced their participation in the community as a real induction into the professional field of CLIL approaches to content and language teaching. During their university course work, they did not have many opportunities to experience such type of approaches since they were rather new in our geographical area. On the other hand, they felt comfortable with the inquiry based science education approaches since they were offered specific science education courses on this topic. However, student teachers have been the most difficult partner to involve since they experienced scheduling difficulties due to course work requirements. Practicing teachers highly evaluated the participation of prospective primary teachers through statements such as bringing new visions to the work, improving the quality of English interactions in the classroom, and increasing primary students' motivations. One of the practicing teachers commented: *"They were very helpful to us since one more teacher in the classroom is very enriching and motivating for the students"*. Student teachers' participation has been considered fundamental since they acted as bridges between the school teachers and the program staff especially during the design and implementation phase. Given the important role played by the student teachers, the staff from the university is planning to develop formal, and thus more stable, undergraduate internships to facilitate their participation in the program.

Conclusions

The triadic partnership was piloted for the first time in the academic year 2014–2015 as a way to promote the integration of inquiry-based science and English in primary education in Catalan schools, and has provided a scenario for successful collaborative and innovative teacher development processes. The first strength of the teacher preparation model has been the explicit theoretical support provided by university and teacher educators from both science and foreign language education. The theory acted as an arena where to create boundary objects that were shared and negotiated among all participants. The boundary objects were supported by the theoretical framework of the program and became materialized as a dynamic tools used by all participants. At the end of the first year, it became evident that these tools could be improved through theoretical refinement, to make them closer to teachers' ways of understanding, and through practical refinement, to better adjust to the particular habitus of teachers' planning. Research work needs to be done in order to investigate the use of these tools by prospective and practicing primary teachers when collaborating in the planning of teaching units.

The second strength of the program has been the composition of the learning community including school teachers, students teachers, university professors and teacher educators from both science and language education disciplines. All participants have developed a positive attitude towards the collaboration within such a diverse community and want to continue the work in the future. The program has included not only teacher development sessions but program staff substantive meetings which have been proven to be crucial. However, the program needs to better articulate the participation of prospective student teachers so that they are offered more systematic scaffolding experiences into the collaborative planning of integrated inquiry based science and English teaching in primary classrooms. It becomes clear that in the next future our particular triadic partnership will need to include three interwoven sequences of professional development appropriate for each partner: student teachers, practicing teachers and program staff.

The third strength of the program points at its particular focus, that is, the attempt to systematically approach the integration of inquiry based science and English teaching in primary classrooms. Once the first trial of the program has finished and evaluated, the staff agreed upon the need to introduce one or two more teacher development sessions, especially in the introduction phase. In doing so teachers would have more opportunities to reflect on the meaning and use of discourse practices in inquiry based science. They would also be better prepared to identify the most appropriate linguistic demands and resources for second language acquisition. The obstacles primary English specialists need to confront when participating in this professional development program are high and lie around the issue of decentralizing their teaching from just promoting English vocabulary learning in school science. Open approaches to science teaching and learning such as inquiry based science education demand students to use a diversity of language modalities, registers, and genres in order to be able to model natural phenomena. Here lies the potential of this type of approaches to act as authentic contexts for both science learning and second language acquisition.

The teacher preparation model has proven to be so promising that two new sites will join in the next academic year: the University of Lleida and the University of Vic both in Catalonia, Spain. These two universities will create two new learning communities with the aim of promoting the integration of Inquiry based science and English in primary classrooms and primary science teacher preparation following the triadic partnership presented in this book chapter.

Acknowledgments The work presented in this chapter has been partially supported by grant 2014SGR1492 from the AGAUR of the Catalan Government and grant EDU2015-66643-C2-1-P from the Spanish Ministry of Economy and Competitivity.

References

Akkerman, S., & Bakker, A. (2011). Boundary crossing and boundary objects. *Review of Educational Research, 81*(2), 132–169.

Azevedo, C., Martalock, P. L., & Keser, T. (2015). The discourse of design-based science classroom activities. *Cultural Studies of Science Education, 10*, 285–315.

Carrillo, N., Valdes-Sánchez, L., & Espinet, M. (2016). Promoting the integration of inquiry based science and English learning in primary education through triadic partnerships. *WFATE Biennial meeting: Innovation in teacher education in a global context,* Barcelona, April 21–23, 2016.

Cummins, J. (2008). Teaching for transfer: Challenging the two solitudes assumption in bilingual education. In J. Cummins & H. Hornberger (Eds.), *Encyclopedia of language and education: Vol. 5. Bilingual education* (2nd ed., pp. 65–75). Boston: Springer Science.

Espinet, M., Izquierdo, M., Bonil, J., & Ramos, S. L. (2012). The role of language in modeling the natural world: Perspectives in science education. In A. K. Tobin, B. Fraser, & C. McRobbie (Eds.), *Second international handbook of science education* (pp. 1385–1405). New York: EEUU: Springer.

European Commission. (2007). *High level group on multilingualism. Final report.* Belgium: Official Publications of the European Community. http://bookshop.europa.eu/en/high-level-group-on-multilingualism-pbNC7807451/downloads/NC-78-07-451-EN-C/NC7807451ENC_002.pdf

Eurydice. (2006). *Content and Language Integrated Learning (CLIL) at school in Europe.* Brussels: Eurydice European Unit. Retrieved from http://ec.europa.eu/languages/documents/studies/clil-at-school-in-europe_en.pdf

Guasch, O., & Nussbaum, L. (Eds.). (2007). *Aproximacions a la competència multilingüe.* Bellaterra: Universitat Autònoma de Barcelona.

LeCornu, R., & Ewing, R. (2008). Reconceptualising professional experiences in pre-service teacher education…reconstructing the past to embrace the future. *Teaching & Teacher Education, 24*, 1799–1812.

Martín, M. J. F. (2008). CLIL implementation in Spain: An approach to different models. In C. M. Coonan (Ed.), *CLIL e l'apprendimento delle lingue. Le sfide del nuovo ambiente di apprendimento* (pp. 221–232). Venezia: Libreria Editrice Cafoscarina.

National Research Council. (2012). *A framework for K-12 science education: Practices, crosscutting concepts, and core ideas.* Committee on a conceptual framework for new K-12 science education standards. Board on science education, division of behavioral and social sciences and education. Washington, DC: The National Academies Press.

Navés, T., & Victori, M. (2010). CLIL in Catalonia: An overview of research studies. In Y. Ruiz de Zarobe & D. Lasagabaster (Eds.), *CLIL in Spain: Implementation, results and teacher training* (pp. 30–54). Newcastle upon Tyne: Cambridge Scholars Publishing.

Osborne, J. (2014). Teaching scientific practices: Meeting the challenge of change. *Journal of Science Teacher Education, 25*(2), 177–196.

Ramos, S. L., & Espinet, M. (2013a). Una propuesta fundamentada para analizar la interacción de contextos AICLE (Aprendizaje Integrado de Contenido y Lengua Extranjera) en la formación inicial del profesorado de ciencias. *Revista Enseñanza de las Ciencias, 31*(3), 1–22.

Ramos, S. L., & Espinet, M. (2013b). Expanded agency in multilingual science teacher training classrooms. In A. N. Mansour & R. Wegereif (Eds.), *Science education for diversity: Theory & practice* (pp. 251–271). New York: Springer.

Star, S. L., & Griesemer, J. R. (1989). Institutional ecology. 'Translations' and boundary objects: Amateurs and professionals in Berkeley's museum of vertebrate zoology. 1907–39. *Social Studies of Science, 19*(3), 387–420.

Stoddart, T., Pinal, A., Latzke, M., & Canaday, D. (2002). Integrating inquiry science and language development for English language learners. *Journal of Research in Science Teaching, 39*(8), 664–687.

Valdés, L., Espinet, M., Aguas, M., & Dallari, L. (2013). La evolución de la co-enseñanza de las ciencias y del inglés en educación primaria a partir del análisis de las preguntas de las maestras. *Enseñanza de las Ciencias. Número Extra, 2013*, 3588–3594.

Wenger, E. (1998). *Communities of practice: Learning, meaning, and identity.* Cambridge: Cambridge University Press.

Chapter 17
Processing Science Through Content and Language Integrated Learning (CLIL): A Teacher's Practicum

Margaret Rasulo, Anna De Meo, and Maria De Santo

Introduction

Non-language Subject Teaching and Reform Processes

In the last 40 years, the Italian education system has been swept by a deep reforming process aimed at innovating curricular requirements and in-service teacher training provision. In compliance with the latest reform of the upper-secondary school sector[1], the Italian national curriculum for high school students now includes dual-focused education that goes by the name of CLIL (Content and Language Integrated Learning) aimed at improving the acquisition of subject content through a foreign language. Among the CLIL subjects chosen by the schools the most popular is Science. This is easily explained as many upper secondary schools with a long-standing tradition in Italy are science-based and go by the name of *Liceo Scientifico*, and those who teach Science in the senior year are required to attend a training course in order to obtain CLIL teaching qualifications.

The role of English is important in CLIL provision as it takes the lead over other European languages that are equally accredited as CLIL vehicular languages such as French, German and Spanish. However, the spread of English in these Expanding Circle countries (Kachru 1985), in which English is a foreign language rather than a second language, has taken different directions. In Italy, for example, the demands of globalization and internationalization processes have given a renewed impetus to

[1] Decree 88/2010, Decree 89/2010.

M. Rasulo (✉)
Second University of Naples, Caserta, Italy
e-mail: mags1@libero.it

A. De Meo • M. De Santo
University of Naples L'Orientale, Naples, Italy
e-mail: demeoanna@gmail.com; marilenads@gmail.com

A.W. Oliveira, M.H. Weinburgh (eds.), *Science Teacher Preparation in Content-Based Second Language Acquisition*, ASTE Series in Science Education,
DOI 10.1007/978-3-319-43516-9_17

the teaching of English from primary to tertiary level education. Learners are now required to study it for specialized purposes and focus on the kind of English they actually need in order to compete with their European colleagues and communicate professionally and/or academically in the workplace or study environment. CLIL provision is Italy's reform solution to prepare students for this challenge and such is the case of English for science education which is the focus of the present research study.

CLIL is a comprehensive methodology as it delivers subject content by making use of the more familiar approaches of foreign language teaching such General English Instruction and other more specialized language education approaches such as English for Specific Purposes (ESP). This approach, based on the teaching of special uses of language in such professional genres as the medical, scientific, economic or legal, typically characterizes the English language curriculum of many Italian university departments and is also part of the general language curriculum of upper secondary schools. As both universities and schools are now faced with the need to increase their content-based language provision, and many are doing so by adopting CLIL, it is important to specify that this methodology tackles instruction from the standpoint of the 'foreignness' of the language, which means that it relies substantially on scaffolding techniques and accessibility processes that are part and parcel of general foreign language teaching and learning. It follows that science teachers need to be equipped to teach the language from different methodological angles and linguistic stances, which implies acquiring knowledge about how to exploit language teaching approaches and select the most suitable according to immediate student needs and requirements. The setback to this is that science teachers do not always possess an adequate level of English to teach CLIL classes as they have most likely learned the language at school or in other contact situations such as travelling, and this means that they need to improve their own language proficiency while attending CLIL methodology training courses.

Setting and Background

The setting of this research study is the CLIL-for-Science Diploma Course held at the Linguistic Center of the University of Naples "L'Orientale". Although the practicum model presented in the following sections of this paper complies with the general CLIL course design guidelines issued by the Italian Ministry of Education, University and Research, the core pedagogical principles underpinning the model are the result of research findings of former training courses and continuous observation of teaching practices conducted by the authors who are also the University's CLIL experts. The participants of this study, selected on a voluntary basis among the 20 teachers attending the course, are 5 Italian science teachers from 5 different science-based high schools of the Campania Region situated in Southern Italy.

In order to implement the model, these teachers were asked to collaborate with the English as a Foreign Language colleagues who are not involved in the Ministry's CLIL training initiatives, albeit one of the CLIL skills to be acquired according to the same Ministry guidelines is that subject teachers should establish 'fruitful partnerships with the language experts'. It is precisely this paradoxical aspect of the requirements that our study wishes to address as we firmly believe that the integration between the C (content) and L (language) of the CLIL acronym can only happen through the deliberate collaboration of the two professionals. In fact, the data collection tools employed in this project such as lesson plans, observation and reflection grids, co-analysis debriefing notes and course materials are implemented by both practitioners. Indeed, the main goal of the practicum model encourages the collaboration of both the Science and English teachers in order to ensure that student learning is scaffolded and supported by both educators.

The Collaborative Nature of the Practicum Model

Collaboration is embedded in CLIL as both Science and English teachers need to familiarize with this new approach to the same extent as their students. This is a necessary condition in Italy where, as mentioned above, the majority of science teachers do not possess a high level of fluency in the target language in order to teach a subject in a foreign language. What is more, one of the tenets of CLIL methodology is the focus on subject discourse in order to improve comprehension of the subject delivered in the foreign language (Marsh 2002), and this understandably requires the intervention of the English teacher, whose role in this partnership is that of the ESP expert. For the sake of simplicity, the acronym ESP will be used henceforth in the article to refer to the language practitioner. This collaborative partnership is strongly advocated by the authors as it also strengthens the science teacher's methodological repertoire with the wealth of strategies, techniques and activities that characterize language teaching.

The study's research questions are representative of the three main threads that run through the entire work and are therefore discussed in the various sections of the paper:

1. How can a CLIL professional training course best satisfy the needs of science educators in their quest for improved student understanding of the subject matter delivered in a foreign language?
2. How can the professional relationship between science and foreign language teachers be encouraged as a result of a training course?
3. How can these teachers effectively contribute to the target language development and proficiency of their students?

Teacher Preparation Model

The Diploma Course

The CLIL-for-Science Diploma Course was designed according to the general structure of the Italian Ministry of Education which includes three main components. The first is an advanced level general English language course, the second is a more specific English-for-Science course accompanied by video-recorded lessons, and the third is a CLIL methodology course consisting of workshops and online discussion forums. The staff members consisted of ESP and science teacher trainers as well as a team of CLIL experts.

For the purpose of this article, only the English-for-Science course is described in detail due to its specific focus on the selected eight functions and discursive strategies (Dalton-Puffer 2007) of academic English which, according to the authors, are necessary to achieve deep understanding of scientific content and materials. What is more, this course component constitutes one of the strongest links between the CLIL science teacher and the ESP teacher.

The CLIL-for-Science Practicum Model

The CLIL-for-Science Practicum Model (henceforth the practicum model or the model) is a training framework with phases, strategies and approaches, activities and materials explored during the workshop lessons and implemented by the trainee teachers in their own classrooms to prepare their final course project. The model's backbone is Tessa Woodward's Loop Input Approach (2003) based on action research methodology. This approach afforded the opportunity to systematically implement the experience taking place within the training context and feed it back into the classroom environment for immediate trial. Figure 17.1 illustrates the practicum's design with the three components of the model placed at the left of the diagram, and the main approaches at the right. The academic functions of English occupy the center of the diagram as this position highlights the crucial role they play as a language-based vehicle through which teachers learn to explore scientific content delivered in a second language during training sessions and in their own classrooms. It is important to remind the reader that since language teachers in Italy

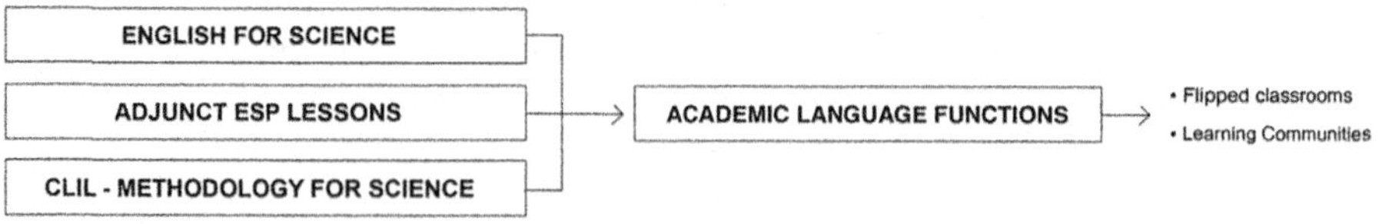

Fig. 17.1 The CLIL-for-science practicum model

are not the direct recipients of training initiatives for CLIL teaching, ESP expert trainers worked with the science teachers during the practicum sessions on the various phases of the model so as to prepare them to do the same with the language colleagues in their own schools (Rasulo 2013). The following sections will provide a full description of the model's components.

English-for-Science Course

The eight language lessons of the English-for-Science component were supported by additional video-recorded segments which were delivered online on the Moodle platform of the University's self-access center. Each 20-min video introduced the academic functions that the teachers explored during their face-to-face lessons as used in scientific contexts and scientific text types. The same functions were then discussed in the workshop lessons backed up by other language activities. This flipped learning lesson arrangement proved to be quite successful as teachers had the possibility to reflect on the content of the videos and practice the functions with their students before sharing their views with the other fellow trainees. The framework below lists these functional categories and their main purposes as well as the discursive strategies that these functions generate in terms of linguistic evidence.

Academic functions of English (Coyle 1999) such as those in Table 17.1 provide clarification of the concepts of language *easification* which, according to Bhatia (1983), means facilitating content accessibility for deep understanding rather than oversimplification of content stripped of any sort of cognitive challenge. Particularly interesting is the argument put forth by Dalton-Puffer (2007) that these discursive strategies have an underlying recognizable structure that is naturally embedded in language. It is precisely this structure that needs to be made explicit to students through what Walqui (2006) describes as a continuing scaffolding cycle of presentation and re-presentation of content. Teachers need to devise a series of activities in order to embed these support structures in the engagement, exploration and elaboration stages of their lessons (Rasulo 2014). The practicum model as developed by the authors requires science educators to work with their language colleague in the exploitation of these functions in delivering their adjunct ESP lesson.

Adjunct ESP Lessons

The success of the model largely depends on the effectiveness of adjunct ESP lessons. As these lessons support instruction from a more language-based approach to the study of scientific content, they are delivered by the ESP language teacher, but are co-planned with the subject teacher and, depending on the overall design of the CLIL syllabus, they can precede or follow the subject lessons. Adjunct lessons zoom in on the academic functions, but they also explore more specific items of the

Table 17.1 Academic functions of English

Functions of English for science	Linguistic evidence
Classifying	How many kinds of… are there?
lustering techniques (mind maps, brainstorming)	How do we classify…?
	Why are these grouped in this way?
outlines with headings and subclasses	How would you classify…?
	Can you rank these according to…?
lists of items in each subclass with details and characteristics	Can you put these into the correct groups?
organization of information in a logical way	This can be classified as… (because…)
Comparing and contrasting	Let's make a comparison between … and …
comparison of data, processes, tables contrasting different subjects	… compared with/to …
	Similarities/differences in …
	In comparison with …
	… Likewise/Similarly, …
	Not only … but also …
	This is actually the same as ….
	This is similar to …
	Is … different from …
	What are the differences between … and?
	Do they share any features?
Explaining and exemplifying	Simplifying:
naming and defining objects, parts, purposes, etc.	That is to say…
	…, meaning that…
	Let me put this another way
describing how something works	I'm going to explain…the reason for…the cause/result is…
	Cause and effect:
	Due to …, …
	Owing to …, …
using sequencing phraseology	Since/As ….
	Because…/Because of…,
transitional words and signposts	Owing to …, …
	Since/As ….
Giving details	Asking for explanations:
topic words	Analyze
reasons	Assess
	Choose
	Decide
Choosing manageable topics	Describe
	Discuss

(continued)

Table 17.1 (continued)

Functions of English for science	Linguistic evidence
clear support examples	Identify
steps of the exemplification process	Illustrate
details arranged coherently	Prove
Hypothesizing and predicting	What if…? (What if we could predict earthquakes?)
	Supposing…? (Supposing Germany had won WW2…)
before and now	Assuming… (Assuming that there is no energy loss, calculate the electrical energy used by the motor.)
making predictions	
talking about probability	Suggest what might happen if…? (Suggest what might happen if we put salt in the water.)
	What is the probability that + clause…?
	What is the probability of '-ing'…?
	How likely is it that + clause…?
	Do you think + sub + will + verb…?
Questioning	Can/could/might:
	What can you see?
Engaging students through	How might this change?
	What might have happened if things had been different
dialogic involvement to think about the topic, or acknowledge alternative views	Present continuous:
	What are we studying?
	Which exercise are you doing?
	Future (going to):
non-rhetorical and	How are you going to begin?
challenging structures	Where are you going to start looking?
	Present perfect simple:
	How long has this happened for?
	Look at these two maps. What has changed?
Sequencing	First of all
	Thirdly
Identifying and defining	Afterwards
	Following that
	At first
the components of an event,	Lastly
process, procedure, theory, etc.	After that
	In the end
	During
	To start with

(continued)

Table 17.1 (continued)

Functions of English for science	Linguistic evidence
the purpose or function within event, process, procedure, theory, etc.	By…
Retelling	
the events, steps, processes within a given text in the o order in which they occur	
Generalization	Qualifying words or phrases:
providing evidence, samples, examples to support topic	among those studied
	are likely to
	one of the reasons
Using marked lexis	Employing discourse markers to draw attention:
creating emphasis	'*Yes, what you said is true*, DNA is a very long and simple double helix molecule'
strategic foregrounding of important meanings	
Paraphrasing	layout (note form, outlines, mind maps, clusters) employing Synonyms/antonyms to vary, compare, simplify, explain specific terminology and concepts
re-organizing the structure of the subject in a more familiar forms	

content language which are typical of ESP methodology such as lexis, aspects of syntax, phraseology or formulaic expressions, and aspects of text genre.

The adjunct lessons are the point of intersection where content and language meet and it is through the delivery of these co-planned lessons that the partnership between the content and language teachers is nurtured.

The CLIL-Methodology for Science Workshops

The practicum model represents one of the possible responses to teaching language-driven content courses as it aims to provide trainers and trainees with a valuable professional course template aimed at developing course materials and activities to ensure accessibility and comprehensibility of science content to students with limited proficiency in the language. At the same time, it aims to improve collaborative partnerships between science teachers and foreign language practitioners in order to strengthen the mutual recognition of roles and contributions. A phase-by-phase description of how the model was implemented is provided in Table 17.2.

At a glance it can be noticed that the partnership between the CLIL and the language teacher is strongly advocated as both teachers are involved in each phase albeit the science teacher is obviously the one who acts as the major *liaison* between

Table 17.2 The CLIL-for-science practicum model implementation phases and activities

Phases	Activities
1. Preparation phase: science teacher with language teacher	co-planning the CLIL science lesson
	co-planning the ESP adjunct lesson
	agreeing on aims and purposes
	planning for observation
2. Implementation phase: science teacher; language teacher	delivering and observing the CLIL science lesson
	delivering and observing the ESP adjunct lesson
3. Extension phase: science teacher with language teacher	analyzing data collected through observation
	re-formulating/re-directing aims and objectives through co-analysis of practice

the training environment and the school classroom. During the Preparation phase (1) the teachers co-plan their lessons by providing their professional expertise for both components using the same lesson template. In so doing, they are both accountable for the success of each other's part and are equally informed on the content of the lessons. The science and the ESP lessons are part of the Implementation phase (2) during which teachers have the opportunity to reflect on the outcome of their part by using a self-observation tool. The Extension phase (3) is dedicated to debriefing activities as the model heavily relies on pre and post co-analysis of lessons in order to encourage reflection practices. To this purpose, the trainees posted their CLIL teaching lesson plans in the virtual community environment on the Moodle platform in order to receive continuous feedback from their trainers and, once back in the training classroom, the teachers had the opportunity to discuss outcomes. The formation of a learning community arranged with the support of flipped methodology was strongly encouraged by the set-up of entire course. The teachers had the opportunity to interact asynchronously on the online forum with other participants and share their individual experiences as well as discuss their self-observation results, and this was accomplished under the moderation of the practicum experts who challenged the teachers by expanding on the issues or questioning their understanding of the implementation tasks. This ongoing and scaffolded activity encouraged both the trainers and trainees to be innovative while reflecting on the question "What is CLIL fluency in teaching Science"? This is the rationale behind the model's existence, firstly because the concept of fluency is generally associated with language, and in the CLIL environment this applies both to the language of *subject content* as well as to the language of *communication*, and secondly because it is understood that acquiring subject content in CLIL depends on the fluent use of the foreign language in order to perform the complex cognitive abilities that are required for knowledge acquisition. It follows that CLIL teacher training relies on teacher partnerships and on the catalyzing effects of both languages of instruction in order to create input that is accessible and comprehensible (Krashen 1985).

Table 17.3 Observation tools and the three phases of the practicum course

Phase	Data collection tool(s)
1. Preparation phase	Checklist of questions
2. Implementation phase	Lesson plan template
Science lesson	Self-observation tool
ESP lesson	
3. Extension phase	Co-analysis of practice observation field notes

Materials

The materials presented in this section capture the essence of the training model and are therefore only a partial sampling of the ones actually employed by the teachers during the action research sessions. A very substantial role was played by observation and reflection tools as the authors support the view that self-inquiry provides insight into the nature of quality teaching and works towards professional development. The Table 17.3 lists the tools used by the science and ESP teachers during the three phases. A brief description of each as well as a sample tool are also provided.

The checklist of questions (Table 17.4) that was used during the teacher preparation phase was designed according to Schön's 'reflection-on-action' theory (1987). It is a simple tool which uses 'thinking-aloud' strategies to discuss what teachers aim to accomplish.

The answers to these questions and the subsequent discussion contributed to the actual implementation phase of the model and the creation of the lesson plan template below which was used to coordinate activities and select the CLIL aspects that both teachers wanted to focus on during their lessons (Table 17.5).

During the implementation phase, as teachers were also putting to practice the newly-acquired concepts in their own classrooms, it was necessary to use a self-observation tool co-designed by trainees and trainers during the workshop sessions to record the lesson events as they were actually experienced. The items of the tool inquired about the strategies whose outcomes the teachers were asked to notice and record after the lesson by simply ticking one of the adjectives that best described the extent to which they had been marginally, sufficiently or totally successful in performing them. By looking at the overall results, the only issue which received a totally successful response from all 10 teachers was related to the safe learning environment of their classrooms which encouraged students to express their needs. Transitioning back to the CLIL lesson or back to the ESP lesson by creating deliberate links through the use of materials or activities received a more hesitant response. The same dubious results were reported by the teachers regarding materials selection and adaptation. Both language and science teachers were indeed unsure about the amount of exposure they were giving students through the content and language information contained in the text types in order to maximize both scientific concepts and ESP language features. The nature of this reply confirms the fact that the availability of adequate science CLIL materials on the market is limited in Italy, leaving teachers with the arduous task of creating in-house materials which, on one hand,

Table 17.4 Checklist of questions

1. Preparation phase	which content objectives, which language objectives?
	how much time do we need?
	what resources do we need?
	what do we expect from students?
	how do we develop awareness that both lessons are working towards the same objectives?
	do we need to give feedback/assessment? when and how?
	how should we observe the lesson? tools?
2. Implementation phase	what teaching materials do I need?
	should I create a handout for students to study from?
	what should students do during the activities?
	how do I create a link with the other lesson? (content-language)
	do they need to produce something?
	how do I assess them? what should I assess?
3. Extension phase	what are the positive aspects?
	what are the weaknesses?
	what needs to be changed?
	what is still missing?
	when would be the best time to discuss experience and results?
	what is the way forward????

are more interesting and effective, but on the other, are extremely time-consuming to put together. As regards the use of strategies for text readability and accessibility such as using English frequently to check for understanding, clarifying language and reformulating concepts, as expected, the language teachers seemed to be more familiar with them, differently from the science teachers, who were at that moment being trained in language-based methodology. Continuous assessment also seemed to concern all 10 teachers who admitted that their present strategies mainly consisted of measuring student performance through written tests and oral questioning, albeit the language teachers were aware of other ways of assessing student learning. The extension phase was characterized by a post lesson debriefing discussion between the two teachers. This co-analysis of practice (Banks et al. 1995) afforded the opportunity to compare notes form the self-reflection tool and discuss strengths, weaknesses and the way forward. This tool guided thinking processes towards evaluating the extent to which they were able to create a content and language integrated lesson. Both the language teachers and the science teachers believed that the most significant strength was the ability to generate increased dialogue among the students due to the discussion sparked by the presence of specific lexical items and syntax structures in the variety of text types belonging to the scientific genre. The weaknesses were identified by both teachers in the areas of materials availability, adaptation and creation, and in the coordination between the teachers in carrying out inclusive assessment procedures that would take into account both content and language.

Table 17.5 Lesson plan template

Title of lesson		
Class		
Time frame		
Content learning objectives (the science teacher)		Objectives that the students will need to achieve
		Specific content/topics
Language learning objectives (the language teacher)	Specific lexis	Subject-specific lexis that characterizes the lesson/s
	Syntax structures/ language chunks	Subject/text specific grammar items and structures
	Skills focus	Language skills
		Research skills
		Study skills
CLIL academic functions		Cognitive skills that the students should execute in order to complete the activities
CLIL teacher and ESP language teacher coordination		Contribution that is expected to link the content lesson to the language lesson (specific academic function(s), specific skill needed to explore the readability or simplification of a written or oral text, grammar support work, etc.)
Personalization of content (if pertinent)		Connections that can be made between the content and distinctive cultural aspects (society, everyday activities, etc), particularly those that involve student life and everyday environment.
Teacher activities		Description of teacher activities such as:
1.		1. Warm-up elicitation activity
2.		2. Presentation of content
3.		3. Asking for feedback
Student activities		Activities that are expected from students that correspond to teacher activities
1.		
2.		
3.		
Group dynamics		Student grouping should be specified for each activity: group work, individual work, pair work, whole class activity, etc
Expansion/final product		Projects/activities that students should complete, usually as a culmination of what they have learned during the unit.
Resources/materials		Resources and materials prepared by the teacher or by someone else that will be used throughout the unit. All will include a reference as to where the materials were found.
Assessment		Specific skills to be evaluated/assessed and the form of evaluation/assessment.

Setting Up the CLIL-for-Science Diploma Course

Any decision to implement a highly complex training activity such as that required for CLIL teaching involves monitoring the feasibility, the effectiveness and the impact of the course on trainees, on the training institution and on the team of experts. Discussing course set-up towards the end of the paper was the authors' intentional choice so as to dedicate enough space to this component without being influenced by the complexities of the model's implementation phases.

Testing Teacher Beliefs

Although this course focused on teaching Science through CLIL, it was necessary to tap into the teachers' prior knowledge of CLIL methodology in general terms. Thus, the first tool was a simple Likert format questionnaire with three-scale adverbs stating whether the teachers strongly disagreed, agreed, strongly agreed with the issues presented by the items. The same questionnaire was administered by the science teachers to their ESP colleagues in their individual schools.

Results from this tool show that both science and language teachers believe that in CLIL teaching there should be a reduction in the use of the first language whose interference can cause confusion and slow down the process of learning the subject through the foreign language. The teachers' attitude towards who should focus on meaning and who should deal with aspects of form reveal a 'role claiming' attitude by affirming that these two aspects should be kept separate and each practitioner should deal with their own area of expertise. The science teachers, in addition, also claim that it is their responsibility to teach content-specific terminology while the contribution from the language teachers should be the teaching of grammar.

As concerns the materials and activities section of the questionnaire, the responses show that ready-made materials and recipes are always preferred to more in-house preparation and adaptation of content. Working towards increasing accessibility while preserving content authenticity takes away far too much time and, according to all of the teachers, it is not their responsibility but that of textbook editors. The answers to these items were precious as they led to the presentation and exploitation of discursive strategies to be used by the teachers during the various phases of the practicum for the benefit of making content both comprehensible and meaningful. Other items questioned the possibility of envisioning a role for language teachers in CLIL provision. The response provided by all the teachers express a general confusion linked to the identification of a specific area of intervention for each practitioner, but assessment seems to be one of the most disorienting and weakest links. Responses also reveal that CLIL fails to be seen as a joint effort supported by whole-school recognition, and it was this response that led to the presentation of teacher partnerships as one of the tenets of the practicum model.

Assessing Feasibility of Course Set-Up

The implementation phase of the course was also observed throughout the entire process as the authors conducted a continuous progress check that consisted in inquiring about course dynamics by using a simple list of questions:

- Are the teachers expressing frustration in terms of constraints of time and content load, coordination between the training environments (training room, the classroom, the online forum) and each individual phase of the model?
- Are the teachers comfortable in expressing their individual needs? Is the climate of the training experience pleasant and flexible? Are conflicts between the science teachers and the language colleagues discussed and dealt with during the course?
- Are the trainers responsive to the teachers' needs by providing regular feedback?
- Are the trainers providing examples of practice?
- Are the teachers reacting to trainer feedback?
- Are the course objectives being met by the activities the teachers are asked to carry out?
- Are the teacher trainers expressing difficulties themselves in terms of constraints of time, content load and the coordination of the three course components and each individual phase?

The answers to these questions led to the creation of a course assessment checklist which was used regularly to observe the individual phases of the entire diploma course. The results of the observations have provided useful insight into the feasibility of the practicum model in terms of teacher knowledge acquisition and other significant aspects which include the model's weaker and therefore critical areas. These aspects are presented as outcomes in the following section.

Outcomes

CLIL methodology offers an ecological framework that facilitates purposeful learning (Marsh 2011), which means that when students are exposed to the information they are required to learn such as their science content, and when they are provided with opportunities to use this information, their learning is cognitively more complex and challenging. Thus, the recall of the information itself is also facilitated (Anderson 1993; Mehisto et al. 2008). It is the authors' view that, along with the more specific outcomes that are discussed below, the general outcome of the study is the meaningful learning that is encouraged by offering a solution that revisits the role of foreign language methodology in terms of what it can do to support science education delivered in second language environments.

Teacher Knowledge Gain and Appraisal

As a form of objective final assessment used to award marks at the end of a diploma course, the writing of a thesis and a thesis discussion with the CLIL experts was not an option as it is a requirement in Italian universities. Thus, the science teachers wrote a thesis on their individual action research project structured as follows: Project description, Aims, Methodological background, Materials, Implementation, and Results.

The authors remind the reader that although these projects were carried out with the contribution of the language teacher and the adjunct ESP lesson component, the thesis was discussed by the science teacher alone through the preferred mode of power point presentations of all the materials and texts created to teach the co-planned lessons. By examining both the individual projects and the oral presentations, the teachers' level of knowledge gain was assessed by the CLIL experts according to the following set of criteria:

- degree of integration between content and language as manifested by co-lesson planning and transforming plans into action;
- degree of awareness of second language teaching as manifested in the implementation of a variety of approaches and activities;
- degree of change occurring in teacher discourse as manifested by the use of the language of teaching (academic functions, specialized terminology deriving from CLIL methodology and foreign language teaching);

The teacher's project was then awarded a final mark based on a scale consisting of the following adjectives: fail, average pass, pass, good pass, pass with merit. The marks awarded to the five science teachers participating in this study are positioned between four teachers with a good pass and one teacher with a pass with merit.

A form of continuous assessment was also implemented during the course. This was the analysis of the teachers' forum contributions which were assessed by the moderators according to the following set of behavioral criteria: frequency of participation, clarity of information asked and given, evidence of critical thinking, and ability to focus on issue. Each criterion was given a score based on a 1–4 scale (1 = not satisfactory, 2 = adequate, 3 = good, 4 = excellent). The moderators then submitted a written description of each teacher's online contribution which became part of the overall mark.

Partnership-Building Capacity

The model enforced and brought to the surface the collaborative effort between science and language educators that would have otherwise remained unexplored. Teacher partnerships encourage the students' use of authentic language as experienced through content, thus resulting in a renewed need to learn the language itself

(Linares 2012), and a scientific text has an increased chance of being interiorized as the concepts themselves are under closer scrutiny while being discussed from a more linguistic point of view (Van Dijck 2008). The students' exposure to both language and content is therefore doubled by approaching the scientific material in different but complementary ways.

Accessibility and Comprehensibility of Scientific Concepts

The academic functions presented in Table 17.1 improve scientific content accessibility through language scaffolding devices rather than through translation from the first to the second language. These devices, if properly exploited, do not oversimplify the texts by stripping them of their fundamental scientific explanations, but afford students the opportunity to work with the texts by manipulating them in various ways.

Transferability to Second Language Environments

Although the language context of this study is English as a foreign language, the authors believe that the model can also be applied to second language contexts. In fact, the model's scaffolding techniques such as reinforcing study skills, working with language functions and text adaptation to enhance comprehensibility as well as preserving the conceptual load of the subject can also help second language students who are not necessarily underachievers, but are experiencing content learning difficulties due to the language barrier, which is also the foreign language student's main obstacle.

Critical areas have also been identified and therefore need to be addressed in future applications of the model especially considering that university language centers are the only institutions that have the Ministry's approval for CLIL teacher training initiatives.

Complex Relationships

The first and foremost critical area dealt with the process of encouraging interdisciplinary relationships between science teachers and language teachers. Indeed, this process was not always smooth due to role sensitivity and the exclusion of the language teachers from CLIL training and therefore from the basic information about CLIL. However, as this standard is part of ministerial CLIL policy and therefore not under the control of those who implement the courses, it is even more pressing to encourage collaboration through practicum activities.

Overload of Course Materials

One of the most demanding aspects of the practicum was the teachers' difficulty in keeping up with all the course materials, especially with those dealing with foreign language methodology, which is not the science teachers' familiar area. A reformulation and a different organization of materials is an issue to deal with in future courses.

Assessment Procedures

Conducting assessment activities in conjunction with the language teacher was a challenging activity as argued throughout the article and expressed many times by the teachers themselves, and should therefore be given more space. If we consider that CLIL methodology is being progressively introduced in Italian education from primary to tertiary levels, the ability to collaborate with language teachers in assessing students so as to encourage a continuum of language learning practices rather than separate experiences is now a crucial asset for science teachers.

Summation

European internationalization processes, especially those concerning the dissemination of scientific knowledge, have progressively been introduced in Italian educational settings in the last decade. In tertiary education, for example, students now have the option to obtain medical degrees entirely delivered in English, and primary and secondary education offers at least one curriculum subject taught in a foreign language. However, CLIL instruction is not without difficulties and the students are the ones who experience its complexities the most because they face the additional challenge of learning subject content in a foreign language. Therefore, as is the case with many other innovative educational approaches, teacher practitioners and researchers must find ways to facilitate the integration of language, content and learning skills so that motivation to learn does not diminish. Teaching partnerships in the CLIL science classroom are strongly advocated in this study as one possible strategy to help teachers and students progress towards an authentic and significant learning experience, characterized by smooth transitions from one lesson to the other, and from one instructional language to the other.

The authors suggest that notwithstanding the overall optimistic results obtained from the practicum model, in order to confirm its validity, the involvement of an increased participation from the teachers is necessary in order to collect a larger corpus of data. Further research especially involving student voices is also required.

To this purpose, it is necessary to investigate student interest levels, cognitive development issues as well as persistent difficulties concerning the achievement of CLIL fluency in both their science and language classes.

References

Anderson, J. (1993). *Rules of the mind.* Hillsdale: Erlbaum.

Banks, F., et al. (1995). Mentoring E8JO. *Master of arts in education.* Milton Keynes: Open University.

Bhatia, V. K. (1983). Simplification v. Easification. The case of legal texts. *Applied Linguistics, 4/1*, 42–54.

Coyle, D. (1999). Theory and planning for effective classrooms: Supporting students in content and language integrated learning contexts'. In J. Masih (Ed.), *Learning through a foreign language.* London: CILT.

Dalton-Puffer, C. (2007). *Discourse in Content and Language Integrated Learning (CLIL) classrooms.* Amsterdam/Philadelphia: John Benjamins.

Kachru, B. B. (1985). Standards, codification and sociolinguistic realism: The English language in the outer circle. In R. Quirk & H. G. Widdowson (Eds.), *English in the world: Teaching and learning the language and literatures* (pp. 11–30). Cambridge: Cambridge University Press.

Krashen, S. (1985). *The input hypothesis: Issues and implications.* California: Laredo Publishing Co Inc.

Linares, A. T. (2012). *The roles of language in CLIL.* Cambridge: Cambridge University Press.

Marsh, D. (Ed.). (2002). CLIL/EMILE- The European dimension: Actions, trends and foresight potential. *Public Services Contract DG EAC: European Commission.*

Mehisto, P., Marsh, D., & Frigols, M. J. (2008). *Uncovering CLIL – Content and language integrated learning in bilingual and multilingual education.* Oxford: Macmillan Publishers.

Rasulo, M. (2013). Co-analysis of practice: A mentoring strategy. *Lingua e Nuova Didattica.* Edizioni Studio Calabiana, 4/13–24.

Rasulo, M. (2014). Popularizing the CLIL classroom. In C. Williams (Ed.), *ESP Across Cultures.* University of Foggia, EDIPuglia, 11, 121–134.

Schön, D. (1987). *Educating the reflective practitioner.* San Francisco: Jossey-Bass Publishers.

Van Dijck, M. (2008). The science and journalism of the Belgian economist Gustave de Molinari. *Science in Context, 21*(3), 377–402.

Walqui, A. (2006). Scaffolding instruction for English language learners: A conceptual framework. *The International Journal of Bilingual Education and Bilingualism, 9*(2), 159–180.

Woodward, T. (2003). Loop input. *ELT Journal, 57*(3), 301–304.

Webliography

CLIL: An interview with Professor David Marsh (2011). International House Journal of Education and Development (Retrieved November 1 2015).

http://ihjournal.com/content-and-language-integrated-learning

Chapter 18
Science Workshop: Let Their Questions Lead the Way

Sara E.D. Wilmes

Introduction

Science Workshop is an integrated science and language literacy program piloted and implemented by primary school teachers in multilingual classrooms in Luxembourg. Grounded in theories supporting the integration of inquiry-based science education and language learning, Science Workshop consists of a teacher professional development program and instructional approach that engages students in inquiry arising from their questions in meaningful learning contexts. In this chapter I detail the strategies and resources used in Science Workshop, a science program which is attuned to student's voices as they question and conduct science investigations, and show how the program supported teachers in implementing integrated science and language literacy instruction at the primary level. Specifically, I discuss how Science Workshop supported the formation of heteroglossic language learning spaces within the confines of a system guided by monoglossic language policies.

Project Context

The Science Workshop Teacher Professional Development (TPD) Project arose from a multi-year research program supported by Luxembourg's national science funding body (Fonds National de la Recherche (FNR)) and the University of Luxembourg. The project details I share in this chapter arose from our team's work during the full-year program pilot (2013–2014), and the first full year of program

S.E.D. Wilmes (✉)
The University of Luxembourg,
2, avenue de l'Université, Esch-sur-Alzette L-4365, Luxembourg
e-mail: sara.wilmes@uni.lu

A.W. Oliveira, M.H. Weinburgh (eds.), *Science Teacher Preparation in Content-Based Second Language Acquisition*, ASTE Series in Science Education,
DOI 10.1007/978-3-319-43516-9_18

implementation (2014–2015). During the pilot phase, our University-based TPD coaching team co-taught several Science Workshop units in a multilingual classroom in a mid-sized city in Luxembourg. In this chapter I draw from both our own classroom experiences piloting the project, and the data resources from the 20 plus teachers who successfully implemented Science Workshop in their primary school classrooms to show how this model successfully supported teachers as they adapted and used the program.

Luxembourg's Language Landscape

The language landscape in Luxembourgish schools, as in many countries, is complex. Teachers are fluent in at least the three national languages; Luxembourgish, German, and French. The student population in Luxembourg is increasingly diverse and multilingual with 49 % being non-Luxembourgish (MENJE 2016). This means that many students speak at least one additional non-national language at home, while being schooled in the three national languages. The teachers participating in this program implemented Science Workshop at a range of grade-levels (ages 5 through 12 years-old) in public primary schools that utilize Luxembourg's national competency-based curriculum. This national curriculum supports a trilingual program with students conducting classroom business in Luxembourgish, and learning German and French literacy as isolated school subjects at the primary level. In this sense, the Luxembourgish language approach is monoglossic, which, as described by García (2009), are educational settings in which "each language is carefully compartmentalized" (p.115). Additionally, from 7 years of age students learn science *through* the German language. Since German is a second or third language for many students (Luxembourgish or another language being their first), this means that almost all of Luxembourg's student population learns science at the primary level through a second or third language. These contextual factors taken as a whole underscored the need for a program that supported teachers in a) integrating science and language instruction, and b) teaching in ways that flexibly support students who are learning both science and an additional language (in this case German) within what can be considered a 'traditional' trilingual education system.[1] In the sections that follow, we detail the approach we used to prepare teachers to address both language learning and science learning needs in synergistic ways.

[1] While the linguistic landscape of Luxembourg is unique in the trilingual demands it places on students, it is similar in many ways to those found in varying degrees in every country. Whether a student is learning multiple languages, or one language (keeping in mind that science is also a culture and language itself (Cobern and Aikenhead 1997)), Science Workshop can support teachers in the integration of science and language in ways that lead to both language and science learning, even in schools with apparently 'homogeneous' language landscapes.

Project Development

Science Workshop is a TPD model and science instructional program developed by our science education team at the University of Luxembourg. The team consisted of a university professor and two doctoral students, all three of whom had extensive experience working with primary age students and developing science education programs. The development of Science Workshop took place following what we learned about science and language instruction in Luxembourgish primary schools during the implementation of the Fibonacci Project (www.fibonacci-project.eu) in Luxembourg from 2010 to 2013. During this time our team assisted in the support of inquiry science instruction in several schools in Luxembourg, and gained first-hand experience through observations and discussions with teachers that led us to identify central facets of a science instructional program that could address key language and science instructional needs. These facets, aligned with the needs described by teachers, formed the backbone of the program and provided the theoretical foundations for the program we developed, Science Workshop.

Theoretical Foundations

Science Workshop consists of a teacher professional development program and instructional support grounded in three key theoretical features arising from practices science and language education research has shown can be effective, namely:

(i) The use of students' questions to drive the inquiry-based science learning process (Exploratorium 2006; Gallas 1995; NGSS Lead States 2013),
(ii) The integration of language literacy and inquiry-based science instruction (Cervetti et al. 2012; Lee and Fradd 1998; Stoddart et al. 2010; Varelas and Pappas 2013), and
(iii) The construction of informal heteroglossic language spaces (Flores and Schissel 2014; Garcia 2009) as places for students to flexibly learn both communicative competencies and science both in and through a second or third language.

I elaborate the theoretical underpinnings of each of the three key features in the sections that follow and describe how each plays a role in the Science Workshop professional development and instructional support program.

Students' Questions Drive the Inquiry Process

Over a decade of education research has established the merits of inquiry-based science education (IBSE) and its positive influence on student science learning (Minner et al. 2010; Rocard et al. 2007). IBSE is rooted in socio-constructivist theories of learning and engages students in actively making observations, posing questions, planning investigations, interpreting data and communicating results to diverse audiences (Minner et al. 2010; National Research Council 2012; Worth et al. 2010). Research has established that IBSE is particularly valuable in multilingual contexts, such as encountered in Luxembourg's schools, because students construct meaning as they experiment, question, and communicate, and thus IBSE provides a rich and authentic context for language development (Haneda and Wells 2010; Lee and Fradd 1998).

The inquiry instructional model employed by Science Workshop first engages students with a phenomenon, and from this initial engagement students are asked to generate questions. These questions are then used to lead students into deeper inquiry (Exploratorium 2006; Gallas 1995; vanZee et al. 2001). Eliciting students' questions in this way accomplishes two goals simultaneously. First, it engages students in a key process of scientific inquiry (NGSS Lead States 2013). Second, it opens the door for students to voice their wonderings and interests. This is valuable for students in that their voices – and their unique interests, worldviews, and perceptions – are revealed. This is at the same time valuable for teachers in that they are afforded a view into their students' curiosities and are then able to use this information to tailor science and language instruction.

Instruction Integrates Inquiry-Based Science and Language Literacy

Research has established that inquiry-based science instruction that integrates language literacy skills (reading, writing, speaking and listening) presents synergistic opportunities for students to learn both language literacy and science (Cervetti et al. 2012; Stoddart et al. 2002). IBSE supports this in that students construct meaning as they experiment, question, and dialogue, and write about their inquiry experiences. Thus they are provided with a rich and authentic context for literacy development (NRC 1996; Lee and Fradd 1998). Science Workshop utilizes this approach in that it incorporates complex and appropriate literacy learning tasks embedded within context-rich inquiry science lessons.

Heteroglossic Spaces Allow Students to Flexibly Utilize Communicative Resources as They Engage in Inquiry-Based Science

Today, many bilingual and multilingual primary school programs in Western cultures remain monoglossic in nature. This means that multilingual students are, more often than not, required to use one language at a time in learning settings (Garcia 2009). This view of multilingualism, Garcia explains, views each language as a discrete separate entity and values the fluent speaker of the language as the desired learning goal. For example, in Luxembourg students who speak Portuguese at home and who attend Luxembourgish primary schools will find that at the age of five they are expected to speak only Luxembourgish in their daily classroom learning routines. When they reach first grade, they begin to learn German literacy as a subject, and one year later French literacy. Rarely are French, German, and Portuguese used in systematic synergistic ways to help the students learn any of the other languages. They are judged throughout their schooling relative to fluent French and German speakers. This is a monoglossic school environment. The work of several education and linguistics scholars including Garcia (2009) have drawn attention to the fact that this monoglossic approach to multilingualism does not reflect the fluid, flexible ways in which students draw on a mixture of semiotic resources during their time outside of school. In contrast to monoglossic learning spaces, heteroglossic learning spaces, in which students are able to flexibly use semiotic resources fluidly without regard to bounded languages as they learn, are more beneficial for the multilingual students' educational career (Flores and Schissel 2014; see also Blackledge and Creese 2014).

The concept of heteroglossia is derived from Bakhtin's (1981) notion of multi-voicedness. At its core it means the valuing of more than one 'voice' speaking at the same time either within a single language or a text.[2] In our work we take heteroglossic to mean spaces within which students and teachers are able to incorporate more than one voice, more than one language, or more than one semiotic resource for expression, communication and learning. We designed Science Workshop to support teachers and students in engaging in learning that draws upon, and makes intentional space for multi-voicedness (Bakhtin 1981) as students synergistically learn both science and languages. For this reason, Science Workshop is built upon a pedagogical stance that incorporates instructional approaches that allows students to draw from their full linguistic repertories (Otheguy et al. 2015). Additionally, as was shown in the work of Flores and Schissel (2014), we feel that the inclusion of heteroglossic spaces in instruction is a way for educators to carve out spaces that value students diverse resources as they learn, and that push back at national education schemes aimed at producing monoglossic speakers of nationalized languages. In the sections that follow we describe the Science Workshop TPD program and its use in classrooms in Luxembourg.

Teacher Professional Development Model

Based on practices that have been shown can be effective in language and science education professional development (Lee 2004; Loucks-Horsley et al. 2009) the TPD program involved a combination of three key features; a teacher workshop

[2] For a more complete discussion of Bakhtin's concept of 'heteroglossia' see Blackledge and Creese (2014).

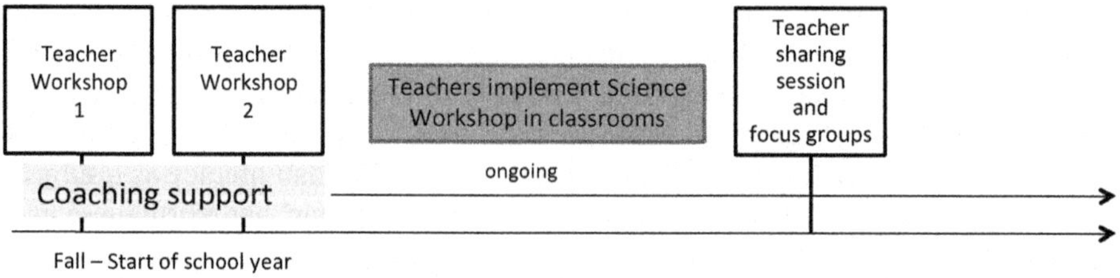

Fig. 18.1 Science workshop's year-long teacher professional development (TPD) program

series, material support, and ongoing coaching support. An overview of a single year of the TPD program is shown in Fig. 18.1. Next we describe each of these features.

Teachers participated in two half-day sessions in the beginning of the school year over the course of 1 week. The purpose of the workshops were to familiarize teachers with the theoretical underpinnings of the program, to allow them to experience inquiry rooted in questions first-hand, and to prepare them to use this learning approach in their classrooms. A brief overview of each of the two three-hour days is shown in Fig. 18.2. The two topics of inquiry we used in the workshops were worms and snails. These subjects were chosen as they are typically taught in Luxembourgish primary school classrooms, and are easy to obtain in our environment. We chose to work with topics the teachers typically teach using transmission-learning based activities in order to provide teachers with the opportunity to compare how they typically teach these topics with the Science Workshop approach.

Inquiry Science Driven by Students' Questions

The science inquiry instructional model that serves as a foundation for the Science Workshop was derived from a conceptual model developed for The Fibonacci Project (www.fibonacci-project.eu), conducted Europe-wide from 2010 to 2013. After its implementation in Luxembourgish schools, we adapted it based on instructional guides for inquiry based on students' questions developed at the San Francisco-based Exploratorium (2006). We emphasized to the teachers that inquiry is driven by the interests and questions of the students and as such focuses specifically on their situated, contextual view of the topics being explored.

Integrated Inquiry-Based Science and Language Learning

Within each stage of inquiry, literacy tasks were integrated that served as instructional opportunities. Each was strategically selected to start with a more context-embedded less-linguistically demanding task. These were then followed by less

Half-day Workshop 1

1. Introduction to TPD Team and ***Science Workshop*** Foundations
2. Using students' questions to guide scientific investigations
   - What does teaching this way look like?
   - Why teach this way?
   - How can I use this in my classroom?
3. Student Science Journals
   - How, why?
4. Inquiry about worms
5. Sharing resources
   - Worm background information, care of worms
   - Literacy strategies
   - Material support

Half-day Workshop 2

1. Inquiry about snails
2. Sharing resources
   - Snail background information, care of snails
   - Additional literacy strategies
   - Material support
   - Coaching support

Fig. 18.2 Teacher workshop program

context-embedded, more demanding tasks. For example, teachers were asked to first discuss verbally with a partner what they were wondering about worms. Next, they were asked to individually write a list of questions in German about worms in their science journal. In this way the person doing the questioning is first provided with social support (the ability to speak informally with a partner) and then is asked to complete the more linguistically demanding task (writing the questions in German in the science journal). This approach, derived in part from the theoretical foundation described by linguist Jim Cummins and illustrated in Cummins' Matrix (1984) shown in Fig. 18.3, is one way to visualize the complexity of language tasks.

Science Workshop integrates literacy tasks in such a way so that they start out less demanding (in both a language and science sense), and then shift to more demanding tasks as students build language skills. Cummins' matrix provides a way to conceptualize these aspects of integrated literacy tasks, and helps teachers contemplate ways to build and extended students' science and language skills across tasks.[3] Examples of how these tasks can be conceptualized a long a continuum in order to support growth in language ability and science understanding are shown in Fig. 18.4.

[3] For a more in depth discussion of the development of Cummins' matrix see Cummins (1984).

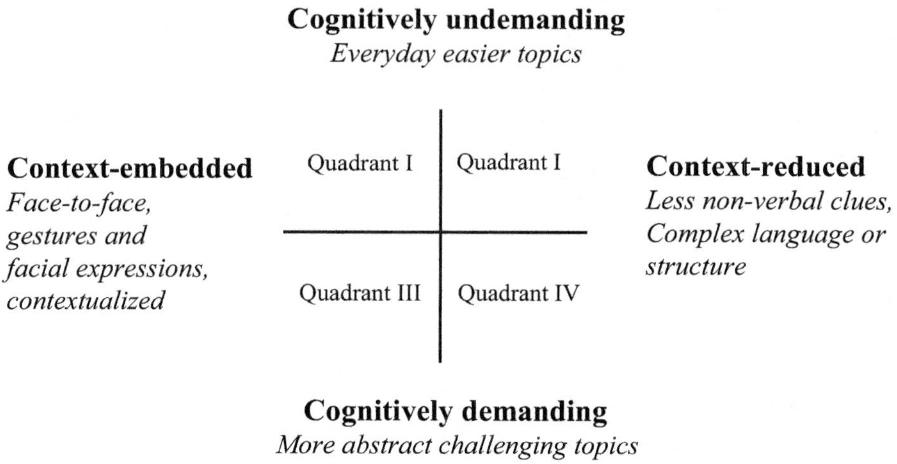

Fig. 18.3 Theoretical foundations of Cummins' (1984) task difficulty matrix

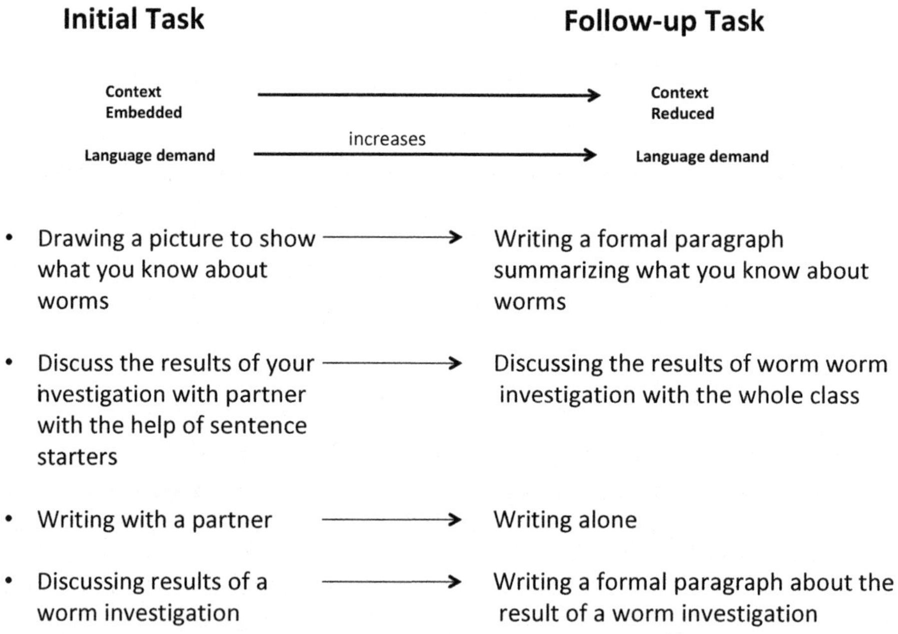

Fig. 18.4 Task sequences embedded in science workshop

Student Science Journals

A key feature of the Science Workshop program is the use of student journals. These journals, when used as an informal writing tool, place value on student voice and allow students to choose how and in which language(s) they record entries. As such, the science journal acts as a space in which students can flexibly and fluidly record questions, wonderings, data, and conclusions in ways that utilize the full complement of semiotic resources in their communicative repertoires. I share examples of how the journals were used in classrooms in the implementation case presented in the following pages. During the TPD workshop we showed teachers how to set up the notebook provided general guidelines for using them with students.

Science Materials

Following participation in the workshops, teachers were provided with kits to support two units of inquiry (one on the topic of worms and one on the top of snails) in their classroom. Each kit contained basic tools – magnifying glasses, rulers, pipettes, spoons, paper trays, cups, containers, colored paper, a class set of blank notebooks, a worm farm and a terrarium in which a snail habitat could be built. The idea behind the kit contents is that first, each student be provided with a science journal, and second, that the kit be stocked with items that are easy to obtain so that they can be used for subsequent inquiry investigations and easily restocked.

Coaching Support

For the duration of the yearlong project, members of the science education professional development team at the University of Luxembourg provided ongoing coaching support for participating teachers. Members of the team have extensive experience teaching inquiry in primary classrooms, supporting teachers in using integrated science and language literacy programs, and using Science Workshop in Luxembourgish primary classrooms. The opportunity for coaching support was introduced to teachers in the initial workshop session. Teachers were invited to contact the PD team when they had questions, or if they would like help – even if this meant just an additional set of hands – while using Science Workshop in their classrooms. Following this invitation, teachers contacted the support team when they desired and asked questions as varied as, 'How do I deal with the noise in my classroom?' to 'Do you know where I can find additional materials?' The TPD team held focus group interviews once all teachers had the opportunity to use Science Workshop in their classrooms for the purposes of checking in with teachers as to how the implementation was progressing, and to support them in further implementation of additional inquiry activities. These focus group interview meetings were key in the sharing of support and resources as teachers shared student work, ideas, and asked questions of each other and the TPD team, reinforcing their resource network for the use of the program.

Project Participants

Over the course of 2 years (two 1-year cycles), we worked with 22 teachers from various regions of Luxembourg. Teacher recruitment was conducted through several established teacher professional development structures. We advertised the project 6 months prior to the start of the first year with the national teacher professional development department, Le Service de Coordination de la Recherche et de l'Innovation pédagogiques et technologiques (SCRIPT), with a science communication website run by the FNR (www.science.lu), and through direct emails to teachers. The result was a group of teachers, 20 % with whom we had previously worked, and 80 % who were new to us and our programs. All twenty-two participating teachers taught in the Luxembourgish public primary school system. This means that they hold Luxembourgish national certification and are fluent in at least the three official languages; Luxembourgish, German, and French. As a group, they varied in they types of training programs they had completed (some had inquiry-based science and project-based learning in their teacher training programs, others had more teacher-centered training) and also in the number of hours they had taken TPD. Participating teachers received continuing education credits for their participation when they attended the full workshop series, taught the program in their classroom, provided documentation from their classroom of use of the program (photos, student work, lesson descriptions), and participated in a focus-group interview/resource-sharing meeting.

While Luxembourg is small in size, the demographics of classrooms across the country vary greatly. This is due in part to the fact that it shares borders with three neighboring countries, Belgium, France, and Germany and partly because of a continuing trend of increasing numbers of non-Luxembourgish families establishing residency in Luxembourg. Therefore each participating classroom had a different linguistic landscape.

Classroom Implementation

Teachers documented their use of Science Workshop through the writing of descriptive lesson logs and photo documentation of students' investigations and science journals. To illustrate how Science Workshop was adapted and used in the classroom, we present below an *implementation case*. An implementation case is a description of the use of Science Workshop in a classroom that weaves together details of instruction along with teacher's reflective comments stemming from surveys and group interviews. Implementation cases allow us to show both, how Science Workshop was used in the classroom, and at the same time represent the implementing teacher's thoughts and impressions about the program's use. Because of space limitations, I present one implementation case in the sections that follow. I share Tristan's[4] case because it shows in general, how he adapted the Science Workshop program to provide integrated science and literacy instruction for his students, and in particular, how heteroglossic language spaces were created through the use of the student science journal.

Tristan's Implementation Case

Classroom Context

Tristan teaches 10–11 year-olds in a primary school on the outskirts of Luxembourg City, the capital of Luxembourg. In the year he participated in Science Workshop, his class consisted of 10 students. Eight of the students spoke Luxembourgish at home. The majority of his students also spoke at least one additional language at home such as Portuguese, Polish, English, French or German.

The first time Tristan used Science Workshop he explained that he implemented the activity sequence exactly as he experienced it during the workshops, and simply adjusted the activity frames to a level appropriate for his students. The lesson sequence used by Tristan was as follows:

- Activity 1 – Quick Write. Write ten words that come to mind when you think about worms. Draw a worm.
- Activity 2 – Observe a small group of worms. Record what you see (Was ich sehe) and questions you have (Ich frage mich).
- Activity 3 – Creation of a class list of questions. Questions marked with 'E' the class believes can be explored using an experiment.
- Activity 4 – Design an investigation to explore a question about worms (Würmer Untersuchung). Record your question, prediction and materials you will need.
- Activity 5 – Document your worm investigation: Explain a. What we did; b. What happened – draw a picture of your investigation, write about your investigation, take a photo.
- Activity 6 – Write to someone you know explaining what you found out from your worm investigation.

Before beginning the inquiry about worms, each student set up a science journal complete with a table of contents, glossary, dated entries, and blank pages to record investigations. As will be explained in the sections that follow, use of the science

[4] All participating teachers and students have been assigned pseudonyms.

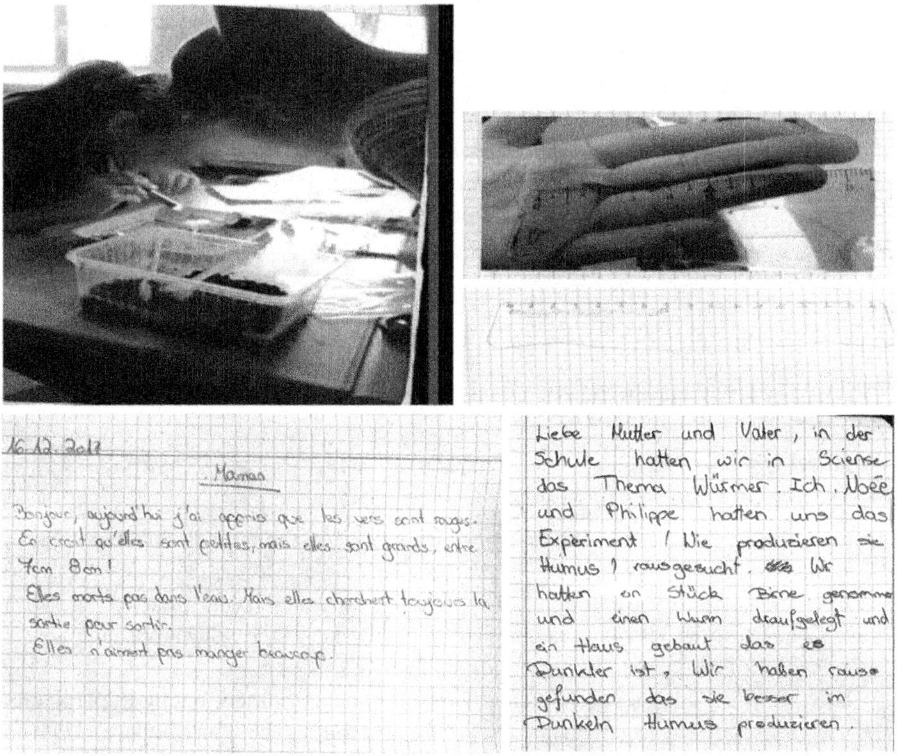

Fig. 18.5 Olivia and Jana's group work

journal was a key tool for Tristan's students with several important implications for student voice, integrated science and language learning, and the creation of heteroglossic spaces.

For their first activity, students wrote and drew what they knew and thought about worms. Then, while working in pairs they observed a small container of worms selected from a larger worm farm set up in their classroom. In their science journals students recorded, "Was ich sehe" (What I see) and "Ich frage mich" (What I wonder). It is during these first two activities that the voices of the students begin to emerge. Figure 18.5 presents sample work from two students, Olivia and Jana, who worked as a pair. We share the work of Jana and Olivia because of their interesting use of languages. As the lower half of Fig. 18.5 shows, Jana (right) consistently used German when writing in her science notebook. Olivia, on the other hand, alternated between French and German. Tristan explained to us that Olivia is a student who has lived in Luxembourg for 2 years, and who identifies as Portuguese. As far as Tristan is aware, she speaks Portuguese at home. In class, she expresses herself in French, and will not converse in Luxembourgish nor in German. But, she is able to comprehend all German and Luxembourgish that is spoken to her.

The first two activities are structured in such a way as to draw upon the contextualized, grounded-in-experience, personal observations and wonderings of each student. In this way, we begin to see the unique perspectives, resources, and understandings (Siry et al. 2016) each student brings and how these emerge in their science journal entries. It is additionally interesting to note, which languages they choose to record different aspects of their multiple entries. Jana records both Activities 1 and 2 in German, while Olivia records the activity title, "Würmer" and directions "(10 Wörter)" in German, while recording her observations in French. Most likely this is because the title and directions were provided by the teacher, while the remaining entries are her original thoughts arising from her inquiry experiences. In comparing these two sets of entries, we see how students' perspectives and language choices within the same activity differ, and thus heteroglossic language spaces were created within the activity structure Tristan provided.

Next, Tristan led a whole-class discussion (Activity 3). In collecting students' individual question in this way, the voices of each student are shared and recorded as a collective class document. This individual-collective dance that occurs as the unit progresses plays a role in supporting students in using the language of their choice, while transforming their responses into the language targeted for instruction – in this case German. Following this activity, the class identified which questions they might investigate using a scientific experiment. Tristan asked each pair of students to investigate a different question. Using an investigation template first shared in the teacher workshop, students designed an investigation to, as Tristan explained, "help them find answers to their questions". In Activity 4, students developed an investigation plan, complete with materials needed, and next (Activity 5) documented using different modes (written, drawn, photo documentation) what happened when they conducted their investigation

We return to a comparison of Olivia and Jana's science notebook entries. In Activity 4, Jana records her question, "Wie lang kann er warden?," (How long can they – the worms - be?) in German. This same question is also in Olivia's notebook, in German. The language(s) Olivia chooses to use for the rest of her entries differ from Jana's choices. As her entries for Activities 4 and 5 reveal, scientific terms provided by the teacher "vorhersagen" (predict) and "beobachten" (explain), are written by both students in German. In the concluding exercise (Activity 6), students are asked to write a letter to a friend or a parent explaining what they learned, Olivia chooses to write in French. Thus, Olivia has effectively utilized her student journal as a heteroglossic space in which she freely alternates her use of French, and German, as well as sketches and photos to document her science experience. In this way, she draws upon her diverse linguistic competencies (Otheguy et al. 2015) and several different modes (Varelas and Pappas 2013) to represent her inquiry experience.

When asked about the integrated language and science approach used in Science Workshop, Tristan shared, "It's quite nice to use the language in a different way." For (the students) it's not really doing work but more like having an experience with worms. Tristan went on to explain that "because the questions, and the development of the experiments and their results are recorded, you can promote linguistic knowledge in science teaching." Therefore, "linguistic learning can quite easily be integrated into science classes."

Regarding the use of students' questions to guide inquiry, Tristan explained that he finds this to be a very authentic way of learning science, as it calls students to think about their own questions and to find solutions to their questions. He noted that his students were very motivated to turn to additional sources beyond the classroom to answer their questions, and asked him if they could do their next science unit about nutrition in the same way, starting with their questions.

Program Outcomes

In this section, I elaborate program outcomes relative to the instruction teachers implemented in their classrooms, and relative to the teachers' perspectives on both question-driven inquiry and integrated science and language learning. In elaborating these outcomes I will highlight both the successes achieved through the use of Science Workshop, and the associated challenges teachers faced.

Classroom/Instructional Outcomes

One of the most apparent features of Science Workshop was the relative ease with which it was adapted for use in primary classrooms. I feel confident saying "ease" in that the participating teachers worked in a wide range of grade-levels (K-6) and each was able to immediately adapt and use the workshop in their classrooms. These adaptations took different forms in each of the participating classrooms. For some, this meant partnerships were formed between pairs of primary teachers who co-taught, meaning one teacher focused on the language instruction while the partner focused on the science instruction. In others, the adaptation took the form of one teacher planning both language and science integrated lessons for the first time. Teachers working in classrooms with younger students focused on ways to introduce students to basic literacy through their questions and investigations (Fig. 18.6), while teachers working with older students adapted the resources provided to meet their students' literacy levels. This speaks to the ability of Science Workshop to provide a framework for integrated science and language instruction adaptable to varying levels of instruction.

Teacher Outcomes

Concluding their participation in the program, teachers completed open-ended surveys and participated in focus-group interviews. These revealed teacher perspectives after their use of the program.

	6 year-olds classroom	10 year-olds classroom
Formulating questions	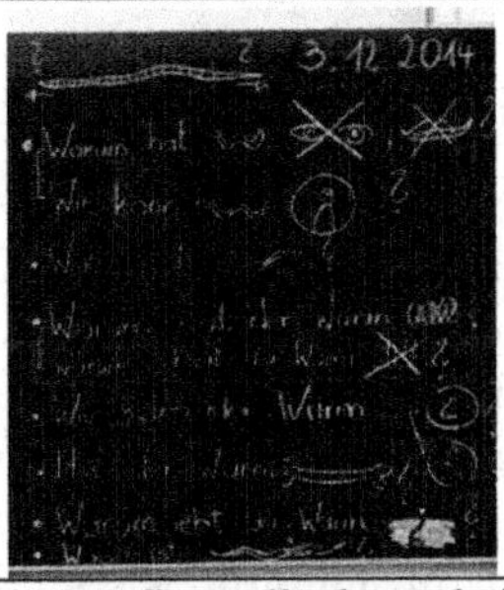	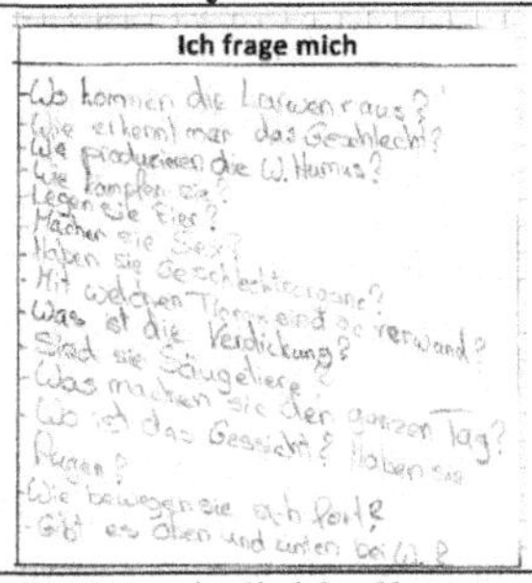
Adaptations	Students first talk about their questions in small groups and then share questions in a whole-class discussion. The teacher records questions on the blackboard integrating images for nouns so that the literacy focus is on question verbs (warum, wie, was/ why, how, what)	Students individually record questions in their science notebook while working in pairs. Next they share questions in a whole class discussion and discuss which can be investigated.

Fig. 18.6 Adaptations of a science workshop activity across grade levels. A common activity, 'Formulating questions', was adapted for use at two different grade-levels

Relative to teachers' perspectives of integrating science and language/literacy learning, teachers shared that the integration of language and science instruction occurred smoothly as it arose from the design of the embedded tasks that were rooted in students' experiences while investigating. For example, when asked, "Did you integrate science and literacy learning?" one teacher responded…"Of course, it was a natural fit…". This shows that the use of this program demonstrates how language learning and science learning go hand-in-hand.

Relative to the use of students' questions as the driver of inquiry, teachers revealed that this method of instruction, increased student engagement in science lessons and motivation. This was apparent in students' enthusiasm in conducting the experiments, in students' increased rate of asking questions when conducting investigations, and their investment in finding answers to their questions. Teachers also shared that this approach gives student's voice a place in the science curriculum in meaningful ways.

Relative to the formation of heteroglossic instructional spaces, analysis of students' science notebook entries (Fig. 18.5), coupled with teachers' comments during focus group interviews, showed that students self-selected to use languages other than the language of instruction at different points during different activities. Teachers also reported, and it was observed during classroom observations, that different groupings of students utilized different languages – other than German – at different points in the inquiry process. This indicates that heteroglossic spaces, while small and intermittent, existed and were used by the students.

Conclusions and Implications

Based on the pilot and field-testing, the following key conclusions can be drawn about the use of Science Workshop:

Science Workshop Supports Locally Contextualized Adaptable/Flexible Integrated Science and Language Literacy Instruction As we saw in the representative implementation case, through the use of students' questions, teachers in Luxembourg across a range of grade-levels were able to design and implement integrated science and language instruction that was locally contextualized and connected to students' personal contexts in meaningful ways. This is particularly notable in an era of science education that tends to promote the adoption of science programs that require significant material investments and the purchase of programs produced and sold by major publishing houses. Science Workshop, in contrast, provided for instruction across various grade-levels in flexible, locally contextualized ways. It accomplished this as it tapped into and utilized the cultural and linguistic resources of the students, and used these "funds of knowledge" (Moll et al. 1992) to provide interesting, engaging instruction.

Science Workshop Helps Create Heteroglossic Spaces Within Traditional Monoglossic Learning Environments Monoglossic multilingual programs that allow one 'voice' or that value the use of one national language at a time can harm a students' language competency development (Otheguy et al. 2015). Through our work in Luxembourg we have seen how open integrated programs, Science Workshop being one of many, provided ways for teachers to help students make connections between the languages skills they already posses and those they are learning in primary classrooms. Even if a school, school system, or national curricula dictate that a programs approach be monoglossic in nature, as it is in Luxembourg, we found traces of heteroglossic approaches to language use within science instruction. This type of instruction, promoted by scholars such as Flores and Schissel (2014), pushes back at monoglossic multilingual programs in ways that position students to use their multilingual competencies as resources for deeper and meaningful learning.

This work has important implications for science and primary school teacher preparation in that it is one example of a program that supports the implementation of an integrated science and language instructional approach that honors the language and cultural resources that students bring with them to the classroom. It additionally shows how teachers working within more traditionally oriented programs can create spaces that are heteroglossic in nature, and thus honor students' communicative strengths in ways that support both content and language learning, even if they are operating within monoglossic educational systems. And perhaps most importantly, it provides a way to listen to our students' questions and to let their voices lead the way.

Acknowledgements I would like to express my gratitude to Chris Siry for her support and guidance during the development, and implementation of Science Workshop, and also for her feedback during the development of this manuscript. I would also like to thank Jana Haus for her support and input during the piloting and the workshop series.

References

Bakhtin, M. M. (1981). *The dialogic imagination: Four essays* (C. Emerson & M. Holquist, Trans.). Austin: University of Texas Press.

Blackledge, A., & Creese, A. (2014). *Heteroglossia as practice and pedagogy*. Dordrecht: Springer.

Cervetti, G. N., Barber, J., Dorph, R., Pearson, P. D., & Goldschmidt, P. G. (2012). The impact of an integrated approach to science and literacy in elementary school classrooms. *Journal of Research in Science Teaching, 49*, 631–658.

Cobern, W. W., & Aikenhead, G. (1997). Cultural aspects of learning science. *Scientific literacy and cultural studies project*. Paper 13. Online at http://scholarworks.wmich.edu/science_slcsp/13

Cummins, J. (1984). *Bilingualism and special education: Issues in assessment and pedoagogy*. San Francisco: College-Hill Press.

Exploratorium. (2006). Workshop III: Raising questions. *The fundamentals of inquiry facilitator's guide*. The Institute for Inquiry. San Francisco. Available online at: http://www.exploratorium.edu/sites/default/files/pdfs/ifi/Raising_Questions.pdf

Flores, N., & Schissel, J. L. (2014). Dynamic bilingualism as the norm: Envisioning a heteroglossic approach to standards-based reform. *TESOL Quarterly, 48*, 454–479.

Gallas, K. (1995). *Talking their way into science: Hearing children's questions and theories, responding with curricula*. New York: Teachers College Press.

García, O. (2009). *Bilingual education in the 21st century: A global perspective*. West Sussex: Wiley-Blackwell.

Haneda, M., & Wells, G. (2010). Learning science through dialogic inquiry: Is it beneficial for English-as-additional-language students? *International Journal of Educational Research, 49*, 10–21.

Lee, O. (2004). Teacher change in beliefs and practices in science and literacy instruction with English language learners. *Journal of Research in Science Teaching, 41*, 65–93.

Lee, O., & Fradd, S. H. (1998). Science for all, including students from non-English-language backgrounds. *Educational Researcher, 27*(4), 12–21.

Loucks-Horsley, S., Stiles, K. E., Mundry, S., Love, N., & Hewson, P. W. (2009). *Designing professional development for teachers of science and mathematics*. Thousand Oaks: Corwin Press.

Ministère de l'Education national, de l'Enfance et de la Jeunesse (MENJE). (2016). *Les chiffres clés de l'Éducation nationale: statistiques et indicateurs - Année scolaire 2014–2015*. Available online: http://www.men.public.lu/fr/actualites/publications/themes-transversaux/statistiques-analyses/ens-lux-chiffres/index.html

Minner, D. D., Levy, A. J., & Century, J. (2010). Inquiry-based science instruction – What is it and does it matter? Results from a research synthesis years 1984 to 2002. *Journal of Research in Science Teaching, 47*, 474–496.

Moll, L. C., Amanti, C., Neff, D., & Gonzalez, N. (1992). Funds of knowledge for teaching: Using a qualitative approach to connect homes and classrooms. *Theory Into Practice, 31*, 132–141.

National Research Council (NRC). (1996). *National science education standards*. Washington, DC: National Academy Press.

National Research Council. (2012). *A framework for K-12 science education: Practices, crosscutting concepts, and core ideas*. Washington, DC: National Academies Press.

NGSS Lead States. (2013). *Next generation science standards: For States, by States*. Washington, DC: The National Academies Press. Retrieved from http://www.nextgenscience.org/

Otheguy, R., García, O., & Reid, W. (2015). Clarifying translanguaging and deconstructing named languages: A perspective from linguistics. *Applied Linguistics Review, 6*(3), 281–307.

Rocard, M., Csermely, P., Jorde, D., Lenzen, D., Walberg-Henriksson, H., & Hemmo, V. (2007). *Rocard report: Science education now: A new pedagogy for the future of Europe*. Technical report, European Commission. Available online: http://ec.europa.eu/research/science-society/document_library/pdf_06/report-rocard-on-science-education_en.pdf

Siry, C., Wilmes, S. E., & Haus, J. M. (2016). Examining children's agency within participatory structures in primary science investigations. *Learning, Culture and Social Interaction*. doi:10.1016/j.lcsi.2016.01.001

Stoddart, T., Pinal, A., Latzke, M., & Canaday, D. (2002). Integrating inquiry science and language development for English language learners. *Journal of Research in Science Teaching, 38*, 664–687.

Stoddart, T., Solis, J., Tolbert, S., & Bravo, M. (2010). A framework for the effective science teaching of English language learners in elementary schools. *Teaching science with Hispanic ELLs in K-16 classrooms*, 151–181.

vanZee, E. H., Iwasyk, M., Kurose, A., Simpson, D., & Wild, J. (2001). Students and teachers questioning during conversations about science. *Journal of Research in Science Teaching, 38*, 159–190.

Varelas, M., & Pappas, C. C. (Eds.). (2013). *Children's ways with science and literacy: Integrated multimodal enactments in urban elementary classrooms*. New York: Routledge.

Worth, K., Saltiel, E., & Duque, M. (2010). *The fibonacci project starting package: Implementing inquiry-based science education.* Retrieved online March 5, 2011, at http://www.fibonacci-project.eu/

Chapter 19
Deaf Students Using Sign Language in Mainstream Science Classrooms

Audrey Cameron, Rachel O'Neill, and Gary Quinn

Introduction

In this chapter, we examine how science teachers can be prepared to use British Sign Language (BSL) in more effective ways than previously with the development of new technical signs for scientific terms. We review the linguistic and pedagogic issues relating to teaching science to deaf signing students and discuss the complexities of working in a mainstream science class with a sign language interpreter.

In the United Kingdom, approximately 4900 children of school age are deaf and use BSL or Sign Supported English (Consortium of Research in Deaf Education 2014/2015). These children are mostly taught in mainstream settings where their teachers facilitate communication in class, although very few teachers are qualified sign language interpreters. The group of sign language users is currently poorly served by professionals. Teachers of deaf children often provide backup lessons in resourced schools to reinforce ideas learnt in the mainstream class. Communication support workers often interpret in the mainstream classroom, although very few are qualified and registered as BSL/English interpreters. Therefore, teachers of deaf children, communication support workers, and the science teachers who have deaf children in their class all need awareness of successful ways to teach science to deaf children and support with sign language skills to teach and communicate effectively. In this chapter, we focus on the needs of those deaf children who use sign language in particular, because successful support for this group depends on a good

A. Cameron (✉) • R. O'Neill
University of Edinburgh, Edinburgh, Scotland
e-mail: Audrey.M.Cameron@ed.ac.uk; rachel.oneill@ed.ac.uk

G. Quinn
Department of Languages and Intercultural Studies, Heriot-Watt University, Edinburgh, Scotland
e-mail: G.A.Quinn@hw.ac.uk

A.W. Oliveira, M.H. Weinburgh (eds.), *Science Teacher Preparation in Content-Based Second Language Acquisition*, ASTE Series in Science Education,
DOI 10.1007/978-3-319-43516-9_19

understanding of linguistic issues. We explore ways in which the various professional groups can receive effective preparation and support to fulfil these roles.

The Scottish Sensory Centre is funded by the Scottish government and provides a programme of continuing professional development for teachers of deaf and visually impaired children, classroom assistants, and mainstream teachers. Since 2007, the Centre has hosted the STEM in BSL Glossary, which has now provided over 1400 technical terms and definitions in BSL online (Scottish Sensory Centre 2015). The subjects represented in the glossary are astronomy, biology, chemistry, geography, mathematics, and physics. The BSL technical terms have been both collected and devised by subject groups of Deaf[1] professionals working as teachers, sign linguists, scientists and in STEM-related occupations. Unlike earlier approaches to standardisation led by hearing teachers who did not use sign fluently, the BSL Glossary project has been Deaf-led and the group has been careful not to impose the suggested signs on the wider Deaf community. Through Facebook groups and public engagement events, the group has brought the new signs to the wider Deaf community and engaged many in science education. For example, the Stargazing events run by the BBC and the Royal Observatory Edinburgh across the UK in 2014 attracted 15,000 people to its events at 14 venues, including many Deaf people. Dr. Audrey Cameron who co-ordinates the Glossary project at the Scottish Sensory Centre and Gary Quinn, the Glossary's sign linguistics consultant, introduced astronomy signs developed by a group of Deaf astronomers and physicists. Many hearing people also found the conceptual and iconic nature of the signs for planets helpful for all students to remember important physical features of each planet.

Teacher Preparation Model

Science teachers in the UK usually complete a science degree and a 1-year initial teacher education course, but they do not spend much time considering inclusive education or the needs of particular groups such as deaf children (see Fig. 19.1). Teachers of deaf children must first qualify to teach all children and then go on to specialise with a Master's level course focusing on deaf education. However, science education is rarely the focus of this course. In addition, these specialist teachers may not have fluent BSL skills since signing students may make up only a minority of their caseload and the UK governments all specify that only level 1 in BSL needs to be held.[2]

In addition, communication support workers who interpret for deaf children in mainstream classes often do not have a degree before starting to work in educational

[1] Following the convention suggested by Woodward (<CitationRef CitationID="CR33" >1972</ Citation Ref>) and now commonly accepted, the use of the uppercase "Deaf" denotes deaf people who use sign language and regard themselves as members of a linguistic and cultural minority, while the lowercase "deaf" refers to those with hearing loss.

[2] Level 1 in BSL can be gained after 50 h study of the language.

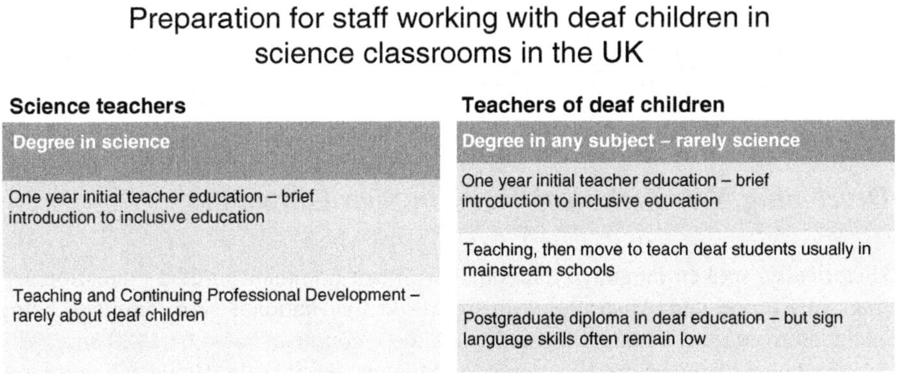

Fig. 19.1 Comparing the usual education of science teachers and teachers of deaf children

settings with deaf children. The Scottish Sensory Centre is an example of an organisation in the UK that can offer continuing professional development for teachers and support staff; but people in these roles sometimes do not have enough scientific background knowledge to adequately support or communicate effectively in science. They need to build on both their own scientific understanding, their use of specialised BSL vocabulary, and their use of fluent BSL explanation. This is a very demanding set of tasks: the teachers and support staff need more support.

In resourced schools, there are larger concentrations of deaf children so that science teachers and specialist staff build good working relationships, which leads to higher quality support for deaf children. The specialist teacher of deaf children in the school usually provides deaf awareness sessions aimed at particular teaching situations. A close relationship develops as the specialist teacher and the science teacher work with different deaf children through the science curriculum; some of these children are likely to be BSL users. In this chapter, we explore ways we have found to support teachers of deaf children and support staff in relation to science teaching using sign language, and science teachers in relation to working with deaf pupils who use BSL as their first or preferred language. First, we explain the literature base that informs our way of working. We then describe how we have provided staff development sessions for teachers of deaf children and interpreters in relation to science and the linguistic needs of deaf children who use BSL. Most of these children attend their local school, so they do not learn alongside other BSL users.

Review of the Literature

Four areas of research literature have informed the way the BSL Glossary team has developed this model of teacher education: research into developing new technical terms in sign languages, how specialist teachers of deaf children work in science contexts, research about how sign language interpreters and other staff work bilingually in science classrooms, and finally the way science teachers can learn to work more effectively with deaf signing pupils.

Developing New Technical Terms in Sign Languages

Historically, spoken language methods have been dominant in deaf education settings in Europe and Australia, with the respective national sign language being excluded from most educational settings in these countries between 1880 and 1975. Sign languages have been only known to be languages since the 1960s, with research first starting into American Sign Language (Adam 2015). With the introduction of Total Communication approach in the 1970s, some hearing teachers of deaf children attempted to extend and regularise signed vocabulary for curriculum subjects, although this was not welcomed by Deaf sign language users (Adam 2015). In the US, the National Technical Institute for the Deaf established one of the first examples of technical vocabulary building in American Sign Language on linguistic principles, although it also aimed for deaf students to be able to pronounce the words (Caccamise et al. 1981). More recent examples of technical vocabulary building in sign languages (Johnston and Napier 2010; Lang et al. 2007) have included discussion of the results amongst wider groups of informants, sign language users, interpreters and sign linguists, to build consensus on the best sign terms to use for a concept. Our own work has been Deaf-led, drawing on the scientific expertise of 24 Deaf participants who work as teachers or in science related occupations (Cameron et al. 2012). Terms use the productive lexicon in BSL (Brennan 1990; Sutton-Spence and Kaneko 2016) using metaphor, iconicity, depicting signs and shared morphological characteristics for groups of signs that are semantically related. Definitions of the terms are given in BSL then translated back to written English so that Deaf students can access the glossary independently. Definitions often include etymological notes in BSL explaining why the sign is made a certain way. We introduce teachers and sign language interpreters to the process of collecting, coining and defining new technical terms in BSL on the course we describe below.

Teachers of Deaf Children in Science Classrooms

In richer countries, the predominant model used in education for deaf children is to encourage the use of speaking and listening, using hearing aids – radio FM systems that reduce the effect of classroom noise and cochlear implants. As deafness is a low incidence disability, this usually means the deaf child does not have deaf peers at school. American Sign Language (ASL) has found wider acceptance in the US education system where the place of ASL has been officially acknowledged since 1988 (Rosen 2006) and residential and day schools are widely available for severely and profoundly deaf children. IDEA, the US education equality legislation, was amended in 1999 to include registered sign language interpreters being expected in educational contexts (Rosen 2006). In the UK and Australia, mainstreaming is the usual approach, and sign languages (BSL, Auslan) have less status. However, in poorer and developing countries, sign language is widely used in day and residential schools; countries such as Brazil, for example, have a rich tradition of bilingual bicultural education. Our project started in Scotland in 2006; this country now has the British Sign Language (Scotland) Act 2015, which may improve the status of the language in educational settings. We focus in this chapter on the education of deaf children who sign, because facilities for this group are often underdeveloped.

There are difficulties that arise from having specialist teachers of deaf children who teach subjects such as science. Graham (2012) outlines the way specialist teachers' views change as they go through their initial teacher education programme. Their conception of science was linear, which led to teaching approaches that prioritised visual information and learning of vocabulary rather than inquiry or discussion of unknowns. Additionally, science teaching was too teacher-directed. Yore (2000) used successful examples of science teaching to hearing students (English learners and low literacy) to encourage teachers working with deaf students to prioritise reading and writing in real communicative contexts in science classrooms. He suggested ways in which less confident readers could develop metacognitive awareness about what they read in science, ways to activate prior knowledge and conscious choice of reading strategy, and ways to improve the students' monitoring of their success as they read. He proposed two strategies for deaf science learners: the joint construction of mind-maps and encouraging students to read science texts then write reflection pieces to improve their summarising skills. Several research projects have illustrated ways in which teachers of deaf students can foster inquiry to reduce the amount of control and structure in the deaf school science classroom, and introduce more autonomy so the children behave more like scientists. Kahn et al. (2013) argue that teachers should allow deaf students to: (1) develop their own science questions, (2) 'fail' in the sense of not stepping in too quickly to correct them, and (3) link classroom science to real-world scenarios and careers. An alternative approach from Lang and Albertini (2001) emphasized more creative reading and writing approaches in science courses such as asking deaf students to write imaginative pieces about scientific phenomena, guided free writing about predictions and consequences, and reflections on what they had learned.

Two ethnographic studies from Sweden and the US explore ways in which teachers connect the signed and written languages in science classrooms. In the Swedish deaf school, collaborative meaning-making was observed as teachers used metadialogue to engage pupils in science, weaving in technical terms (Lindahl 2015). Depicting signs were a mediating resource that allowed students and teachers to express and understand science concepts. Rapid and confident translanguaging shifts occurred between Swedish Sign Language, Signed Swedish and fingerspelling. More importantly, this classroom offered deaf students the chance to use all their bilingual resources. In the US study, teachers also had many opportunities to move between English and ASL in science discussions (Lane-Outlaw 2009), but the teacher input was in ASL and the student output mainly in written English. There were few opportunities to read science texts and older students had less hands-on inquiry than younger pupils. The teachers often translated science texts to sign language for the students.

With a focus on developing the deaf children's conceptions of science through sign language discussion and inquiry, Pinto Silva et al. (2013) took the learning away from a Brazilian deaf school into a university science lab and encouraged the pupils to set their own questions, plan their own experiments and report back at the end of each day in an intensive week-long summer school. Teachers from the deaf school worked in a group alongside the children, participating as equals. In a parallel project with primary pupils, Deaf trainee teachers used a discovery approach and technical signs developed by the university project. They did not tell the children the answers, but asked more questions, an approach which differed from that of their usual class teachers (Da Fonseca Flores and Rumjanek 2015).

Deaf students often show significant delays in literacy skill, which has an impact on their science learning (Vosganoff et al. 2011). Taking a Universal Design approach, and suggesting that perhaps using print literacy may not be necessary, Wang (2011) puts forward the concept of 'performance literacy' which can include using multimedia science resources, subtitles, signed ASL texts and students recording their own signed explanations. Favouring inquiry methods, she suggests that teachers of deaf children must be carefully prepared to introduce these ways of working. Among the practices considered to be successful in science classrooms with deaf children, Easterbrooks and Stephenson (2006) include more hands-on inquiry, teachers who are qualified to teach science, more discussion of real world problems in science, using consistent signs agreed in advance, and encouraging deaf pupils to rewrite texts themselves rather than being given simplified texts.

The language demands of science texts are examined by Fang (2007), who conclude that exposure to science writing is not enough. Science teachers also need to give explicit attention to the terms and structures used in scientific texts. Similarly, the study of a group of Deaf Norwegian teachers reflecting on their own progress in learning physics at deaf school (Roald 2002) concluded that, although their teacher was not completely fluent in Norwegian Sign Language, he was systematic and set high expectations. The students used illustrated diagrams in their lab reports and discussed the textbook closely using sign language. As teenagers, they were the first deaf students to go on to university.

Interpreters in Science Classrooms

In the UK, deaf children who use BSL are often in mainstream schools where a support worker interprets between spoken English and BSL or a more English-like form, Sign Supported English. However very few of these interpreters are qualified, leading to gaps and misconceptions for deaf learners. Brenda Schick has addressed this question systematically in the US, showing that the mean level of educational interpreter is below the minimum recommended (Schick et al. 2006). Mainstream teachers often see the interpreter as taking responsibility for the deaf learner, not treating the interpreter as a partner (Schick 2008). The interpreter, in her view, should be strongly involved in the preparation with the class teacher, providing ongoing support about ways in which deaf children learn.

Schick and colleagues go further suggesting that these poorly prepared educational interpreters are actually creating barriers to learning for deaf children in mainstream science classrooms (Kurz et al. 2015). Their research showed that deaf pupils learning science directly from a Deaf science teacher understood more than via an experienced science teacher working with an interpreter, although this was a small study with only 19 students. Other researchers have compared how much deaf college students understand through a sign language interpreter and a note-taker working to screen (Borgna et al. 2011); results were very similar. These researchers demonstrate that deaf students are much weaker at monitoring their own comprehension in science lectures than hearing students. They propose that there are cognitive differences between deaf and hearing students, probably caused by different early language learning environments. Their recommendations are for teachers to focus on teaching deaf learners to integrate different parts of a question, and to use scaffolding prompts such as summaries of content more effectively as ways of building metacognitive awareness.

Science Teachers Learning About Signing Deaf Pupils

Apart from Schick's (2008) work mentioned above, there are very few studies looking at the attitude of mainstream science teachers to working with signing pupils and an interpreter. To overcome this limitation, we have drawn on studies about other marginalised groups and used them to inform our thinking about science teachers teaching deaf learners.

'Wait time' when asking questions is a crucial issue for science teachers to consider when there is a deaf pupil in the class and a sign language interpreter. Grimes and Cameron (2005) found that the mean wait time following a question from the teacher was 1.4 s, which meant the deaf pupil following a sign language interpreter never got an opportunity to answer; the interpreter lag was always more than the wait time. The researchers also noted that the deaf pupils sometimes became detached from the main class, communicating with a specialist teacher or interpreter and not with other classmates. Naming the pupil who should answer the question will help the deaf pupil to participate. The type of question is important in all science classrooms. According to Lustick (2010), who was addressing science teachers in general, 'focus questions' should address a specific scientific phenomenon and be tailored to the learners' life experiences and environment.

The five essential (5E) features of science inquiry learning put forward by Wilson et al. (2010) are strongly language-based and would be useful for deaf learners too: engage, explore, use evidence, give explanations and evaluate. The researchers used a randomised control at a summer school and allocated hearing students to an inquiry approach or a commonplace approach. They found that non-white students made good gains on the inquiry approach, though the socioeconomic status of the student made no difference. Using high quality pre-prepared materials which focus on discussion activities, they argue, is a motivating way to introduce teachers to the inquiry approach and improve the achievement of all students. An example of the materials is at https://www.nutrientsforlife.org/for-teachers.

Critical Race Theory can also be a useful perspective for considering the achievement of pupils who come from other minority communities, such as Deaf learners who use sign language. Working in New Zealand, Tolbert (2015) used observations of science lessons in schools with a high proportion of Maori pupils, where the schools employed culturally responsive teachers to support the science teachers. The role of the Maori mentor was that of a teacher with insider knowledge of Maori culture who used it to focus on achievement and inclusion in the class through discussions with the science teachers. This approach could be used with sign language users, particularly in schools with larger numbers of deaf children.

Materials Developed

Scottish Sensory Centre's BSL Glossary

The University of Edinburgh, acting through the Scottish Sensory Centre has undertaken projects with a team of 24 scientists, mathematicians, sign linguists and teachers to create, define, catalogue and develop BSL signs for Science and Maths for deaf people and their teachers. The glossaries contain BSL signs for mathematical and scientific terms. Each entry provides a BSL sign in video format for a mathematical or scientific term with a movie that explains the term in BSL, a definition, and an example of how the term can be used in BSL, within the context of a science discussion or experiment. Written English translations also accompany the BSL movies and often images too. The Scottish Sensory Centre's online BSL Science Glossary is an evolving resource. To date we have 1410 signs for different scientific terms along with BSL video clips of definitions and experiments (http://www.ssc.education.ed.ac.uk/BSL/list.html). Currently, the BSL Glossary website receives approximately 8000 hits per month with typically increased usage during the week and less during the weekends and school holiday periods.

Positive feedback has been received from deaf students and professionals using the glossary. They say it is a useful resource not only for professionals, such as teachers of deaf children and sign language interpreters, but for deaf students too:

> The Science glossary website helps me a lot. The website is good because it encourages us [students] to study independently. It is not difficult to use. It is good to read the text to help us to understand questions in English and how to write in English for the exams. It is good to see a resource that is in BSL – my own language. (Deaf pupil, 16 years old)

> When I started teaching deaf students I was quick to notice the lack of science signs, and the absence of standardisation of the limited number. Individual teachers were compelled to make up their own signs with the students or relying heavily on fingerspelling. There is a vast amount of vocabulary within any science syllabus and my level of BSL was not adequate to tackle the creation of new signs, therefore, I was exceptionally pleased to be introduced to the Science Signs Glossary. The science signs make up a valuable part of my lessons, both for my own preparation beforehand and to use as a reference tool during. Using science signs is vital for fluency in explanations and discussions. The glossary also allows students to access information at home which is a necessity as the language in many science textbooks is not appropriately graded. (Chemistry teacher working with deaf children)
>
> I am engaged by Education Queensland (Australia) to develop subject specific signs (Auslan). I look to this site to glean some ideas, as well as comparing our signs with BSL. I found this site very useful. Keep up the good work. (Teacher of deaf children from Australia)

Identification of appropriate words and terms for development focus on school subject syllabuses from across the UK, textbooks and in-depth discussions with subject specialists. The following specific topics and sub-topics are now in the glossary (Table 19.1). The focus of our work often depends on the funders, as the project receives small grants from many bodies. For example, the geography funders had a strong interest in physical geography. While our initial work has focused onto terminology associated with the school curriculum, the terms developed can be applied to more advanced levels, including undergraduate degree levels and professional settings. Our goal is to continue to develop more advanced terminology.

Developmental work is conducted by a working group consisting predominantly of contributors from within the Deaf community with expertise in science, maths – at degree or PhD level – teaching, deaf education and BSL linguistics. The participants discuss each scientific concept, including definitions, applications and contexts to establish how best to represent it in BSL. Often one term in English corresponds to two in BSL.

Table 19.1 Topics and sub-topics in the British Sign Language science glossary

Topic	Sub-topics
Astronomy	Deep sky objects; solar system; stars; star constellations; stargazing
Biology	Adaptation; cells; ecosystems; genetics; humans and the environment; nervous and hormonal systems; photosynthesis and respiration; protein and enzymes; reproduction and inheritance; specialised and non-specialised cells; transport and exchange
Chemistry	Acids and alkalis; reactions of acids; atoms, elements and the periodic table; how atoms combine; carbohydrates; chemical reactions; corrosion; fertiliser; fuels; hydrocarbons; making electricity; metals and non-metals; plastics; properties of substances; separation; solution and solubility; states of matter
Geography	Geology; geographical information systems (GIS); glaciology; maps; rivers; weather
Maths	Algebra; arithmetic; geometry and trigonometry; probability; statistics
Physics	Electricity; electromagnetic spectrum; electronics; energy; force; health physics; heat; light and sight; measurement; movement; nuclear physics; radiation; sound and waves; telecommunications; telescopes; space physics

Fig. 19.2 Signs for MASS, DENSITY and WEIGHT with the closed fist representing MASS. The DENSITY sign has the mass sign as its base with the VOLUME sign. (http://www.ssc.education.ed.ac.uk/BSL/physics/density.html) The WEIGHT sign has the MASS sign with the GRAVITY sign pulling down.(http://www.ssc.education.ed.ac.uk/BSL/physics/weight.html)

The new signs have been developed to be visually representative and to conform to the principles of BSL linguistics. When creating new signs, we often found that the scientific signs are linked together morphologically, that is they are united by a common handshape or movement. In this process, the key concept is identified and a central handshape or movement is linked to it. This common handshape or movement is then evident in many of the related terms. For example, the sign for MASS is used to help create the sign for DENSITY and WEIGHT (see Fig. 19.2). The development team expect that the morphological link between signs for conceptually related terms will assist teachers of deaf students and sign language interpreters to prepare clearer explanations and interpretations for the science class.

Another principle of developing new signs was not to create signs based on fingerspelling – i.e. using the initial letter from the English spelling of the term. However, there are some exceptions; firstly for names of units, so the finger-spelt letter N, for example, is Newton. Secondly, in chemistry we use finger-spelt letters for chemical elements. Also when distinguishing different enzymes in biology, initialisation is used to distinguish *amylase*,[3] *catalase*,[4] *invertase*[5] and *pepsin.*[6] These terms are all hyponyms of the word *enzyme*, in that their meanings are all included in the word *enzyme.*[7] Visually, therefore, they are closely related despite the use of initialisation. Other sign languages, such as ASL, make greater use of initialisation than BSL.

Individual terms are only approved once the group has reached agreement, and often these terms are refined or redrafted following further discussions during and after the working sessions. Once agreed, approved terms are recorded along with

[3] http://www.ssc.education.ed.ac.uk/bsl/biology/amylase.html

[4] http://www.ssc.education.ed.ac.uk/bsl/biology/catalase.html

[5] http://www.ssc.education.ed.ac.uk/bsl/biology/invertase.html

[6] http://www.ssc.education.ed.ac.uk/bsl/biology/pepsin.html

[7] http://www.ssc.education.ed.ac.uk/bsl/biology/enzyme.html

Fig. 19.3 Screenshots of video clip definitions of ATOM, PRISM, and ARÊTE (Geography)

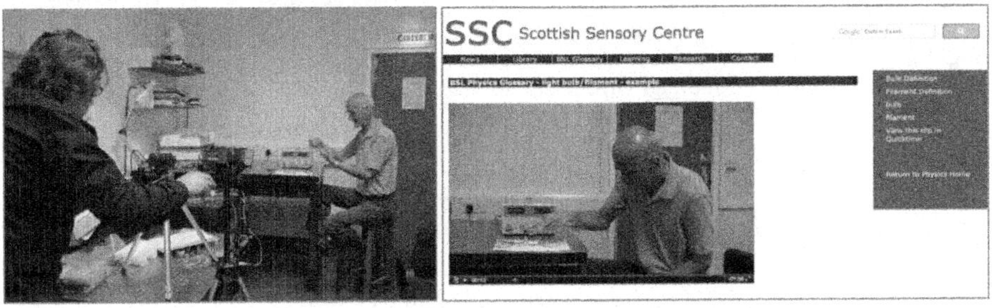

Fig. 19.4 Filming a demonstration in a physics laboratory

clear definitions, explanations and useful material e.g. photographs, film clips and diagrams (Fig. 19.3).

There are also video clips of demonstrations using the signs (Fig. 19.4).

A Deaf pupil and a communication support coordinator confirmed the importance of including video clips of laboratory demonstrations or examples with the developed signs in the glossary when asked for their feedback. At the start of the project, demonstrations and examples were only found in the Chemistry and Maths glossary but after listening to the deaf young people and professionals, we now try to include as many demonstrations and examples as possible on the glossary website:

> You should have more examples and lab movies because I think they will help deaf pupils to understand the definitions better. (Deaf pupil, 16 years old)
>
> The glossary is so much more than lists of new BSL signs; the meanings of complex concepts are made clear and exemplified, making it a vital tool for Deaf students and for a range of educational professionals. (Communication support coordinator at college)

A lecturer of science education also supported the inclusion of definitions and examples in the glossary:

> The glossary of terms provides learners with empowerment to learn about, think about and talk about physics and science. The presenters engage in a dialogue with learners, blending signing, symbolising and explaining processes as they go. I found the explanations of how lenses help us to see and how light behaves as it passes through a glass block to be as good as any I could have mustered as a teacher. This work breaks new ground and provides an invaluable set of resources for deaf learners.

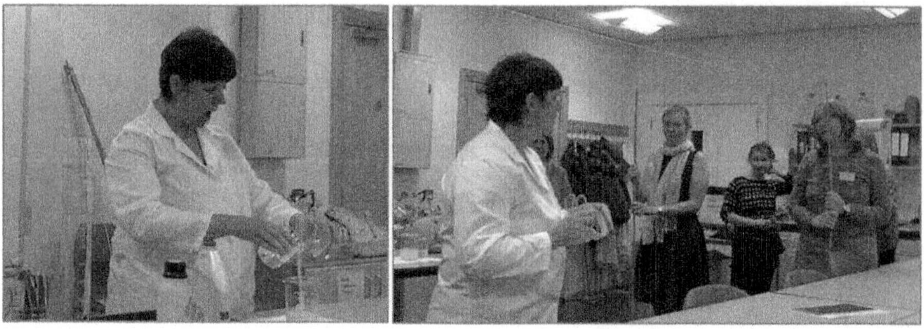

Fig. 19.5 Workshop with teachers of deaf children and interpreters to show how to demonstrate practical activities using the signs from the glossary – mixing of two substances and balloon racing

The work done to publicise and disseminate the new terms has been extensive, involving presentations to the general public and to Deaf people in the form of science festivals, invited lectures and workshops for teachers working with deaf children, thus confirming that the science glossary is being well received.

Implementation

Over the past 5 years, the BSL Glossary team has provided regular 1-day workshops for teachers of deaf children, sign language interpreters, and science teachers working at both primary and secondary levels (Fig. 19.5). On average, there are twelve participants in each workshop. As the Scottish Sensory Centre is based in a School of Education, we have also provided workshops for students in initial teacher education as an enrichment activity. In this section, we provide the plan we use for a typical workshop, along with briefing notes about working with a sign language interpreter in science classrooms and a handout about the practicalities of teaching deaf pupils.

Program Schedule

The following schedule of activities is provided to participants. This template could be adapted by other centres such as universities providing education for specialist teachers of deaf children or introducing science teachers to work with deaf students. Information in the right hand column refers to further details and handouts given below.

	Lesson plan	Objectives	Timing (min)	Information
1	Welcome and introduction	To introduce the participants to each other and find out their aims and motivation for this course to help the tutor to cater the workshop for the participants	10	
2	Practicalities of teaching deaf students in a mainstream class	To introduce good teaching practice with deaf students	60	Handout – practicalities
		Knowledge and experience of deaf students		
		Strengths and issues for deaf learners in mainstream science classes		
		The importance of questioning and wait time		Briefing notes
		How to work with a sign language interpreter		
		Class setting tips		
3	Introduction of the STEM glossary	To introduce the STEM in BSL glossary	60	Glossary website
		Why do we need to develop new BSL signs for STEM subjects?		
		STEM in BSL glossary website		
		Impact of the glossary		
4	Break			
5	Demonstration of practical activities in the laboratory	To demonstrate practical activities using the signs from the glossary to explain the concepts behind each experiment	45	Science demo
		To show how to demonstrate in front of a deaf student to make sure that this student is able to follow and understand		
		Introducing inquiry methods in mainstream class, and in backup sessions where a specialist teacher often works in a support space, not a lab		Everyday materials as specialist teachers may not have access to a lab
6	National assessment arrangements	To discuss how to support the deaf students during public examinations using the signs from the glossary	45	
7	Discussion and summing up	To allow participants to ask questions or discuss issues linked to teaching or supporting deaf children learning science	30	
8	Course evaluation and close	To collect feedback from the participants to help shape the next workshop	5	

Working with a Sign Language Interpreter

During deaf awareness sessions, participants are informed about some of the issues that arise in science classes when working with a sign language interpreter and deaf student(s). A sign language interpreter working in an educational setting is a professional who provides bilingual communication support working for individuals or groups of deaf students. This support generally involves a two-way exchange of information, through sign language and the spoken language of the country, providing access to the curriculum and the wider school environment to meet the needs of the deaf student wherever possible.

It is essential to promote the employment of appropriately experienced and qualified sign language interpreters, ideally to work as part of a team and not as the sole 'specialist'. It is good practice to match the needs of the deaf student with the skills and experience of an individual interpreter, especially with knowledge of science if in science classes.

Mainstream science teachers need to work with interpreters as partners to ensure smooth exchange of information with the deaf students. The ideal location for an interpreter within the classroom is at the front near the science teacher and source of visual information (e.g., electronic whiteboard or when demonstrating) so that the deaf student can see both the science teacher and interpreter easily. It is important to remember that it is not the interpreter's role to explain the lesson content. The deaf student should be encouraged to seek clarification from the science teacher.

There is a danger of deaf students being disengaged in lessons if the interpreter possesses a low level of signing skills or background knowledge of the subject – especially in scientific subjects. It is therefore vital that science teachers try to include deaf students in the lesson with support from a suitably qualified and experienced interpreter.

Before each lesson, the interpreter should have access to lesson plans and materials including the country's Science in Sign Language glossary (see Appendix A) to allow them to prepare before each lesson so that they can construct meaning from the many participants and events in the science classroom, and make sure the deaf student's contributions are voiced with understanding.

Science teachers must be mindful of deaf students when asking questions and should cooperate with the interpreter to ensure the deaf students receive the questions quickly enough to be able to offer answers without undue delay. The signed message is likely to be behind the spoken message, so the deaf students will need time to catch up to enable them to participate fully (Powers 1999).

Increasing wait time when asking questions will be helpful and will give the interpreter sufficient time to translate the questions and relay the answers from the deaf students to the science teacher (Grimes and Cameron 2005). There is a danger that the science teacher and interpreter may be out of synchronisation. The student may receive a reduced message as the interpreter tries to keep up with the lesson and thus abbreviates some of the information. It is important for the science teacher to maintain control of who is answering questions in class so that deaf students have the chance to be asked to contribute to class or group discussions.

The deaf students will need clear guidance about where to look. If there are a number of sources of information, it can be difficult to decide. This applies particularly to classroom discussions. If visual materials are used, the deaf student and interpreter will need time to look before communication continues, as they are using the visual channel for all information and communication. If a science teacher wants to comment on something that is presented visually, they should be careful to allow pauses for the deaf student to check the visual information.

The same is true in relation to lab demonstrations. The science teacher must be aware that the deaf student will need time to watch their explanation of what they are going to do before starting the demonstration. The teacher will need to allow the interpreter time to relay the information to the deaf student before carrying out the demonstration. The deaf student will not be able to watch the interpreter and the demonstration at the same time.

New terms may present problems as the deaf student may be unfamiliar with them or there may not be a sign for them. The interpreter and deaf student will need to time to discuss the new terms and signs, preferably before the start of the lesson. Because the deaf student may be the only child in his/her class with an interpreter, he/she will get undivided attention. This may sometimes get too much and there is a danger that the deaf student could become too reliant on the interpreter. However, having effective communication access will empower deaf students, giving them confidence to participate in the class as equals with their hearing peers.

Handout on Working with Deaf Learners

A guiding handout (Appendix B) is given to participants during the glossary workshops and can be adapted with acknowledgement by other centres.

Outcomes

After each course, participants were invited to complete an evaluation form. Three major themes emerged from their responses. First, teachers reported that they felt more confident about teaching deaf children in mainstream contexts:

> There is a S1 hearing impaired pupil who I feel would benefit from use of this in science classes, and the glossary is useful for explaining concepts. Also discuss areas of concern with the [mainstream] science teacher. (Teacher of deaf children)

The teachers were also very positive about learning why and how the signs were developed:

> As a non-signer with a science background, the new signs based around family words, e.g. chemical change, were simple and made sense! (Mainstream science teacher)
>
> It was a complete eye-opener. I realised how useful signing could be for all visual and kinaesthetic learners in class. (Mainstream science teacher)

Teachers also reported a growing awareness of the inquiry approach introduced on the course and an increased understanding of how to work with interpreters:

> I will learn all of the tips to ensure that I am a great facilitator to learning, letting the children lead their learning. Also more aware of BSL resources which will help me to challenge my kids. (Mainstream science teacher)
>
> The importance of class layout and further awareness of pupils in my class with hearing impairments – be aware of the wait time. I'm going to look into BSL courses. (Mainstream science teacher)

Course Review

Responses about the course were very positive and included comments such as the following:

> We had great discussions and will go back to school to disseminate information and hopefully organise future in-service events for staff, pupils and maybe the local community. (Teacher of deaf children)
>
> The whole course was excellent. One of the best. (SL interpreter)

The team has used evaluations to improve the sessions, reducing the amount of time spent explaining the origins of the glossary and focusing more on working with a sign language interpreter in a mainstream science class.

Summation

The work of the BSL Glossary team over the past 8 years has gradually moved from collecting and creating technical terms in BSL to disseminating and exploring ways in which the signs and definitions can be used. The range of contexts the technical signs have been used for is very broad, from mainstream schools to deaf schools and resourced schools with groups of deaf learners, and at both primary and secondary level. Interestingly, the signs have also been used by mainstream science teachers to teach hearing children (Ashby 2013). The iconic and visual metaphors in the signs can help a wide range of children to understand scientific concepts.

In Scotland, a British Sign Language Act has recently been passed. which may support the use of BSL in educational settings with deaf children. On the other hand, evidence from the rest of the UK suggests that BSL as a language is in rapid decline in the education system. Worldwide, however, there are many contexts where sign languages are being used in bilingual science teaching. Deaf children often choose to move fluently between modes and languages, and where they do we think there is a place for high quality resources, which will stimulate them to find out independently about science, to answer their own questions about the world.

Appendices

Appendix A: Science Sign Language Glossaries from different countries in the world

Glossary	Country	Website
STEM in BSL Glossary, Scottish Sensory Centre, University of Edinburgh	Scotland, UK	www.ssc.education.ed.ac.uk/BSL/list.html
Science Signs, Wolverhampton University	England, UK	www.sciencesigns.ac.uk
Projeto Surdos Glossario (LIBRAS), Universidade Federal do Rio de Janeiro	Brazil	http://projetosurdos.bioqmed.ufrj.br/categoria/produtos/projeto _glossario/
NTID Science Signs Lexicon (ASL), National Technical Institute for the Deaf at Rochester Institute of Technology	US	www.rit.edu/ntid/sciencesigns/
Sign Language for the Classroom (ASL), *embe* Outreach	US	www.needsoutreach.org/Pages/sl.html
Signing maths and science (ASL)	US	http://signsci.terc.edu/SSD/index.htm
First Steps in Physics for the Deaf project (Bulgarian Sign Language)	Bulgaria	www.signlanguage-bg.com/products.php?page=10

Appendix B: Handout on Practicalities of Teaching Deaf Pupils in Science

Knowledge and Experience

- Be aware that deaf children may not have the same experiences or backgrounds as hearing children – they are likely to have less general world knowledge and know fewer technical terms. Encourage all pupils to discuss their experiences related to the topic being taught.
- The English language is often a challenge for deaf pupils, especially when reading scientific text; focus questions are important in class and in relation to reading.

Using Visual Resources Carefully

- Use visual materials to support your lesson – PowerPoint slides, smart boards or the basic white board. All pupils will benefit.
- Only show one source of visual information at a time.
- Use and demonstrate real-life objects that apply to the science being taught.
- Use the classroom walls to build up visual resources and definitions of key terms; use colour for different groups or concepts.

Teaching

- Make eye contact with pupils while teaching. Make sure you have their attention before you start teaching or during discussions/ debates – flicking the lights on and off is a good way of gaining deaf pupils' attention.
- Try to meet all pupils' communication preferences (BSL, Signing with spoken language order, or spoken communication). Work with the sign language interpreter if you are not able to sign yourself. Expect deaf pupils to switch between languages depending on who they are communicating with. This may include writing notes and using speech.
- Write important information on the whiteboard or wall to let the pupils know e.g. dates for tests/ exams, homework, change of venue, etc.
- At the start of the lesson, give an outline of the lesson plan on the whiteboard or wall so they know what to expect.
- When demonstrating, ensure all are watching you before you start the demonstration; explain what you will do first, and repeat this after you have done the demonstration. Deaf pupils are not able to watch the teacher talking and the demonstration at the same time.
- Circulate among the pupils while they carry out their activities. Stand still and in a good light if talking to a deaf pupil. Usually the sign language interpreter will move next to you.
- At the end of the lesson provide a written recap on the whiteboard or discuss what was learned during the lesson.
- Before starting a new topic, make sure that pupils understand what has just been taught -use questioning and homework tasks to check.

Questions

- Use different styles of questioning and ask focus questions which are relevant and will encourage inquiry.
- Encourage deaf pupil(s) to ask or answer questions in the classroom. Allow **'wait time'** to encourage more thorough responses and to allow time for the sign language interpreter to get the questions to the deaf pupils.
- Look at the deaf pupil, not the interpreter when replying to her or his questions.
- Let the deaf pupil know who is talking; pupils asking questions should put their hands up. Repeat the question if the deaf pupil is not able to follow.

Classroom

- In mainstream settings, ensure the deaf pupil is able to watch the mainstream teacher, sign language interpreter and the whiteboard.
- Try to arrange the desks and chairs in the classroom to allow the deaf pupil to see who is contributing more easily, for example try a semi-circle.
- Ensure deaf pupils are safe when doing experiments, but do not over-protect them.
- Do not stand in front of the window because deaf pupils will not be able to see your face properly, important for speech reading.

References

Adam, R. (2015). Standardization of sign languages. *Sign Language Studies, 15*, 432–445.

Ashby, R. (2013). Signing in science. *Primary Science, 127*, 5–7.

Borgna, G., Convertino, C., Marschark, M., Morrison, C., & Rizzolo, K. (2011). Enhancing deaf students' learning from sign language and text: Metacognition, modality, and the effectiveness of content scaffolding. *Journal of Deaf Studies and Deaf Education, 16*, 79–100.

Brennan, M. (1990). *Word formation in British sign language*. Stockholm: University of Stockholm Press.

Caccamise, F., Smith, N., Yust, V., & Beykirch, H. (1981). Sign language instructional materials for speech, language and hearing professionals. *Journal of the Academy of Rehabilitative Audiology, 4*, 33–61.

Cameron, A., Quinn, G., & O'Neill, R. (2012). *Collecting, coining and defining new terms in BSL.* Poster at Reflecting on Deaf Education conference, University of Edinburgh, 12 Mar 2012.

Consortium of Research in Deaf Education. (2014/2015). *CRIDE 2015 survey*. http://www.ndcs.org.uk/professional_support/national_data/cride.html

Da Fonseca Flores, A., & Rumjanek, V. (2015). Teaching science to elementary school deaf children in Brazil. *Creative Education, 6*, 2127–2135.

Easterbrooks, S., & Stephenson, B. (2006). An examination of twenty literacy, science and mathematics practice used to educate deaf students who are deaf or hard of hearing. *American Annals of the Deaf, 151*, 385–397.

Fang, Z. (2007). The language demands of science reading in middle school. *International Journal of Science Education, 28*, 491–520.

Graham, S. C. (2012). *Deaf education preservice teachers' perceptions of scientific inquiry and teaching science to deaf and hard of hearing students*. Ph.D. thesis, University of Tennessee, Knoxville. Retrieved from http://trace.tennessee.edu/utk_graddiss/1297

Grimes, M., & Cameron, A. (2005). *Inclusion of deaf pupils in Scotland: Achievements, strategies and services.* Paper presented at the Inclusive and Supportive Education Congress, Inclusion: Celebrating Diversity? Glasgow. Retrieved from http://www.isec2005.org.uk/isec/abstracts/papers _g/ grimes_m.shtml

Johnston, T., & Napier, J. (2010). Medical signbank: Bringing deaf people and linguists together in the process of language development. *Sign Language Studies, 10*, 258–275.

Kahn, S., Feldman, A., & Cooke, M. L. (2013). Signs of autonomy: Facilitating independence and inquiry in deaf science classrooms. *Journal of Science Education for Students with Disabilities, 17*(1), 13–35.

Kurz, K., Hauser, P., & Schick, B. (2015). Deaf children's science content learning in direct instruction versus interpreted instruction. *Journal of Science Education for students with disabilities, 18*, 23–37.

Lane-Outlaw, S. (2009). *A qualitative investigation of ASL/English bilingual instruction of deaf students in secondary science classrooms*. Ph.D. thesis, Gallaudet University, Washington DC.
Lang, H., & Albertini, J. (2001). Construction of meaning in the authentic science writing of deaf students. *Journal of Deaf Studies and Deaf Education, 6*, 258–284.
Lang, H., Hupper, M., Monte, D., Brown, S., Babb, I., & Scheifele, P. (2007). A study of technical signs in science: Implications for lexical database development. *Journal of Deaf Studies and Deaf Education, 12*, 65–80.
Lindahl, C. (2015). *En studie om dialog i ett multimodalt, teckenspråkigt tvåspråkigt NO-klassrum*. (pp. 126–138). PhD thesis, University of Stockholm, Stockholm. Retrieved from http://bit.ly/1SHcypU
Lustick, D. (2010). The priority of the question: Focus questions for sustained reasoning in science. *Journal of Science Teacher Education, 21*, 495–511.
Pinto Silva, F., Martins, P., & Rumjanek, V. (2013). Rousing interest in science among secondary deaf students. *Scholarly Journal of Scientific Research and Essay, 2*, 104–108.
Powers, S. (1999). Deaf and hearing impaired pupils learning mainly via aided hearing. In L. Watson, S. Gregory, & S. Powers (Eds.), *Deaf and hearing impaired pupils in mainstream schools*. London: David Fulton Publishers.
Roald, I. (2002). Norwegian deaf teachers' reflections on their science education: Implications for instruction. *Journal of Deaf Studies and Deaf Education, 7*, 57–73.
Rosen, R. (2006). An unintended consequence of IDEA: American sign language, the deaf community, and deaf culture into mainstream education. *Disability Studies Quarterly, 26*(2). Retrieved from http://dsq-sds.org/article/view/685/862
Schick, B. (2008). A model of learning within an interpreted K-12 educational setting. In M. Marschark & P. Hauser (Eds.), *Deaf cognition: Foundations and outcomes* (pp. 351–386). New York: Oxford University Press.
Schick, B., Williams, K., & Kupermintz, H. (2006). Look who's being left behind: Educational interpreters and access to education for deaf and hard of hearing students. *Journal of Deaf Studies and Deaf Education, 11*, 3–20.
Scottish Sensory Centre. (2015). *British sign language glossaries of curriculum terms*. Retrieved February 13, 2016, from http://www.ssc.education.ed.ac.uk/BSL/list.html
Sutton-Spence, R., & Kaneko, M. (2016). *Introducing sign language literature: Folklore and creativity*. London: Palgrave.
Tolbert, S. (2015). "Because they want to teach you about their culture": Analyzing effective mentoring conversations between culturally responsible mentors and secondary science teachers of indigenous students in mainstream schools. *Journal of Research in Science Teaching, 52*, 1325–1361.
Vosganoff, D., Paatsch, L., & Toe, D. (2011). The mathematical and science skills of students who are deaf or hard of hearing in inclusive settings. *Deafness and Education International, 13*, 70–88.
Wang, Y. (2011). Inquiry-based science instruction and performance literacy for students who are deaf or hard of hearing. *American Annals of the Deaf, 156*, 239–254.
Wilson, C., Taylor, J., Kowalski, S., & Carlson, J. (2010). The relative effects and equity of inquiry-based and commonplace science teaching on students' knowledge, reasoning, and argumentation. *Journal of Research in Science Teaching, 47*, 276–301.
Woodward, J. (1972). Implications for sociolinguistic research among the deaf. *Sign Language Studies, 1*, 1–7.
Yore, L. (2000). Enhancing science literacy for all students with embedded reading instruction and writing-to-learn activities. *Journal of Deaf Studies and Deaf Education, 5*, 105–122.

Part IV
Science Teacher Preparation

The Project

The Inheritances Books are a collaborative project between Ichabod Crane High School's Illustration and ELL students. It is made possible by a grant from the Berkshire Taconic Community Foundation.

The Student-Artist

Sanford Fels, grade 11, 16 years old.

Chapter 20
Epilogue: Emergent Trends and Threads

Alandeom W. Oliveira and Molly H. Weinburgh

In addition to highlighting the complexities (theoretical and practical) of preparing school teachers to effectively support learning at the intersection of science and second language acquisition, chapters in this volume provide us with a more sophisticated, theory-based understanding of the target pedagogical activity toward which science teacher preparation in content-based second language acquisition is aimed. More specifically, they help us better understand the nature of content-language integration in science as a pedagogical endeavor that involves simultaneous lexicalization and conceptualization of one's experiences in a science classroom. Additionally, the preceding chapters illuminate the specific type of pedagogical expertise (i.e., knowledge and skills) that science teachers need to develop in order to effectively accomplish such an educational endeavor. These emergent themes are discussed below.

The Nature of Science-Language Integration

As indicated in our introductory chapter, integration of science and language learning is a highly complex and dynamic pedagogical endeavor wherein a skillful teacher (or team of co-teachers) dynamically provides learners with linguistic and epistemic scaffolds that enable simultaneous acquisition of both scientific knowledge and an additional language. Such a dual acquisition process involves both

A.W. Oliveira (✉)
State University of New York, Albany, NY 12222, USA
e-mail: aoliveira@albany.edu

M.H. Weinburgh
Texas Christian University, Fort Worth, TX 76129, USA
e-mail: m.weinburgh@tcu.edu

A.W. Oliveira, M.H. Weinburgh (eds.), *Science Teacher Preparation in Content-Based Second Language Acquisition*, ASTE Series in Science Education,
DOI 10.1007/978-3-319-43516-9_20

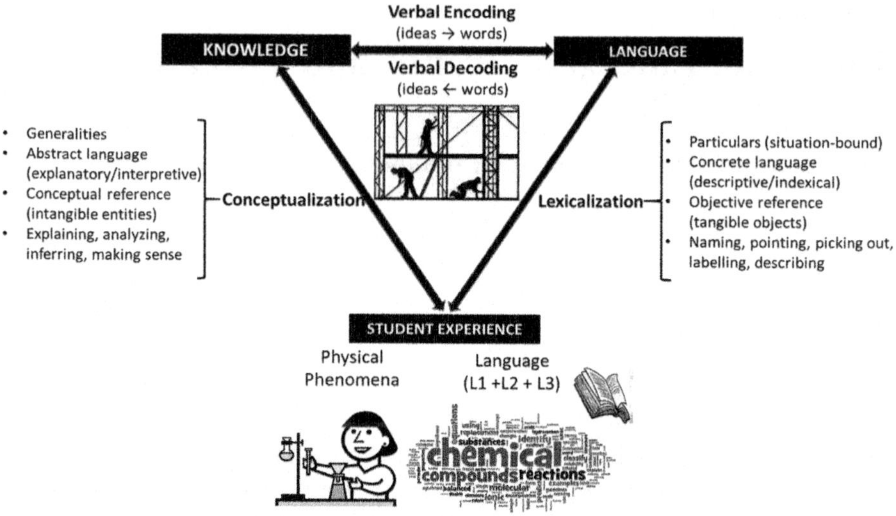

Fig. 20.1 Science-language integration

lexicalization and conceptualization of science classroom experiences (Fig. 20.1). Under the teachers' guidance and in collaboration with peers, students lexicalize classroom experiences (i.e., put them into words) by describing physical actions performed, objects handled, and observations made while engaged in hands-on science activities. New terms are acquired through objective reference and concrete language use such as naming, picking out, labelling, and pointing to real objects in the immediate physical and ostensive context. Words are mapped onto the physical world as speakers show a referent (tangible object) to each other and engage in *naming games* (Tomasello 2001). Additionally, students conceptualize their classroom experiences (i.e., articulate ideas and thoughts) by performing cognitive acts such as analyzing experiences, explaining observations, and inferring interpretations (i.e., making sense of experiences had). Content is acquired through abstract language use and conceptual reference to intangible entities (e.g., gravity, molecules) and by generalizing (deriving general principles beyond the immediate context).

Consistent with the above perspective on science-language integration, teacher preparation programs invariably emphasized the importance of allowing second language acquisition students to experience science firsthand through inquiry-based pedagogy (Chaps. 8, 15, and 17), small-group work, and demonstrations (Chap. 18). These science experiences were either paralleled or followed by various forms of verbal encoding/decoding activities wherein ideas were put into words (lexicalization of science experiences) and words used to construct ideas (conceptualization of science experiences). Learning emerged from the pedagogically scaffolded packing and unpacking of ideas. Common among these activities were teacher-led whole-class discussions (Chap. 9), journal writing (Chap. 17), and visual supports (Chaps. 3, 13, and 18). Only two programs favored use of the traditional science textbooks (Chaps. 2 and 14). These programs sought to prepare science teachers to effectively scaffold second language acquisition students' comprehension of textually encoded ideas, hence prioritizing textual experience over physical experience as a source of lexicalization and conceptualization.

Emergent Trends

A noticeable trend in science teacher preparation in content-based second language acquisition is the concerted effort made to ensure its theoretical coherence. More than merely providing science teachers with a set of "tricks of trade" (disconnected strategies or techniques), programs in the preceding chapters were designed to imbue in science teachers the ability to combine instructional pedagogies more holistically (i.e., went beyond mere adoption of isolated pedagogical strategies). Introduction to linguistic and epistemic scaffolds that could be used to support second language students' science learning was situated in larger, research-based frameworks such as systemic functional linguistics (Chap. 13), translanguaging (Chap. 17), assets/funds of knowledge (Chaps. 4 and 9), GLAD (Chap. 3), and CLIL (Chaps. 15 and 16). Rather than simply serving as a source of superficial "know-how," these programs equipped science teachers with frameworks that enabled them to develop more solid, research-based understandings of the communicative processes underlying second language science teaching practices. Interestingly, several programs had pedagogical frameworks rooted in educational policies and documents such as the NGSS, CSSS, WIDA, CREDE, and CERF. Grounded primarily in widely accepted standards of excellence in teaching and learning, these programs emphasized officially endorsed expert knowledge and skills – content, pedagogies, and practices sanctioned by policy makers at the state, national, and international levels.

Another emergent trend is an apparent tendency among professional developers to extend science teacher preparation above the level of *lexicon* (content-specific words or terminology). In addition to educating science teachers to more effectively deal with the cognitive and linguistic demands of the specialized vocabulary of science, several programs promoted science teachers' expertise at higher linguistic levels such as *utterances* and *genres*. Consistent with genre-based approaches to teaching of a second language (Henry and Roseberry 1998), science teacher preparation programs focused on the *speech acts of science* – the specific types of discursive moves that readers and writers typically need to be able to participate in any science activity such as posing a question, hypothesizing, and generalizing (for a more comprehensive list, see Chap. 16). As described in our previous work (Oliveira and Weinburgh 2016), such a conception of scientific language as social action assumes that the main purpose of language is "doing things," and that what truly matters is what we socially accomplish by uttering something (rather than language structure). For instance, by uttering the words "what is a chemical reaction?," a science teacher performs the speech act of asking a question. Other chapters sought to increase teachers' understanding of the rhetorical organization of written texts such as rheme-theme patterns (Chaps. 2 and 14). Across these programs, we notice a general shift in focus from localized patterns of language structure to larger issues of language functionality, as well as clear potential to help science teachers develop more sophisticated understanding and improved ability to support language acquisition.

EL Science Teacher Expertise

The preceding chapters also allow us to outline the type of pedagogical expertise that science teachers need to develop in order to be able to effectively promote content-based second language acquisition. Looking across chapters, several understandings and skills stand out as being central to the emergence of such a teaching ability. These are listed on Table 20.1.

The list of pedagogical understandings and skills outlined on Table 20.1 begin to illuminate what it means to become an expert in second language science teaching. Despite its potential to inform the design of future professional development programs in this area, it should be noted that previous research on language teacher cognition has shown that effective language teaching is a highly contextual and dynamic pedagogical ability contingent upon a variety of situational factors. Borg (2006) summarizes this body of research as follows:

> More expert language teachers are characterized by cognitions in which different forms of formal and experiential knowledge function as integrated whole and which enable teachers to envision learning potential in instructional contexts, anticipate problems and to respond (often improvisationally) in ways which are both technically skilled and sensitive to learners (p. 328).

Likewise, it can be argued that second language science teacher expertise does not develop in isolation of the social and institutional setting in which instructors have to work. While the above understandings and skills can inform science teacher cognition and practice in content-based second language acquisition, they should not be taken as a foolproof guarantee of success. Rather than developing a simplistic and static set of skills and understandings, science teachers need to ultimately develop the ability to capitalize on the dialect relationship between language and thought through thoughtful and contextually appropriate practice.

Future Work

While the professional development programs in this volume considerably extend our current understanding of how to prepare science teachers to effectively support emergent bilinguals' concept and language acquisition, much remains to be done. Future work in this area will need to continue to explore the effectiveness of

Table 20.1 Outline of second language acquisition science teacher expertise

The expert second language science teacher knows/understands…	The expert second language science teacher has the ability to…
The complex nature of scientific/academic language;	Simultaneously scaffold conceptual development and vocabulary development;
The process of second language acquisition;	Foster student mastery of the disciplinary language of science;
Language experientially and functionally rather than atomistically (mere accumulation of vocabulary);	Combine collaborative science inquiry with comprehensible input;
The areas of convergence and synergy between science and language;	Provide second language students with diverse visual supports (images, gestural signs) and written supports (journals);
That words can serve as concepts;	Go beyond pedagogical "tricks & tools";
That science processes can be conceived in terms of productive/receptive language;	Build on second language students' funds of knowledge through cognitively demanding, differentiated, language-rich tasks;
The key linguistic features of science texts;	Foster contextualized and authentic use of language in science through systematic content-language integration;
Scientific language as sense-making (not just vocabulary);	Adopt linguistically/culturally responsive pedagogy;
Language and content demands of educational policies such as NGSS and CCSS;	Take an asset-based approach (building on the strengths of second language students);
Themselves as "language planners";	Engage students in meaningful classroom interactions;
The pedagogical value of translaguaging and multilingualism;	Make science texts accessible to ELs through language simplification (without loss of rigor in content);
The importance of taking into account language during lesson planning;	Collaborate with language specialists.
Language-intensive pedagogical frameworks;	
A variety of pedagogical supports (curricular materials and technology) for supporting second language students (linguistically and epistemically).	

introducing science teachers to the latest linguistic and epistemic supports at their disposal. One good example is the set of bilingual glossaries recently developed by the Office of Bilingual Education and Word Languages in New York State (OBE-WL 2015). Meant to serve as reference material for teachers when planning lessons and students when taking State examinations, these bilingual glossaries provide direct translations of content-specific terminology from English to World languages commonly spoken among immigrant student populations in New York State (Fig. 20.2). Organized by subject area, these bilingual glossaries provide only direct translations of individual words (without being accompanied by definitions or explanations for each term). Word-to-word translation serves to accommodate second language

a **Elementary School**

Science Glossaries (grades 3-5)	Arabic (346 KB) | Bengali (304 KB) | Chinese Simplified (428 KB) | Chinese Traditional (445 KB) | Haitian--Creole (347 KB) | Korean (338 KB) | Punjabi (244 KB) | Russian (341 KB) | Spanish (274 KB) | Urdu (341 KB) | Vietnamese (325 KB)
Science Glossaries (grades 6-8)	Arabic (310 KB) | Bengali (248 KB) | Chinese Simplified (379 KB) | Chinese Traditional (393 KB) | Korean (318 KB) | Punjabi (252 KB) | Russian (300 KB) | Spanish (233 KB) | Urdu | Vietnamese (364 KB)

High School

Physical Setting Earth Science	Arabic (628 KB) | Bengali (360 KB) | Bosnian (A-L) (552 KB) | (M-Z) (778 KB) | Chinese (Traditional) (528 KB) | Haitian-Creole (427 KB) | Korean (331 KB) | Polish (425 KB) | Russian (405 KB) | Spanish (660 KB)
Chemistry	Bengali (747 KB) | Chinese (Traditional) (631 KB) | Haitian Creole (1,090 KB) | Korean (716 KB) | Polish (1,024 KB) | Russian (862 KB) | Spanish (1,266 KB)
Living Environment	Arabic (A-I) (612 KB) | (I-Z) (616 KB) | Bosnian (A-I) (690 KB) | (I-Z) (701 KB) | Chinese (Simplified) (725 KB) | Russian (A-I) (754 KB) | (I-Z) (752 KB) | Spanish (367 KB)
Physics	Chinese (traditional) (A-L) (525 KB) | (M-Z) (534) | Spanish (2,473 KB)

b

E	
earthquake	ভূমিকম্প
eccentricity	ভিন্নকেন্দ্রী
electromagnetic energy	তড়িৎ চুম্বকীয় শক্তি
element	পদার্থ
ellipse	উপবৃত্ত
elliptical	উপবৃত্ত সংক্রান্ত
energy	শক্তি
epicenter	উপকেন্দ্র
equilibrium	ভারসাম্য
equiox	বিষুব
erosion	ক্ষয়ে যাওয়া
escarpment	বন্ধুর উতরাই বা ঢল
esker	এসকার
eutrophication	ইউট্রোফিকেশন
evaporation	বাষ্পীভবন
evaporite	বায়ু নির্ণায়ক
extrusion	বল প্রয়োগ দ্বারা বহিষ্কার

c

B	
balance	저울, 균형, 균형을 이루다
balanced forces	밸런스(상쇄)하는 힘
bar graph	막대 그래프
barometer	기압계
battery	전지
behavior	행동, 행태
beneficial	유익한, 유용한
best	최고
bird	새, 조류
blink	눈을 깜박이다
block	블록, 막다, 방해하다
blood tissue	혈액 조직
blood vessel	혈관

Fig. 20.2 (**a**) Grades 3-12 bilingual glossaries for science subjects; (**b**) Snapshot from the English/Bengali glossary for physical setting and earth science (high-school subject); (**c**) Snapshot from the English/Korean glossary for science grades 3-5

students linguistically but not epistemically – a pedagogical effort to separate language from content. The effectiveness of introducing these specialized bilingual glossaries to science teachers during professional development as well as their potential impact on EL science achievement remain to be examined.

In addition to these novel forms of curricular support, potentially useful technological devices are constantly being developed and placed at teachers' disposal, often free of charge. One good example is *Rewordify*, an open-source, online program that enables teachers to conveniently and quickly simplify science texts through automatic rewording (http://rewordify.com/). By systematically replacing more difficult words with more comprehensible words based on a large database (called Corpus of Contemporary American English), the program allows teachers to considerably reduce the difficulty level of texts written in English and adapt them with the click of a button to a level that is more comprehensible for ELs to read. Another technological tool of great pedagogical potential is the app *Google Translate*. This free computer program is capable of automatic translation of text, speech, image, or video. Downloadable to mobile devices (e.g., cell phones), this powerful device can be used by second language students to translate oral speech in real time and also pictures taken with their cell phone cameras, instantly. Like bilingual glossaries, these technological tools remain to be systematically integrated into science teacher preparation programs in content-based second language acquisition.

In addition to increasing science teachers' familiarity with the latest translation technologies for linguistically supporting second language students, professional development programs can benefit from the incorporation of recent work on *translation-based pedagogies* – classroom activities wherein learners translate texts (written or audiovisual) to learn an additional language and content. In a recent issue on the pedagogical use of translation, Laviosa (2014) emphasizes that engaging students in translation tasks (e.g., video subtitling, reverse subtitling, oral interpretation, shadowing, meaning-based communication, etc.) can be effective in supporting student acquisition of an additional language as well as developing interlinguistic and intercultural competence, critical language awareness, problem-solving skills, and cognitive flexibility. In the field of science teacher education, more careful consideration needs to be given to effectively developing science teacher expertise in the design and implementation of such translation tasks as a means to promote learning of scientific content. As our previous work has shown (Oliveira et al. 2009a, b; Quigley et al. 2011), translating to learn science is far from being a simple and straightforward endeavor as many complexities often arise (e.g., lack of specialized translation equivalents in the target language).

Lastly, there currently exists a growing need for research on how to effectively prepare science teachers to pedagogically support and meet the needs of students with a refugee background. Also known as *Students with Limited or Interrupted Formal Education* (SLIFE), this particular subpopulation of second language students typically has more extreme needs that go far beyond having limited proficiency in the language of instruction. With little or no formal education due to war, migration, or economic necessity, SLIFE have been shown to commonly face a variety of challenges when trying to adapt to western educational systems, including feelings of isolation, cultural dissonance or incongruity due to very different assumptions about teaching and learning and unique orientation to education. As DeCapua and Marshall (2011) emphasize, "reaching SLIFE entails not only addressing language and content, but also addressing culture… these students have a very different learning paradigm from that of mainstream U.S. schools and face cultural dissonance" (pgs. 4–5). Effectively teaching SLIFE requires more than adoption of culturally responsive pedagogy and provision of culturally relevant materials (Gordon and Yowell 1999). Supporting SLIFE science learning requires teacher awareness of these students' often low levels of literacy in their first language (Miller 2009) as well as familiarity with instructional models such as the *Mutually Adaptive Learning Paradigm* (DeCapua and Marshall 2011).

We would like to conclude this book by emphasizing the importance of effectively preparing school teachers to embrace language and multilingualism as an important part of being an expert science educator. Rather than seeing themselves as professionals whose pedagogical expertise is limited to matters of content and for whom linguistic form does not matter, it is critical for science teachers to become more cognizant of the centrality of language to the human experience. As the Canadian poet Carl Leggo (1998) writes "everything is constructed in language, our experiences of lived time and lived space and lived body and lived human relation are all epistemologically and ontologically worded/lined/known/revealed/disclosed/understood/lived in words" (p.175). Developing such a realization is of paramount importance if science teachers are indeed to become capable of effectively promoting content-based second language acquisition and help all students develop a scientific voice regardless of mother tongue or sociocultural background. This is our hope for this book.

References

Borg, S. (2006). *Teacher cognition and language education: Research and practice*. London: Bloomsbury.

DeCapua, A., & Marshall, H. W. (2011). *Breaking new ground: Teaching students with limited or interrupted formal education in U.S. secondary schools*. Ann Harbor: University of Michigan Press ELT.

Gordon, E., & Yowell, C. (1999). Cultural dissonance as a risk factor in the development of students. In E. Gordon (Ed.), *Education and justice: A view from the back of the bus* (pp. 34–51). New York: Teachers College Press.

Henry, A., & Roseberry, R. L. (1998). An evaluation of a genre-based approach to the teaching of EAP/ESP writing. *TESOL Quarterly, 32*(1), 147–156.

Laviosa, S. (2014). Introduction to special issue on translation in the language classroom: Theory, research, and practice. *The Interpreter and Translation Trainer, 8*, 1–7.

Leggo, C. (1998). Living un/grammatically in a grammatical world: The pedagogic world of teachers and students. *Interchange, 29*(2), 169–184.

Miller, J. (2009). Teaching refugee learners with interrupted education in science: Vocabulary, literacy, and pedagogy. *International Journal of Science Education, 31*(4), 571–592.

Office of Bilingual Education and Word Languages in New York State (OBE-WL). (2015). *Bilingual glossaries*. Retrieve April 20, from http://www.p12.nysed.gov/biling/bilinged/bilingual _glossaries.htm

Oliveira, A. W., & Weinburgh, M. (2016). Coming to terms with language: Editorial for the EJSE special issue on science and language (Special issue editorial). *Electronic Journal of Science Education, 20*(3), 1–10.

Oliveira, A. W., Colak, H., & Akerson, V. L. (2009a). "Who polluted the Potomac?" The translation and implementation of a US environmental story in Brazilian and Turkish classrooms. *Cultural Studies of Science Education, 4*(1), 89–132.

Oliveira, A. W., Colak, H., & Akerson, V. L. (2009b). Curriculum translation and environmental education: Considering issues of discursive intentionality, interpretation, and validity. *Cultural Studies of Science Education, 4*(1), 149–155.

Quigley, C. F., Oliveira, A. W., Curry, A., & Buck, G. (2011). Issues and techniques in translating scientific terms from English to Khmer for a university level text in Cambodia. *Language, Culture Curriculum, 24*(2), 159–177.

Tomasello, M. (2001). Perceiving intentions and learning words in the second year of life. In M. Bowerman & S. C. Levinson (Eds.), *Language acquisition and conceptual development* (pp. 132–158). New York: Cambridge University Press.

Notes on Contributing Authors

Rouhollah Aghasaleh is a PhD candidate and an instructor in the Department of Educational Theory and Practice at The University of Georgia. Rouhollah served as a classroom teacher, curriculum developer, and instructional coach in urban 4–12 grades before starting a doctoral program in 2011. Rouhollah's scholarship and teaching emphasizes the understanding of embodiment in the curriculum and policing and surveillance of bodies through school discipline policy and pedagogy. Rouhollah collaborated with a team of scholars to investigate science teaching and learning in the LISELL-B project, which is focused on science education for emergent bilingual students, professional development for their science and ESOL teachers, as well as their families. Rouhollah is a co-founder of *Feminist Scholar-Activists*, a student-lead organization that aims to consciously merge feminist theories and everyday practice to work for social change through providing a critical atmosphere for scholars to apply their scholarship in solidarity work of activism.

Martha Allexsaht-Snider completed her doctorate in Crosscultural Education at the University of California at Santa Barbara (1991) and is an Associate Professor in the Department of Educational Theory and Practice at the University of Georgia. Her research interests include: family-school-community interactions in diverse settings including Latino communities and rural México; professional development and equity in mathematics and science education; and creative approaches to teacher education in diverse contexts. She has worked with national and international grants in the areas of science and math education for immigrant students and families and teacher education for rural and indigenous teachers in México. Currently, she is collaborating with colleagues at UGA, and in Hall County and Clarke County Schools, in a 4-year National Science Foundation grant titled *Language-rich Inquiry Science with English Language Learners through Biotechnology (LISELL-B)* that involves Latino/a middle and high school students, their families, and their science and ESOL teachers.

A.W. Oliveira, M.H. Weinburgh (eds.), *Science Teacher Preparation in Content-Based Second Language Acquisition*, ASTE Series in Science Education,
DOI 10.1007/978-3-319-43516-9

Jamie S. Baker is an adjunct professor of teacher education at New Mexico State University in Las Cruces, New Mexico. She earned a Ph.D. in Curriculum and Instruction at New Mexico State University (2012). She is a National Board Certified Early Adolescent English/Language Arts Teacher at Gadsden High School in Anthony, New Mexico. Her academic interests include Secondary teacher development, STEM recruitment, retention and teacher preparation, and literacy instruction across the content areas.

Julie A. Bianchini is a professor of science education in the Gevirtz Graduate School of Education at the University of California, Santa Barbara (UCSB). Her research investigates prospective, beginning, and experienced teachers' efforts to learn to teach science to *all* students in equitable and reform-minded ways. She serves as Faculty Director of UCSB's CalTeach Initiative, a University of California System-wide program designed to encourage science, mathematics, and engineering undergraduates to consider K-12 science and mathematics teaching as a career. She is the principal investigator (PI) or Co-PI on three teacher education research and development grants funded by the National Science Foundation: CalTeach Physical Sciences and Engineering (CTPSE); STEM Teachers for English Language Learners: Excellence and Retention (STELLER); and Science and Mathematics Teacher Research Initiative (SMTRI).

Marco A. Bravo is an associate professor at Santa Clara University where he teaches courses in language and literacy development to pre-service teachers. He earned a Master's degree in Human Development and Psychology from Harvard University (1995) and a PhD degree in education with a focus in Language, Literacy and Culture (2003) from the Graduate School of Education at the University of California, Berkeley. He was a post-doctoral researcher at the Lawrence Hall of Science where he was a member of the *Seeds of Science/Roots of Reading* research and curriculum development team. He authored several children's science books as part of his work on this project. He is a former bilingual first grade teacher. His research interests include science and literacy integration, vocabulary development and the process by which English language learners learn to read in English and their native language.

Cory A. Buxton is *Athletic Association Professor of Education* in the Department of Educational Theory and Practice at the University of Georgia. Buxton has a BS in Geology from University of Maryland; an MAT in Earth Science Education from Tulane University; and a PhD in Science Education from the University of Colorado at Boulder. He formerly taught science and English as a second language in Guatemala, New Orleans and Colorado. His current research focuses on designing teacher professional learning experiences and classroom interactions that foster more equitable science learning opportunities for emergent bilingual learners. Other research interests include the language of science, the uses of science to solve social problems, building family-school-university partnerships, and how success in science serves as a gateway to broader academic success.

Audrey Cameron is project manager of the Scottish Sensory Centre's STEM in British Sign Language Glossary at the University of Edinburgh. She earned a PhD degree in Chemistry at Strathclyde University (1996) and PGCE in Chemistry with Science at University of Edinburgh (2004). She has taught science in mainstream schools before joining the SSC glossary team to develop new signs for STEM. With Gary Quinn, she runs professional development courses on using sign language in science lessons for staff working with deaf children. They also present science shows at festivals and schools with deaf children to introduce the new signs for science. Her research interests include access to science education for deaf children.

Lourdes Cardozo-Gaibisso a former middle school teacher in her native Uruguay, is a Ph.D. student in the Language and Literacy Education Department at The University of Georgia. She is a Research Assistant for the LISELL-B project and serves as editor for the Scholars Speak Out (SSO) section of the UGA Journal of Language and Literacy Education. Her research at this time focuses on how Latino Emergent Bilingual Learners can successfully develop science literacy through a model of scientific inquiry and a pedagogy of translanguaging.

Stacey L. Carpenter is a doctoral candidate at the University of California, Santa Barbara (UCSB). She has a B.S. from the University of Hawaii at Manoa and taught high school science for 3 years in Honolulu. She has an M.A. in Education from UCSB and her research focuses on science teaching and teacher learning.

Núria Carrillo Monsó is an English and Catalan language teacher since 1984. She earned a degree in Philosophy and Educational Sciences – Psychology (1990) and a degree in English Philology (2008) at the University of Barcelona (UB). She has worked as an English language teacher and CLIL (Content and Language Integrated Learning) specialist at Vila Olímpica primary school in Barcelona. She also worked on a bilingual project for 3 years (1995–1998) in a school in Lynwood, Los Angeles, California; and she spent one school year (2004–2005) working on a CLIL project in a couple of schools in Ormskirk, UK. She is a co-author of eight Science CLIL modules- "The Thinking Lab", Cambridge University Press, for upper levels in primary school. She has participated as a speaker at seminars and workshops. She has worked in foreign languages teacher education and she is currently a member of CESIRE (Resource Centre for Innovation and Educational Research) at the Ministry of Education in Catalonia, in the linguistic field, and she participates in several interdisciplinary and innovation projects.

Christopher D. Carson is an instructor of culturally and linguistically diverse education at the University of Colorado Denver. He earned a MA in Secondary Education from the University of Mississippi (1997) and an MA in Linguistics from the University of Utah (2002). He was a science teacher for 11 years, teaching multilingual students in a diverse array of classroom settings from upper elementary to high school.

Ana Castillon Pascual is a special needs teacher, and a kindergarten and primary school teacher at Pau Romeva primary school in Barcelona. She earned a degree in Geography and History and a degree in Pedagogy at the University of Barcelona (UB). She is a member of CESIRE (Resource Centre for Innovation and Educational Research), in the natural science field, and she participates in some interdisciplinary and innovation groups. She is co-author of several articles about social and natural sciences. She has been a trainer on science teaching and learning for kindergarten and primary teachers. She has participated as a speaker at seminars and workshops. She collaborates in different working groups about science learning, and innovation.

Anna De Meo is an associate professor of language acquisition at the University of Naples L'Orientale, Italy. She received her PhD in Linguistics at the University of Pisa, Italy, in 1993. She is currently President of the Linguistic Center, Rector Delegate for the Lifelong Learning Program, and responsible for the Student Refugee Program. She is President of the Italian Association of Applied Linguistics and Coordinator of the Spoken Communication Group of the Italian Linguistic Society. She is actively involved in the national programs of teacher education and has developed, coordinated and implemented teacher training courses in CLIL and second language teaching. In 2014, she was named Guest Professor of the Tianjin Foreign Studies University, China. Her research interests include second language acquisition, L2 speech perception and production, prosody, interlanguage pragmatics, and translation.

Luciana C. de Oliveira is Chair and Associate Professor in the Department of Teaching and Learning in the School of Education and Human Development at the University of Miami, Florida. Her research focuses on issues related to teaching English language learners (ELLs) at the K-12 level, including the role of language in learning the content areas and teacher education, advocacy and social justice. Currently, Dr. de Oliveira's research examines the linguistic challenges of the Common Core State Standards for ELLs and their implications for teachers of ELLs. She is the series editor of five volumes focused on the Common Core and ELLs (2014–2016) with TESOL Press.

Maria De Santo is a self-access center manager and language counsellor at the University of Naples L'Orientale. She earned a Master's Degree in Foreign Languages and Literatures (2000) and a Teaching of English as a Foreign Language to Adults Certificate (2002) at L'Orientale University in Naples. She has taught courses for school teachers of English, French and CLIL. She also teaches blended courses for teachers of Italian as a Second Language. Her research interests include learner autonomy, self-access center management, technology enhanced language learning environments and CLIL.

Uyen H. Do is the Program Manager at the University of California, Davis School of Education. Ms. Do has earned a Bachelor's degree in Human Development and

Psychology, Masters Degree in Education, as well as a Multiple-Subject Teaching Credential from the University of California, Davis, and pursuing her Ed.D. in Educational Leadership and Policy. She spent 5 years in the classroom and over 15 years working with administrators, teachers, students, in a variety of educational settings.

Max Vázquez Dominguez received his PhD in Educational Theory and Practice at the University of Georgia (2016) where he worked in several science education projects with middle school science teachers, ESOL teachers, emergent bilingual students and their families. With the Master's degree in early childhood education with an emphasis on science from the University of Georgia (2009), he worked in México to provide pre-service and in-service teachers with professional learning in science education in urban and rural regions of the state of Veracruz. His research interests include using emergent bilingual students' interests and passions in the teaching/learning process, family engagement, science and soccer, the use of alternative concepts of space to enhance science learning, and bilingualism in science teaching and learning.

Mariona Espinet is a professor of science education at the Universitat Autònoma de Barcelona in Catalonia, Spain. She earned a PhD degree in science education at the University of Georgia, Athens in 1990 thanks to a Flubright-LaCaixa scholarship after teaching science in primary and secondary schools. At present she teaches science education university courses offered in pre-school and primary teacher education undergraduate programs, as well as science education research courses offered at the master and doctoral levels. She is the coordinator of the Official Master of Research in Education Specialty in Science Education at the UAB, and the coordinator of two research groups Gresc@ (Education for Sustainability, School and Community) and SGR LICEC (Language and Contexts in Science Education). Her research and innovation interests are strongly interdisciplinary and focus on classroom discourse and critical literacy in multilingual science learning environments, and education for sustainability at the interphase between schools and communities.

Susan Gomez Zwiep is a professor of science education at California State University, Long Beach. She received a BA in biology from the University of California, Berkeley (1993), an MA in education from Whittier College (1995), and a PhD in science education from the University of Southern California (2005). Her research has focused on establishing equitable access to rigorous, inquiry-based science instruction for all students, the integration of science with other disciplines and teacher/faculty professional development. She currently serves as a regional director for the K12 Alliance/WestEd.

Laura Farró Gràcia earned a degree in Hispanic Philology and Catalan Philology at the University of Barcelona. She is a member of CESIRE (Resource Centre to Support Research and Innovation Education) at the Ministry of Education in

Catalonia. Professional Experience: Catalan language and literature high school professor; technical teacher educator at the Institute of Sciences of Education at the University of Barcelona (UB); technical advisor for educational guidance at the Ministry of Education in Catalonia; teacher educator for leadership, communication and emotional intelligence in business and education; teacher educator for practicum psychologists tutors at the Open University of Catalonia (UOC). She has coordinated various working groups for innovation, and has participated as a speaker at seminars, conferences and congresses. She has published articles related to experts training, language and literature didactic books, and about emotional intelligence.

Amy Heineke is the Associate Professor of Bilingual and Bicultural Education in the School of Education at Loyola University Chicago. She holds a Ph.D. in Language and Literacy and M.Ed. in Curriculum and Instruction from Arizona State University. She began her career as an elementary teacher of predominantly immigrant students in Phoenix, Arizona, where she begin developing her expertise and advocacy related to the teaching and learning of English learners (ELs). Dr. Heineke currently facilitates teacher learning for ELs through field-based teacher education coursework and grant-funded professional development focused on supporting students' language development in the Chicago Public Schools. Her research agenda centers on teachers' active roles in language policy and teacher preparation for culturally and linguistically diverse students.

Anita C. Hernandez is an associate professor of language and literacy education and the Don and Sarrah Kidd Endowed Chair in Literacy at the New Mexico State University. She earned a Master's degree in reading education at California State University and a Ph.D. degree in Language, Literacy, and Culture at Stanford University. She has taught bi-literacy courses in Mexico, California, and New Mexico and organizes professional development programs for preservice and classroom teachers. She has co-authored two books: *Theme Sets for Secondary Students* and *Interactive Notebooks for English language learners*. Her research interests include bi-literacies, professional development for teachers of bilingual learners, and language connections through Spanish-English cognates.

Cecilia M. Hernandez is an assistant professor of science education at New Mexico State University in Las Cruces, New Mexico. She earned a master degree of science in biology from Texas Tech University (2002), and a Ph.D. in Curriculum and Instruction from Kansas State University (2011). She has taught courses related to science K-16, and has worked with school districts, community colleges, and university faculty in continuously improving teacher education programs. Her research focuses on diverse learners, culturally responsive teaching in science education, and ESL/Bilingual education, as well as issues related to increasing the participation of underrepresented populations in the sciences.

Yainitza Hernández-Rodríguez completed her bachelor's degree in Microbiology at the University of Puerto Rico-Mayagüez (2004) and her doctorate degree in Fungal Biology and Genetics at the University of Georgia in Athens (2011) where she also completed a post-doctorate in Fungal Cell Biology. Interested in science education and curriculum development, in 2014 she joined the LISELL-B project as the project coordinator in the Department of Educational Theory and Practice at the University of Georgia. The project works in collaboration with UGA faculty, graduate students and Hall County and Clarke County Schools science and ESOL teachers, Latino/a middle and high school students and their families. The primary goal of this project is to test instructional strategies to prompt students' science talk, writing and action with a focus on emergent bilinguals and biotechnology. Yainitza is interested in co-developing meaningful inquiry science curriculum with science teachers and science education researchers that integrates academic and scientific language to meet the needs of emergent bilinguals. She is also interested in family-school interactions to promote and encourage science learning and career readiness.

Sarah Hough is a project scientist at the University of California, Santa Barbara. She earned a Master's degree in Applied Mathematics (1996) and a PhD in Educational Psychology (2002). She has conducted evaluation research for a wide range of teacher education initiatives. Her current research interests include STEM teacher development and inquiry-based teaching and learning in mathematics.

Mehtap Kirmaci is a Ph.D. student in the Department of Educational Theory and Practice at the University of Georgia. She currently works as a research assistant for the LISELL-B project, focusing on science/ESOL teachers, their middle and high school Latino/a students and their families. Mehtap received her Bachelor's degree in elementary education from Celal Bayar University in 2008 and taught elementary grade levels in Turkey. Her current research interests include community-school-family interactions, parent involvement in science education, education of immigrant/refugee students and critical pedagogy.

Amanda M. Latimer is currently a graduate research assistant working on her doctorate in Educational Theory and Practice at the University of Georgia. After earning a Master's of Science in animal science (1989), she worked in obesity research and then for a biotechnology company. She began a biotechnology based teacher outreach and professional development program (2007) while working with a Department of Labor grant at Athens Technical College. She is now part of the Teaching Science to English Language Learners: Language-Rich Inquiry Science with English Language Learners through Biotechnology (LISELL-B) project where she is interested in the professional development of science and agricultural teachers, creative approaches to teaching science, and the agency of science kits.

Jin Sook Lee is a professor of education in the Gevirtz Graduate School of Education at the University of California, Santa Barbara. She earned her MA degrees in

Linguistics at Yonsei University, (Seoul, Korea) and in language education at Stanford, University as well as her PhD degree in language, literacy and culture at Stanford, University (2000). She has taught EFL/ESL courses in both Korea and the US and has served as a Fulbright scholar to develop English language programs abroad. Her research interests include academic language in English language learners, teachers' pedagogical language knowledge, and the development of bilingualism in home, community, and school settings.

Sarah Lord is a doctoral student in mathematics education at the University of Wisconsin-Madison. She has been a classroom and bilingual teacher, a math coach, and a district math teacher leader before deciding to pursue graduate work. She has been leading professional development and graduate courses for in-service teachers for the past 8 years. Her primary research interest is children's mathematical development.

Edward Lyon is an Assistant Professor of science education in the School of Education at Sonoma State University. He received his PhD in science education from the University of California, Santa Cruz, and his BS in psychobiology from the University of California, Los Angeles. Dr. Lyon studies how secondary science teachers engage in instructional and assessment practices that promote authentic science learning and literacy development in multilingual classrooms. He is currently Co-Principal Investigator on the NSF funded project, *Secondary Science Teaching with English Language and Literacy Acquisition* that examines how novice teachers become prepared to teach science to English learners. Dr. Lyon teaches secondary science methods and research paradigms in education.

Rita MacDonald is an applied linguist and academic English language researcher at the 38-state WIDA Consortium at the Wisconsin Center for Education Research, University of Wisconsin-Madison. Since completing an MATESOL degree at Saint Michael's College in Vermont in 2002, she has worked in educational linguistics, first as an ESL teacher, then as a teacher educator for both ESL and content teachers, and now as a member of WIDA's research team. Her work and research has focused on supporting content teachers in building students' academic discourse and academic language. Additional areas of focus include formative language assessment, mentoring of co-teaching teams, and the development of resources to support equitable, cross-state identification and reclassification of English Learners.

Elizabeth Mahon is an assistant clinical professor of culturally and linguistically diverse education at the University of Colorado Denver. She earned a PhD in Educational Equity and Cultural Diversity from University of Colorado Boulder (2004). She was a bilingual school counselor and English language development teacher for 20 years before coming to the University of Colorado Denver in 2011.

Emily Miller is a lead writer for the Diversity and Equity Team on the Next Generation Science Standards and a coauthor of "NGSS: All Standards, All Students and Case Studies", and NGSS for All Students, with Dr. Okhee Lee with NSTA Press. Emily has taught for 18 years as an ESL and Bilingual Resource science specialist in Madison, Wisconsin at a Title I school. Emily serves as a curriculum developer on multiple grants: A Project Based Learning grant with Dr. Joe Krajcik and Dr. Annemarie Palinscar, an NSF grant with Dr. Okhee Lee and Dr. Guadalupe Valdez, and a NSF Teacher Professional Development grant with the Wisconsin Center for Educational Research. Emily has a MS in Bilingual Studies from the Department of Curriculum and Instruction at the University of Wisconsin, Madison, a BS from the same department with a Spanish minor, and ESL and Bilingual teaching certifications. She is pursuing a PhD in Curriculum and Instruction at the University of Wisconsin-Madison.

Renee N. Newton is senior director of Community and School Partnerships at the University of California, Davis School of Education. She earned a baccalaureate degree in Applied Behavioral Science from the University of California, Davis (1989) and a Master of Public Administration degree from the University of Southern California (1990). Over the past 15 years, she has been P.I. for educational programs organized to strengthen K-16 education systems including school and community outreach programs and services. She has led a multidisciplinary team in support of a statewide, expanded learning STEM initiative that cultivated stronger pathways between STEM learning and expanded learning programs.

Susan O'Hara is Executive Director of Resourcing Excellence in Education (REEd) at the University of California Davis. Susan has authored more than 50 publications related to English learners, innovative uses of technology, and professional development for teachers. Most recently she has led a team in the development and testing of Essential Practice Frames that articulate high leverage practices to meet the learning needs of all students. She is a principal investigator on two large federal grants, one investigating best models for improving outcomes for English Learners and one focused on the development of Integrated Professional Learning Systems. Prior to her current position, Susan was an Associate Professor and Executive Director of the Center to Support Excellence in Teaching at Stanford University, as well as a professor of education at Sacramento State University.

Alandeom W. Oliveira is an associate professor of science education at the State University of New York at Albany. He earned a Master's degree in science education at Southeast Missouri State University (2002) and a PhD degree in science education at Indiana University Bloomington (2008). He has taught science education courses to teachers in Brazil and the US and has coordinated multiple professional development programs for school teachers, including Science Modeling for Inquiring Teachers Network, and Technology-Enhanced Multimodal Instruction in Science and Math for English Language Learners. His research interests include cooperative science learning, inquiry-based teaching, and classroom discourse and language use.

Rachel O'Neill is a lecturer in the School of Education at the University of Edinburgh. Qualifying first to teach deaf children, then to teach English as an Additional Language, Rachel's Master's degree at the University of Manchester (1993) was in linguistics. She taught a wide range of deaf students in mainstream secondary schools and colleges for 25 years before moving to the University in 2006. There she is programme director for the MSc in Inclusive Education, teaching on the deaf education pathway. Rachel has supported the Scottish Sensory Centre's BSL glossary since 2006. Her other research interests include the attainments of deaf children, language policy, and linguistic access strategies for deaf children at school.

Deborah Pitta is Vice President of Professional Learning for *EPF for teaching*, was an elementary and secondary teacher as well as Assistant Superintendent of Curriculum and Instruction where her primary responsibility was to provide professional development for teachers. Debi supports the professional learning of teachers and administrators on topics ranging from implementation of common core state standards, professional learning communities, and differentiated instruction.

Robert Pritchard is a language and literacy specialist who has worked extensively with schools, school districts, and state departments of education on a wide range of professional development projects. A classroom teacher and reading specialist for 13 years, Dr. Pritchard also worked internationally as an ESL teacher and curriculum specialist. He has authored numerous publications related to English learners, innovative uses of technology, and professional development for teachers.

Gary Quinn is Assistant Professor in the Department of Languages and Intercultural Studies at Heriot-Watt University. He worked at the University of Central Lancashire between 2000 and 2005 as a lecturer on the Deaf Studies degree, also researching deaf people with minimal language skills. His MA in Language Studies was from Lancaster University in 2004. At Heriot Watt University Gary is the head of the BSL section and programme director for the MA BSL Interpreting degree. His PhD study is on features of BSL discourse. He has been an active member of the BSL glossary project as the sign linguist in the development teams, and has presented many science shows with Audrey Cameron.

Margaret Rasulo is a researcher of English language and linguistics at the Second University of Naples, Italy. She earned a Master's degree in education at the Open University (2002) and a PhD degree in English for Specific Purposes at the University of Naples 'Federico II' (2007). She has worked with Italian local and national education authorities in coordinating and implementing university and school-based foreign language projects for teacher education and requalification. She is currently involved in CLIL methodology and second language education and training for primary and high school teachers. Her research interests include learner autonomy, second language acquisition, computer-mediated communication and web-based popularization genres.

Núria López Rebollal is a primary school teacher at Llacuna primary school at Poblenou. She is a specialist in natural science. She teaches science education university courses offered in primary teacher education undergraduate programs at Universitat Autònoma de Barcelona (UAB) in Catalonia. She is co-author of the book *Química a infantil i primària. Una nova mirada*, and she has collaborated with other authors in other text books dealing with science education. She has participated as a speaker at seminars and workshops. She is a member of CESIRE (Resource Centre for Innovation and Educational Research), in the natural science field, and she participates in some interdisciplinary and innovation groups such as creative technologies. She is a member of several teacher working groups about science learning (Kimeia).

Christine M. Reyes is a visiting clinical assistant professor and interim director of the ESOL program in the department of linguistics at the University of Texas at El Paso. She earned a Master's degree in ESL/Bilingual education at Texas Tech University, Lubbock Texas (2002) and a Ph.D. in curriculum and instruction at Kansas State University (2013). She has been involved in teaching ESL/EFL methods courses to visiting teachers from China and Ecuador. Her current research interests include the role of Latina culture in distance education, discourse patterns in language development and improving methods in language acquisition.

Sarah A. Roberts is an assistant professor of mathematics education in the Gevirtz Graduate School of Education at University of California, Santa Barbara. She earned a Master's degree in Education at Stanford University (2000) and a PhD degree in Mathematics Curriculum and Instruction at University of Colorado at Boulder (2009). She is a former middle school and high school mathematics and science teacher. Her research interests include equity in mathematics education, with attention to supporting English learners in mathematics classrooms, providing professional learning experiences for mathematics teachers of English learners, and developing preservice teachers' of ideas of equity.

Lara K. Smetana is the Associate Professor of Science Education in the School of Education at Loyola University Chicago. She holds a Ph.D. in Science Education from the University of Virginia. She began her career as a middle grades physical science teacher in Alexandria, VA and has been supporting elementary and secondary teacher candidates in a variety of educational settings since. Her research agenda centers on field-based teacher preparation for culturally and linguistically diverse students and better understanding school-level factors that promote equitable science education in urban schools.

Jorge L. Solis is an Assistant Professor in the Department of Bicultural-Bilingual Studies at the University of Texas at San Antonio. He holds a PhD degree in Language, Literacy, and Culture from the University of California, Berkeley. He teaches undergraduate and graduate courses addressing bilingual science methods instruction, sociocultural contexts of literacy, and bilingual/second-language research methods. He is currently collaborating on projects integrating science-language pedagogy with elementary and secondary school pre-service teachers working with English Learners and bilingual students. His research interests include science literacy practices, preparing novice bilingual teachers, tensions/adaptations in classroom learning activity, and academic transitions of English Learners.

Trish Stoddart is Professor of Education at The University of California, Santa Cruz. She has a Ph.D. in Educational Psychology from the University of California, Berkeley and MA and BA degrees in psychology from the Universities of Birmingham and Leeds in the UK. Her research focus is effective STEM pedagogy for English Learners and teacher preparation. In her program of applied research, she uses mixed methods quasi-experimental research at multiple sites to investigate the impact of pre-service teacher education on the development of novice teacher's expertise in integrating the teaching of academic language and literacy and science instruction for English Learners. She is the founding Director of the TEEL (Teacher Education and English Learners) Center and Principal Investigator of the ESTELL, ELLISA and SSTELLA projects funded by the NSF and USDOE.

William J. Straits is a professor of science education at California State University Long Beach. He has earned Master's degrees in Biology at California State University Fullerton (1995) and in Curriculum and Instruction (1997) at The University of Texas at Austin, and a Ph.D. degree in science education also at The University of Texas at Austin (2001). Throughout his career, as a science teacher and as a teacher educator, he has emphasized the natural and important connections between science and language literacies. He currently serves as the director of the National Center for Science in Early Childhood and focuses much of his scholarly work on early childhood science education.

Lisa Sullivan is Senior Evaluation Analyst at the University of California, Davis School of Education. Dr. Sullivan has extensive experience providing evaluation services to education partners – including program design, developing data management and data collection tools, and providing formative and summative assessment of program activities. Dr. Sullivan completed the doctoral program in Learning and Mind Sciences at the School of Education at UC Davis in 2010. She was a classroom teacher in Los Angeles for 5 years. Her main area of interest is supporting educators to translate research into practice that will improve outcomes for all students.

The eCALLMS Team is a large group of scholars, students and teachers who collaborated to create the online content featured in this chapter. The project began in 2011 under the four Co-PIs for the eCALLMS grant, Kara Mitchell Viesca, Jacqueline Leonard, Honorine Nocon and Cindy Gutierrez, and has included instructional technologists, culturally and linguistically diverse education experts, mathematics education experts, science education experts, graduate and undergraduate students and practicing teachers. Significant collaborators on the eCALLMS project currently and historically have included: Christopher Carson, Nancy Commins, Anne Davidson, Susan Detrie, Helen Douglass, Polly Dunlop, Carlos Garcia, Cindy Gutierrez, Bonita Hamilton, Colin Hueston, Kim Hutchison, Nicole Joseph, Joy Barnes-Johnson, Jacqueline Leonard, Elizabeth Mahon, Cheryl Matias, Sally Nathenson-Mejia, Roberto Montoya, Naomi Nishi, Honorine Nocon, Ozyer Aysenur, Luis Poza, Sheila Shannon, Valencia Seidel, Lee Tran, Gama Viesca, Kara Mitchell Viesca, Geeta Verma, and Brent Wilson. Many local practicing teachers and other collaborators have also tested our work and provided extremely valuable feedback. Without all of these collaborators, the success of eCALLMS and the work presented in this chapter would not be possible.

Sara Tolbert is Assistant Professor of science education in the College of Education at The University of Arizona. She is a former secondary science and ESOL teacher who received her Ph.D. in education from the University of California, Santa Cruz. Dr. Tolbert investigates how teachers can be prepared to create more contextually authentic science learning experiences that leverage students' lived experiences through engagement with sociopolitical and community issues in science. She is Co-Principal Investigator on Secondary Science Teaching with English Language and Literacy Acquisition (SSTELLA). She has received several awards for her work, including the National Academy/Spencer Postdoctoral Fellowship Award in 2015. Dr. Tolbert teaches elementary, secondary, and bilingual science methods courses as well as graduate courses on equity, social justice, and traditional indigenous knowledge in math and science education.

Lida J. Uribe-Flórez is an associate professor at Boise State University. Previously, she was an associate professor of mathematics education and research methodologies at New Mexico State University. She earned a Master's degree in applied mathematics at University of Puerto Rico (2001) and a Ph.D. degree in mathematics education at Virginia Tech (2009). Her research interest includes teacher education (including teacher candidates and in-service teachers) as well as the use of tools to support learning in mathematics classrooms. Her work has been presented and published in several national and international venues.

Laura Valdés-Sánchez is an associate professor and a research assistant at the Universitat Autònoma de Barcelona in Catalonia, Spain. She earned a degree in Biology at University of Barcelona (2008) and a Master's degree in science education at Universitat Autònoma de Barcelona (2012). She has worked as an educator at the Natural Science Museum of Barcelona and at the Botanic Garden of Barcelona and she teaches science education university courses offered in pre-school and primary teacher education undergraduate programs. At present she is finishing her PhD degree (2016) in which she analyses co-teaching as a tool for professional development and as a strategy for the collaborative construction of Content and Language Integrated Learning projects. Her research focuses on classroom discourse in multilingual science learning environments.

Kara Mitchell Viesca is an assistant professor of culturally and linguistically diverse education at the University of Colorado Denver. She earned a Master's degree in education from Stanford University (2002) and a PhD in curriculum and instruction at Boston College (2010). She has taught multilingual learners from grades PreK-PhD across several countries and states including Mexico, China, California, Virginia and Massachusetts. She has led various projects including the National Professional Development project e-Learning Communities for Academic Language Learning in Math and Science (eCALLMS) and the state-wide collaboration of higher educators in Colorado focused on improving the education of multilingual learners and their teachers called Higher Educators in Linguistically Diverse Education (HELDE). Her research focuses on advancing equity through the policy and practice of educator development, particularly for multilingual learners.

Roser Martínez Vila earned a degree in Catalan Philology at the University of Barcelona (1989). She is a Catalan language and literature secondary school teacher at Valldemossa high school. She is currently a member of CESIRE (Resource Centre for Innovation and Educational Research), in the linguistic field, and she participates in several interdisciplinary and innovation groups. She has worked on languages teacher education at the Ministry of Education in Catalonia. She has been coordinator of the language, multiculturalism and social cohesion program in her high school. She has coordinated and participated in working groups of teachers for educational innovation. She has participated as a speaker at seminars and workshops.

Molly H. Weinburgh is the William L. & Betty F. Adams Chair of Education and Director of the Andrews Institute of Mathematics & Science Education at Texas Christian University. Her honors include the *Chancellor's Distinguished Achievement as a Creative Teachers and Scholar, Piper Professorship,* and Fellow of the *American Association for the Advancement of Science.* Her scholarship focuses on equity issues in science education and inquiry-based instruction. Her most recent research has centered on academic language acquisition and conceptual understanding in science by emerging multilingual students. She is an active member of numerous science education societies and editor of the Electronic Journal of Science Education.

Sara E.D. Wilmes is a doctoral student at the University of Luxembourg. She earned a Master's degree at the State University of New York at Buffalo in Education (2000). She has taught middle-school science in Western New York, has developed issue-based science curricula with SEPUP at the Lawrence Hall of Science, University of California, Berkeley, and has worked with teachers through various teacher professional development programs in both the US and Luxembourg. Her research interests include science education in multilingual spaces and the exploration of self-critical research methodologies such as collaborative autoethnography.

Wei Zhang is an Assistant Professor of linguistics and TESOL at The University of Akron. She earned a Master's degree in TESOL at Grand Valley State University (2004) and a Ph.D. in Linguistics at Purdue University (2010). She has taught courses in linguistics, second language acquisition, second language teaching methods, and second language writing. She has also coordinated and taught in-service teachers in professional development programs. Her research interests include academic language development of English learners, TESOL teacher training and program design, and second language phonology.

CPSIA information can be obtained
at www.ICGtesting.com
Printed in the USA
LVOW02*1030020417
529325LV00001B/48/P

9 783319 435145